Slither

Slither

*How Nature's
Most Maligned Creatures
Illuminate Our World*

STEPHEN S. HALL

GRAND
CENTRAL

New York   Boston

Grand Central Publishing
Hachette Book Group
1290 Avenue of the Americas, New York, NY 10104
grandcentralpublishing.com
@grandcentralpub

First Edition: April 2025

Grand Central Publishing is a division of Hachette Book Group, Inc. The Grand Central Publishing name and logo is a registered trademark of Hachette Book Group, Inc.

The publisher is not responsible for websites (or their content) that are not owned by the publisher.

The Hachette Speakers Bureau provides a wide range of authors for speaking events. To find out more, go to hachettespeakersbureau.com or email HachetteSpeakers@hbgusa.com.

Grand Central Publishing books may be purchased in bulk for business, educational, or promotional use. For information, please contact your local bookseller or the Hachette Book Group Special Markets Department at special.markets@hbgusa.com.

Print book interior design by Jeff Stiefel.

Library of Congress Cataloging-in-Publication Data

Names: Hall, Stephen S., author.
Title: Slither / Stephen S. Hall.
Description: First edition. | New York : GCP, 2025. | Includes bibliographical references and index.
Identifiers: LCCN 2024048341 | ISBN 9781538741337 (hardcover) | ISBN 9781538741351 (ebook)
Subjects: LCSH: Snakes—Behavior. | Reptiles—Evolution. | Snakes—United States—Identification. | Herpetology—Fieldwork—Anecdotes. | Herpetologists—Anecdotes.
Classification: LCC QL666.O6 H325 2025 | DDC 597.96—dc23/eng/20250124
LC record available at https://lccn.loc.gov/2024048341

ISBNs: 978-1-5387-4133-7 (hardcover), 978-1-5387-4135-1 (ebook)

Printed in the United States of America

LSC-C

Printing 1, 2025

For Del and Bob,
who didn't love snakes,
but loved and encouraged the child who did

Contents

Slither

"Snakes, Ardency Of"

W hat does wildness, a true animal life-force, feel like in your hands? What is it like to hold a sinuous form of lightning? I was lucky enough to find out, more than half a century ago. I've never forgotten the feeling.

At the time, I was walking near the end of a single-file gaggle of young boys, perhaps 10 of us, traipsing through a grassy patch of wetland in southern Michigan. It was a Saturday morning, late March or early April, the dull wash of clouds overhead more like winter than spring. The landscape, damp underfoot and the color of muddy straw, was not exactly wilderness, but not inhabited, either. It was one of those patches of undeveloped exurbia that surround every population center in every country on the planet, yet to feel the bite of theodolite and bulldozer. But if you paused, you could hear the machines coming.

The blur happened after most of the boys had walked ahead. Something on the ground shot shockingly fast into my peripheral vision. Dark and rectilinear, yet curvy and shiny. I did not know it at the time, but a part of my brain's visual system (and yours, too) likely evolved in our primate ancestors tens of millions of years ago

1

to detect just such a shiny blur. (That neural innovation also helps humans read fine print, such as this very text.)

The object sped across the path, right in front of me, darting toward the salvation of thick underbrush to the right. My brain processed this visual information almost subconsciously, overriding the warning—uttered a few minutes earlier by Mr. Fowler, the elder in the group—to be careful because this was a perfect environment for the eastern massasauga (*Sistrurus catenatus*), a dark and chunky and, more to the point, venomous rattlesnake. Before I had time to process a proper thought, I lunged. Down on my knees, reaching out, I grabbed at the contrail of motion.

Got it.

Wild desperate thrashing. So thin, yet surprisingly muscular. So *alive*.

That was the first time I experienced the gravitational tug, the psychological push–pull of the Snake: a relationship built of equal parts fascination, fear, admiration, and perhaps a secret sentimental affinity for a particularly wild and despised form of Otherness. I had felt the existential squirming of other creatures in my hands, fish and frogs and salamanders and turtles, but this was a life-force like nothing I had held before.

I felt a little unnerved by the animal's continuing struggle yet couldn't help but keep holding on. The snake was beautiful. Dark but glistening brown skin, brilliant yellow stripes down the back and side. At one point, I felt a strange bump-bump-bump against my hand. The snake was striking, trying to bite me. But its aggression felt like being poked by an oversize eraser, the kind we used to slip over the end of number two pencils. Blunt, painless, no chance of venom (or so I thought). Most of all: harmless. My left hand squeezed behind the head. Wildness contained. But not, not ever, extinguished.

One thing I did not do is wonder what the animal felt. Its cold-blooded heart was pumping wildly. The nerves fanning out of its dorsal ganglia were aflame with the insult of an alien grip. No one teaches us to imagine what it might be like to be grabbed by the nape of the neck and jerked roughly into the sky in the grip of a predator. That's not part of our imperial imaginations.

I held the prey up like a trophy, for all to admire. A clot of adolescent boys gathered around the catch.

"That's a nice little ribbon snake," said James Fowler. There was an unmistakable bit of "ho-hum" in his voice.

A ribbon snake (*Thamnophis sauritus septentrionalis*). The better-dressed cousin to the garter snake. Handsome as a bow, a chocolate-brown slither machine with lightning-yellow racing stripes. *Dainty* is the word Raymond Ditmars, a demigod in the world of 20th-century herpetology, used to describe it. On a scale of macho herpetology, the ribbon snake barely ranked above a worm, but no matter. I slipped the snake into a soft bag, cinched it tightly, and rejoined the walk. No one spoke of catch and release in those days.

It was 1965, somewhere north of Detroit, where leafy suburbs bled into cropland. The new subdivision where our family lived ("The Dells of Bloomfield!") had only recently been an apple orchard, and an ancient Native American trail that appears on an 1817 survey map, about which we were typically oblivious, cut across the backyard; the acreage of the Martin farm began about a quarter mile from our home. Very few people back then connected the dots of our little suburban disturbances of the environment—asphalt driveways with their basketball hoops and concrete sidewalks leading to faux-Colonial homes and paved streets stilettoed with mailboxes—with habitat destruction. I was 13 years old. My parents had parked me in a Saturday-morning "Herpetology for Juniors" class at the nearby Cranbrook Institute of Science, and this was our one and only field trip.

There is book learning, and there is fingertip knowledge. Like many children, I had devoured books about snakes—the Golden Guide book on reptiles and amphibians, Raymond Ditmars's *Snakes of the World*. I had seen the animals in zoos and in the wild. But it wasn't until that spring morning in Michigan that I had felt their innate, unique life-force in my fingertips. That was the true beginning of my fascination with snakes. It was also the beginning, as for so many other people, of my misunderstanding of them.

I took that ribbon snake home and unknowingly committed every misdemeanor you could in animal husbandry. It lived in a repurposed aquarium located in the garage; I did not know that I should have provided a heating element. The aquarium contained a random piece of wood, but no proper place for a secretive animal to hide. The light in the garage was dim; herpetologists would not know for another decade or two that snakes need a regular dose of ultraviolet B radiation to maintain good health. I assumed the snake was harmless; I did not know that the ribbon snake and the garter snake, commonly considered innocuous, are technically venomous—they just can't hypodermically deliver a huge gush of venom through fangs. And I contained the ribbon snake, and several later additions to the collection, with a wire-mesh top held down by heavy ingots of lead; we did not know what good escape artists snakes are until my mother's air-raid-quality screams alerted us to the fact that my eastern milk snake (*Lampropeltis triangulum triangulum*) had squeezed out of its cage and was sunning itself on the one patch of sunlight on the garage floor. It was trying to tell us something we were incapable of understanding, as snakes often do.

~~

Decades later, I found myself thumbing through the index of one of Charles Darwin's opuses, *The Descent of Man*, looking for anything the great naturalist may have said about reptiles, and came across an entry that read "snakes, ardency of." *Ardency* is a quaint, old-fashioned 19th-century noun you rarely hear these days, but as soon as I saw the word, I thought back to the Saturday-morning field trip. It was exactly the right word to describe that snake. *Ardency* means being alive in some hot-wired, zealous, visibly fervent way. It means thrashing and writhing and darting and slithering and bursting with life; it means eluding and camouflaging and ambushing and fending off with a vital life-force. (It also, according to the *Oxford English Dictionary*, means "warmth of feeling"—an *OED*-certified blast of irony aimed at a cold-blooded creature.)

You can feel ardency in your hands if you're lucky enough to grasp it. I wish everyone could know that feeling. And I now wish I had let the snake go. But I couldn't. That, too, was something I couldn't understand.

Snakes don't let go of our collective imagination. Of all the creatures on earth, they seem to incite more awe and fear, more earthly fascination and philosophical musing, more veneration and more hatred than any other animal with which humans interact. They slither through the Bible, Hebrew and King James; denote letters in ancient Egyptian papyri; decorate the holiest sites of Buddhism, the sanctuaries known as stupas, which house relics touched by the Enlightened One; and defend the most majestic temples of Mesoamerican cultures. They are a drearily predictable trope of feral terror in popular culture (*Anaconda, Snakes on a Plane*), yet have inspired the design of "spooky, but adorable" mass-market glassware and candelabra at West Elm retail outlets and the $172,000 Serpenti

Viper Necklace, in 18-karat white gold and diamonds, sold by Bulgari, whose snake-inspired Serpenti line has become the luxury jewelry maker's bestselling style since its introduction in 1948. In the hands of classic poets, snakes are foundational storytelling characters in mythology, from Medusa's writhing hair in Ovid to the healing arts of Asklepios (learned, according to multiple accounts, from a snake); they infiltrate the poems of moderns, too, like Mary Oliver and D. H. Lawrence and Louise Glück, to mention just a few. And who knew that Andy Warhol adopted the persona of a serpent in probably his least-known series of drawings, collected in *The Autobiography of a Snake*? To the creative imagination, snakes become irresistible metaphoric vehicles to explore human limitation, human fallibility, human mortality, human otherness.

Snakes have found comfortable habitats in our cultural, scientific, and ecological landscapes, even as they trigger paradox and contradiction. Underneath it all is an innate psychological reaction so deep and so primal that, as D. H. Lawrence observed of snakes in *Apocalypse*, his last book, "a rustle in the grass can startle the toughest 'modern' to depths he has no control over." But it wasn't until I met Matthew Holding, a scientist studying snake venom at the University of Michigan, that I began to understand how to make sense of these profound contradictions.

All these zero-to-one emotional switches, buried like primeval neural bytes, may be hardwired in the human brain. Holding referred to a provocative hypothesis by Lynne Isbell, a professor at the University of California–Davis. Isbell has pulled together a wealth of compelling evidence to suggest that early in the evolution of modern primates, at least 60 million years ago and perhaps much earlier, the ability to *see* snakes—that is, quickly detect them in the visual field—was crucial for primate survival; in fact, she argues that snakes influenced the evolution of the mammalian visual system in a way that allows (indeed, *forces*) modern humans to spot serpents rapidly. As Isbell put it, "We primates uniquely evolved the combination of excellent vision and large brains in response to snakes."

Riffing off Isbell's notion of a "detection module" in the brain related to serpents, Holding went on to map out humankind's unique neural plumbing when it comes to snakes this way: "I think snakes trigger an innate excitement factor in humans, and then that reaction gets channeled one way or another. When those neurons start to fire, it gets converted to fear or wonder." In other words, we're biologically hardwired to react to snakes, yet culturally wired to channel that reaction in a particular emotional direction. The *nature* of that reaction (and how it plays out) is colored by learning, experience, risk tolerance, and other distinctly human overlays to that initial, instantaneous excitability in the "detection module." We can't help but react to them, but we might be able to control *how* we respond to them. Perhaps even bend the arc of that reaction more in the direction of awe and wonder, if we pause to consider their remarkable qualities.

Consider personality. "I've spent hours and hours and hours of my entire adult life in the field with rattlesnakes," said Emily Taylor, biologist and author of the recently published *California Snakes and How to Find Them*, "and I will 100 percent tell you, they absolutely have personalities. They are utter individuals, and it's absolutely amazing how unique they are. There are ones that will never rattle, no matter what you do. There are ones that will track you with their eyes and their heads as you walk by. There's ones that will rattle you from the bushes from 10 feet away every time. They have personalities, just like we do."

They may also, in a biological sense, have *attitude*. Daren Card, who studied the evolutionary genomics of reptiles at Harvard University, put his finger on it in both a scientific and a metaphoric way: Snakes and lizards (known collectively as squamates) are animals that refuse to stay in their phylogenetic lane. And snakes seem to break rules that most other vertebrate animals (birds and mammals, amphibians, other reptiles) follow. Some lay eggs; others bear live young. Some live in trees; others live underground. Some eat frequently; others eat once a year. Some snakes have sex chromosomes that resemble those of birds; others have sex chromosomes that resemble mammals. Birds and dogs use a very particular molecular mechanism to "recombine" the DNA in their sex cells during meiosis, which is essential to the process of genetic diversity, while most other mammals use a completely different mechanism. "Snakes seem to do both," Card said. "There are just no rules anymore. They seem to meander. They have different sex chromosome systems. They have things like parthenogenesis. They just seem to do everything." They've also evolved some pretty nifty innovations, like the sea snakes that figured out a way to vent nitrogen gas through their skin so they don't get the bends. As Card detailed all the ways snakes break the rules of biology that define virtually all other vertebrate animals, I began to think of them as the Renegades of Chordata— the animals, unique among vertebrates, who seemed to adopt whatever means of reproduction, genetics, metabolism, or movement was necessary to survive.

Rick Shine, an eminent herpetologist in Australia, urged a bit more humility. "I'm not sure that we know enough about them to have a really clear grasp of what some of those rules are," he said— which was saying something, coming from a scientist who has published more than 1,100 papers (mostly on snakes), authored the recent *So Many Snakes, So Little Time*, and coined some of the wittiest journal titles in the scientific literature ("Are snakes right-handed?" and "May the (selective) force be with you"). That said, he immediately added, "I think most ecologists that work with snakes end up being astonished by their flexibility and plasticity, their ability to change some of the things they do in response to novel challenges."

Shine went on to say something that came as close as anything I've encountered to suggesting that snakes have not only a worldview, but in fact an *admirable* worldview. He was talking about his fieldwork in the mountains of Tasmania, with its notoriously miserable climate. "There's probably only 20 or 30 really nice warm days a year," he said. "But the reptiles that live up there sit under the rocks during the cold days, and they only come out on the warm days, so as far as they're concerned, they're living in a tropical paradise. The only time that they're out there doing anything, it's warm and sunny and it's lovely! And the rest of the time basically doesn't exist. So we walk around as these sort of constant-rate, warm-blooded creatures thinking, *This is a god-awful, horrible environment. How can anything live here?* And as far as they're concerned, they're living in the villa by the sea in a warm climate, because that's the only time they're active." Shine paused and then ventured a thought that landed somewhere between Buddhist-like empathy and reptile envy. "I really want to know what it's like to crawl out from under the rock on a cold morning and feel your body warm up and feel your organ systems start to switch on and your brain start to work," he said. "That must be an amazing sensation."

All those biological "transgressions"—reproductive, chromosomal, metabolic, and of course skeletal—perhaps also explain the special appeal of snakes to a distinct sliver of humanity. James Murphy, a longtime zoo curator and herpetologist, refers to this simply as the Madness—a fanatic, all-consuming fascination with snakes that changes one's behavior, habits, curiosity, acquisitiveness, sense of risk. Part of this allure is the sheer Otherness of snakes. It's not just that they're different, but that they're different *and* loathed. I can't help but wonder if a lot of human outliers (including alienated adolescents growing up in Michigan) identify with that Otherness.

But it's more even than that. Snakes, seemingly so constrained by biological limitations—so "primitive"—are incredibly successful *despite* all those supposed limitations. It's not just that snakes lack limbs; it's that they had them and then renounced them at least 80 million years ago. Bipedal chauvinists may consider that a limitation, but snakes have turned that limitation into just another way of breaking the rules. Henry Astley, an expert on snake locomotion at the University of Akron, has pointed out that virtually every other terrestrial animal forced to traverse a messy, complex, obstacle-strewn landscape has to slow down to pick its way through; snakes by contrast use obstacles as an accelerant, speeding up and moving faster. They lack eardrums and thus cannot hear in any conventional sense; yet they can detect and interpret extremely low-frequency vibrations as if they were sonic booms. They lack eyelids, can never close their eyes, and are anatomically locked in an eternal state of vigilance, even though they are thought not to see very well; yet they can detect tiny chemical differences on either side of a blade of grass and measure temperatures to several thousandths of a degree Celsius. The rigidity of their skin and scales freezes their mouths into oddly sinister grins or chillingly indifferent gapes; yet the plasticity of the bones in their skulls allows them to eat more forms of prey-food than any other vertebrate. They lack the metabolic machinery to warm their own blood and generate bodily heat—and are thus supposedly fated to chase sun and shade for their very survival; yet some of them have concocted an out-of-the-box biochemical wizardry that humans—especially humans with metabolic disorders such as diabetes—can only dream about. Their nervous system is so puny, so supposedly primitive, that *reptilian brain* has become an enduring, all-purpose cultural insult; yet they may be capable of a kind of synesthesia—a blending of multiple sensory inputs, such as "seeing" thermal radiation or "visualizing" odors—that creates a sense of the world we can't even begin to imagine. The venoms cooked up by deadly snakes induce widespread terror and terrible human mortality in places like Asia and Africa; yet venomous snakes are masters of genetic mixology, creating three-dimensional tweaks in the atomic structure of their toxins that are the envy of medicinal chemists, who have developed lifesaving drugs based upon serpent ingenuity.

All those "yets" derive from recent scientific research on snakes, which is ultimately what reignited my interest in them as a subject. I don't consider myself a "herper," those impassioned snake-lovers who think nothing of wading through swamps or careening down desert highways at night in search of serpents, and I haven't kept a snake as a pet since those days in Michigan. But I do write about science, and about 15 years ago, I began to notice some very unusual scientific papers about snake biology in the most influential scientific journals, including major articles in *Nature* and *Science*. It wasn't until the advent of modern scientific technologies—radio telemetry, genomics, wildlife videography, sophisticated genetic manipulations—that we began to realize that the most amazing traits of these creatures were qualities we'd never seen before, often microscopic, molecular, hidden from view. Many of those discoveries flipped the phylogenetic bias of humans on its head—supposedly inferior qualities (limblessness, cold-bloodedness, and the like) turn out not to be so inferior at all, just different. Those Renegades of

Chordata have been a revelation of all the ways that biological rules can be broken in the name of survival.

And snakes are, above all, survivalists. Despite all their limitations, they've been around for some 130 to 150 million years, when the first proto-snakes peeled off from the lizard lineage, elongated their bodies, added ribs, and subtracted limbs. Their signal trait is diversity. Triggered by "one or more unknown and unknowable singular events" deep in the evolutionary past, perhaps 65 million years ago, snakes experienced an explosion in biodiversity, according to a recent research report in *Science*, and became radiant masters of all environments. New species are identified almost monthly. At last count, there were approximately 4,000 species in roughly 20 families, the best known being the colubrids (more than 3,000), the vipers, the elapids (cobras, mambas, kraits), the pythons, and the boas. They inhabit every continent except Antarctica. They range in size from three inches to 30 feet. They may eat once a day or once a year. They may perish in the jaws of predators in the first week of life or live for decades. They've survived dinosaurs, droughts, asteroids, ice ages, and all manner of predator. In an age of climate change, they are prodigies at adjusting to thermal variations in the environment. If we stop loathing (and killing) them, we might learn something from them.

~

Just as catch and release was not the norm many decades ago, the world—and our way of sharing it with serpents—has dramatically changed. The swashbuckling naturalists of a century ago appear to us differently today; the first popularizers to capture animals on film, once viewed as innovative pioneers, look different to contemporary eyes, which detect episodes of exploitation and insensitivity. Decades ago, keeping snakes as pets was a hobby reserved for weirdos, loners, and people with perhaps an unseemly hunger for attention, like that guy (and it always seems to be a guy) who's walking down the boardwalk with a boa constrictor on his shoulders. Now the retail reptile market extends to ubiquitous franchises like PetSmart and serves millions of hobbyists, creating cottage industries of captive breeding and reptilian meals-to-go, while unfortunately still sustaining the legacy trafficking in illegal wildlife that began long before my Michigan field trip.

But some things have not changed. Habitat destruction has relentlessly continued, and safe havens for snakes have conversely shrunk, as have their odds of surviving. The signs of stress are both evidence-based and anecdotal. A 2022 study in *Nature* suggested that 21 percent of the world's reptiles were in danger of extinction, and an esteemed backwoods herper-writer named the Weazel said you can barely find snakes anymore in the wilds of North Florida. "In the late Sixties, I would routinely find a dozen or more of various species while making a night time 20-mile loop south of Gainesville to Micanopy and back," he told me. "I drive those same roads today on my way home from town. On a good night in the late 60s I saw more snakes of various species than I have seen in all the years since. Now I see none." And, he added, "The first predictor of extinction is rarity." It is a refrain, anecdotal but consistent, that I heard again and again, in different environments, on different continents, from different hemispheres—there aren't as many snakes as there used to be.

Given how many different ecosystems they inhabit, it would be tempting to call snakes the true canaries in the global coal mine, but that's a role they'll never be invited to play. Like many, I constantly get solicitations for donations from conservation organizations with their calendars and return-address stickers showing fuzzy, beloved, heartwarming creatures: owls, bees, hummingbirds, elephants, foxes, lemurs, even bats. When was the last time you saw a snake on a fundraising pitch?

How do we reconcile that widespread cultural antipathy with the need for protection? This ultimate paradox gets at the most unique ecological niche inhabited by snakes: the human psyche. This symbolic slithering mirrors their agility at infiltrating the narrowest, most inaccessible, most secretive physical spaces. Maybe a lot of people don't like snakes because they have always highlighted existential tensions of the human condition. Beyond the obvious poles of fascination and dread, we waver between temptation (sin) and avoidance (abstinence), reverence and demonization, benevolence and vengeance, regenerative life—or sudden death. Snakes inspire aesthetic rapture, but also rabid serpenticide. Perhaps the ultimate test, for all of us, is to aspire to a kind of spiritual form of conservation, where we meet the psychic and ecological challenge of protecting Otherness. When it comes to vertebrates, no class of creatures is more different, more abhorrently Other, than snakes. If there is room for them in our psychological ark, there's space for everyone.

Evolutionary biologist E. O. Wilson, who was an avid snake hunter during his Alabama childhood, noted in his 1998 book *Consilience* that serpents are "the wild creatures most frequently conjured around the world in dreams and drug-induced hallucinations." The pioneering cultural historian Aby Warburg, after studying indigenous rituals in the American Southwest, went even further: He argued that snakes represent the single most important visual symbol of the human relationship with nature—an interaction he traced from antiquity to Renaissance art, from pre-Christian pagan cultures to the ubiquity of snake iconography in religious art. Snakes have always been symbolic to and of the human condition:

temptation in the Garden, power in the Egyptian tradition, healing in the Greek tradition, serpents entwined like the double helix around the staff of Asklepios, good and evil, and of course mortality.

Of course.

That dainty ribbon snake of mine shed its skin several times, trading its dull outerwear for a brilliantly colored new coat—a modest, garage-band version of the symbolic immortality that humans have always associated with snakes. But acknowledging the symbolic immortality of snakes is also a way of acknowledging our own mortality.

In 1968, Louise Glück published a beautiful, haunting, deliciously cold-blooded poem called "Cottonmouth Country" as part of her first collection. The poem is dark and furtive, like the namesake serpent, a venomous creature of piney woods and waterways in the American South, its black outer skin shockingly contradicted by the glistening white of its open mouth, ominously framed by two fangs. The snake in the poem lies "uncurled" on a bed of moss, indifferent to the mortality that an incidental disturbance or accidental misstep might provoke. In eight spare, unsentimental lines, the unfurled snake symbolizes in Glück's rendering all things dangerous, all things unheeded, all the unpredictable but inevitable calamities set in motion and eventually unleashed by the mere act of being born and being alive. Whether we know it or not, it is clear from Glück's penultimate line—"Birth, not death, is the hard loss"—that we all live in Cottonmouth Country.

Snake Road
Catskill, New York

On April 1, 1841—when John Muir was a three-year-old child in Scotland, more than a decade before Henry David Thoreau would publish *Walden*—the landscape painter Thomas Cole gave a public talk to his neighbors in Catskill, New York, extolling nature, wilderness, and beauty. Cole is rarely considered a conservationist; he was "an unusual combination of writer, poet, philosopher, observer, and painter," according to art historian John Howat, and most people know him as the founder and earnest philosopher-king of the Hudson River School of artists, a group of American landscape painters active in the mid-19th century who celebrated nature for the sake of nature. Their paintings captured both the majesty of untrammeled wilderness and, if you look closely at the canvases, the petty aspirations of tiny humans who encroached upon its margins.

Cole's talk at the Catskill Lyceum, titled "Lecture on American Scenery," was part celebration, part cautionary exhortation. Though he viewed the American landscape through the eyes of an artist, not a scientist or environmentalist, he reminded his audience that "it would be well to cultivate the oasis that yet remains to us, and to cherish the impressions that nature is ever ready to give, as an antidote to the sordid tendencies of modern civilization." To Cole, the American landscape embodied not only wildness, beauty, and the sublime, but also what he called "futurity." At the very end of the lecture, with that future already imperiled, Cole departed from his prepared text to speak of matters closer to home. The words *habitat destruction* and *environmentalism* were yet to become common coin in the national conversation about conservation, but that was exactly what he chose to talk about.

Cole understood what was at stake because he had spent his formative years under the permanent human-made eclipse of smoke and soot produced by the Industrial Revolution in northern England. Having experienced firsthand the degradation of nature in the name of progress and prosperity, he recognized the first signs taking place in 19th-century America. One of his favorite spots to paint the slope-shouldered, brooding Catskill Mountains was in a grassy meadow near the intersection of Kaaterskill and Catskill Creeks, just outside the town proper. It was a beautiful valley, but a local sawmill—and the felled cedar trees that fed local commerce—was already converting its wildness to human deforestation, human industry, human encroachment. That's what prompted Cole to deviate from his prepared text. After apologizing for keeping his audience a bit longer, Cole said, "I cannot but express my sorrow that much of the beauty of our landscapes is quickly passing away; the ravages of the axe are daily increasing, and the most noble scenes are often laid desolate with a wantonness and barbarism scarcely credible in a people who call themselves civilized."

What does Cole's 1841 lecture have to do with snakes?

Part of the connection lies in a research article that appeared nearly two centuries later in the journal *Nature*. That's the article, mentioned earlier, that assessed the global extinction risk to reptiles at 21 percent—significantly more than similar surveys predicting extinctions for birds (13.6 percent) and slightly less than for mammals

(25.4 percent). Among the leading causes of these extinction risks? Precisely the activities that Cole identified (and lamented) way back in 1841: agriculture, logging, and urban development. What we now call habitat destruction.

Another part of Cole's snake connection lies about 180 miles east of Catskill, in Boston. One of Cole's most famous works is a dark (in every sense) oil painting, given pride of place in a gallery of the Museum of Fine Arts, called *Expulsion from the Garden of Eden*. Completed in 1828, the allegorical painting depicts Adam and Eve, tiny figures bent over in shame, as they exit paradise and slouch into a world of mortality, sin, menace, and uncertainty. Cole was a deeply religious and reverent man, so he surely knew the opening lines of Genesis chapter 3, where the serpent tricked Eve into eating the apple from the Tree of Knowledge. Words matter, especially in the Bible. *The New Oxford Annotated Bible* described the serpent as "more crafty than any other wild animal that the LORD god had made," but there's no unanimity in the damning adjectives applied to the snake. Robert Alter, in his famous translation of the Five Books of Moses, chose the word *cunning*. The Torah of the American Hebrew Congregation opted for *shrewdest*.

Whether crafty or cunning or shrewd, all hint at duplicity and deceit. The snake in the Garden of Eden was more than crafty. It could speak and converse with humans (like only two other animals in the Bible). It was bold enough to contradict God, assuring Eve that, contrary to what God's warning had implied, she would not die immediately if she ate the fruit from the Tree of Knowledge. And it spoke the truth when it said if humans partook of the fruit, their eyes could be open to the knowledge of good and evil. When God learned that his instructions had been disobeyed, of course, and confronted Eve, she threw the snake under the bus (what might be considered the Original Roadkill) by saying, "The serpent tricked me..." "Because you have done this," God thundered at the serpent, "cursed are you among all animals and among all wild creatures; upon your belly you shall go, and dust you shall eat all the days of your life." And there is unanimity about the word God used in punctuating this curse: There would always be "enmity" between the snake and all of Eve's offspring, which is to say all of humanity, for all time. Then God "drove" Adam and Eve out of the Garden and stationed a winged beast with flaming sword to bar their reentry. All three expelled creatures hit the road.

The temptation in the Garden of Eden as well as Adam and Eve's expulsion from the Garden are among the most dramatic and fateful episodes in Genesis, in all of Christian mythology. In the history of Western art, the roster of painters who have taken on the challenge of depicting that fateful scene is pretty much an all-star lineup: Albrecht Dürer, Pieter Brueghel, Lucas Cranach, Benjamin West, Rubens, Michelangelo, Massaccio, Tintoretto, Holbein, Titian, William Blake, and Paul Gauguin, to name a few. In many depictions of the Expulsion (but not all, including Cole's), the scene includes not just the two disgraced humans but, over there on the ground, one particular limbless fellow traveler: the dust-eating snake. In some of these paintings, the serpent proffers the apple; in others, it coils around a nearby tree, as if in ambush, a patient spectator awaiting the maelstrom it has unleashed; in yet others it slinks along the ground, carving a serpentine trail of shame in the dust for the exiles to follow. Not every snake is depicted with naturalistic fidelity. Michelangelo's bloated serpent, floating on the ceiling of the Sistine Chapel, looks more like an overinflated balloon than anything you'd see in a herpetology book. But that's not the point.

The visual climax of the Expulsion scene is not only a timeless Judeo-Christian parable of sin and mortality but also an ecological metaphor that assumes special resonance in the 21st century. Only three creatures usually take the exit ramp from eternal peace, ignorance, bliss, immortality: Adam and Eve, their heads bowed and bodies bent with the burden of knowledge, which is shameful, and the Snake. The obvious message is disgrace. Perhaps the hidden message, however, is that whatever fate awaits them, the destiny of all three creatures is forever linked by the Expulsion.

Thomas Cole explicitly connected his version of unspoiled nature to the Garden. As he told the audience in Catskill, "Nature has spread for us a rich and delightful banquet—shall we turn away from it? We are still in Eden; the wall that shuts us out of the garden is our own ignorance and folly." Viewed through both Cole's sentimental lens and our latter-day taxonomic one, it's hardly a stretch to see the Expulsion as a conservationist metaphor—a particularly thorny and modern take on conservation because the animal to be conserved in this picture is the object of collective, eternal enmity. The serpent, like it or not, is our fellow traveler on this secular and ecological road. The fate of the snake, Other and loathsome though it may be, is our fate, too.

I hadn't thought much about this confluence of painterly images, biblical allegory, and conservationist omens until I took a short hike in Catskill that retraced the same ground Thomas Cole traversed on the way to one of his favorite spots to set up his easel. A trail sign along the path quoted an excerpt from Cole's 1841 lecture, the part he added at the end, in which he confessed his regret over how his beloved local landscape was already being desecrated. "Among the inhabitants of this village, he must be dull indeed, who has not observed how, within the last ten years, the beauty of its environs has been shorn away," he said. "...Year by year the groves that adorned the banks of the Catskill wasted away; but in one year more fatal than the rest the whole of that noble grove by Van Vechten's mill, through which wound what is called the Snake Road, and at the same time the ancient grove of cedar, that shadowed the Indian burying-ground, were cut down...Where once was beauty, there is now barrenness." The signage, and indeed Cole's lecture, marked an early entry in our ledger of desecrations.

Snake Road. "Futurity." The Expulsion, and its suggestion that the fates of humankind and the snake are improbably entwined. It was an unexpected mingling of signs. It occurred to me that perhaps we might think about the future survival of snakes as a proxy for our own environmental, ecological, biological, and perhaps even spiritual future. And it got me wondering about other "snake roads"—roads that tell us something about conservation, snakes, us.

1

Female 21 and the Black Mamba

~~~~~~

Two Snake Tales

W here do most people encounter snakes? One might reasonably think it is at the zoo, on a hike, in a farm field, at a reptile expo, perhaps even in the backyard garden. But the most likely habitat, especially when it comes to scary snakes, is in the media, particularly on the internet. A sampler of recent headlines includes "Maryland man with 124 snakes in his house died of snakebite, autopsy finds" and "Snake on a plane! Live reptile discovered in overhead cabin on Bangkok flight," along with YouTube titles like "Fatal Black Mamba Bite!!! RIP Ryan" (on a video with more than 4.4 million viewers). These nuggets of "news," these bonbons of dread, are like junk food for the amygdala. And leave it to Charles Darwin to capture this perverse fascination with a brilliant turn of phrase. He distilled the paradoxical behavior of young monkeys, terrified of snakes but compelled to peek repeatedly into a basket containing them, as "satiating their horror in a most human fashion." The key word is not *horror* but *satiating*. Attraction to fear is a kind of hunger.

I could not resist a bit of binge-reporting on my own when I ran across another macabre headline from a few years ago: "Black mamba pet snake suspected in death of Putnam woman." I was curious to learn more.

The tale began with a 911 call into the Putnam County Sheriff's Office in upstate New York around 7:00 p.m. on June 14, 2011. A male voice on the line explained that he had returned home from work to the small bungalow that he shared with his girlfriend in the community of Putnam Lake, according to the official incident report, only to discover her body on the bed—face up, unresponsive, cold to the touch. The sheriff's department dispatched emergency responders at 7:06 p.m. At some point, fairly early on, the first responders learned about the snakes.

Putnam Lake is a bucolic, tree-enshrouded enclave about 65 miles north of New York City. Once a weekend getaway spot, it is now more suburban than wilderness, more modest than ostentatious. The houses, dotted along the winding streets and wooded hills that descend to the lake, seem utterly ordinary, with front-yard flower beds, basketball hoops, children's toys strewn on the lawn. It is not what you would consider a typical habitat for a species of venomous snake whose natural range is limited to sub-Saharan Africa. Yet given our global commerce in exotic animals as well as exotic cars and foodstuffs these days, any home in any neighborhood in any town can house a covert menagerie of exotic—and potentially lethal—animals. And when things go wrong locally, everyone quickly learns about it globally.

That was exceptionally true of what went wrong in the house at

21 Hopewell Drive, a short, narrow street on a hill that rises above the eastern shoreline of the lake. When first responders arrived at the house, they encountered the distraught boyfriend, identified as Vito Caputo in the police incident report. He told officers that there were approximately 50 venomous snakes kept in double-locked cages in the residence. He also mentioned that one of the locks on a cage housing an especially deadly snake known as the black mamba (*Dendroaspis polylepis*) had been removed and was on the floor, although he claimed the snake was still in its cage. The responding officer entered the house just long enough to confirm the motionless body, then quickly retreated. More officers were dispatched to the scene. The provisional reason for the police activity was "Unattended"—as in unattended death.

The lead investigator, Nicholas DePerno, arrived soon after. When he heard about the black mamba, he went up to the boyfriend and said, "I'm not pulling any punches. Are you sure that snake is in the cage? Because if it isn't, and it bites me, I'm going to be really pissed!" He would be more than pissed; prior to the advent of antivenom, black mamba bites were considered 100 percent lethal. Assured that the snake was in its enclosure, he entered what was still a potential crime scene with another officer. His initial assessment was, "Holy shit!"

It was immediately apparent that this wasn't your typical suburban interior decorating scheme. Heavy curtains covered the entryway to most of the rooms. There was a weird background sound that filled the house, almost like steam leaking from a pipe or a chorus of domesticated cicadas. DePerno walked into the living room. One entire wall, floor to ceiling, where normally there might be an entertainment center or bookshelves, housed custom-built, zoo-quality, wood-framed containers holding snakes. Lots of them. "It was a very, very slow search," he admitted. "You get a 'pucker factor' when you're in a house full of venomous snakes." Every time he walked in front of a large enclosure in the kitchen, there was loud hissing, followed by the sound of a snake striking the glass. It contained, investigators soon learned, an agitated king cobra (*Ophiophagus hannah*). Every time the snake struck, one official noted later, "You could see the plexiglass bend a little."

DePerno withdrew from the house and called Ken Ross Sr., chief of Putnam County's Society for the Prevention of Cruelty to Animals. In New York State, the SPCA has law enforcement status when it comes to captive animals. Ross took one quick look at the cages after he arrived and immediately called his "snake guy." This was Christopher Cooper, a self-described critter consultant who worked at the Katonah Bedford Veterinary Center in nearby Bedford Hills.

By the time Cooper arrived, a large contingent of first responders and onlookers had clotted the narrow road leading up to the house—local police, county sheriff's department, state police, paramedics, curious neighbors. When Cooper—the son of an NYPD cop and a former biker (though you'd never guess from his slender build, tame beard, and wire-rimmed glasses)—strode up to the group, the officer in charge turned to the SPCA chief and, voice dripping with skepticism, asked, "Is this him? Is this the guy?" Cooper was used to hearing that. It took him back to the days when he was a gawky adolescent growing up in the Bronx, with a bowl haircut and buck teeth, "mercilessly" picked on by schoolmates. He realized early on that owning and handling animals, especially snakes, gave him a cachet, an authority, and, in his opinion, a way to prevent even more bullying.

Despite first impressions, Cooper had been handling reptiles, including venomous snakes, for more than three decades. He had once been called on to tangle with a runaway six-foot Nile monitor lizard on the main street of Beacon, New York. When it came to exotic animals, especially reptiles, he described himself as the "sort of go-to person for really bad stuff." No one wanted to accompany him back into the house at first. Because it was a potential crime scene, however, someone had to do it. That was DePerno.

The moment Cooper set foot in the house, he heard "a tremendous amount of hissing." It is not unusual for snakes to hiss, he knew, but after their routine had been disturbed by the comings and goings of strangers, this sounded different. "Overt loud hissing is unusual," he said. "And over this overt hissing comes a chorus of rattling. So now I'm like, 'Okay, rattlesnakes!'"

On his first short reconnaissance into the house, Cooper pulled away the heavy curtain covering the first door to the left off the central hallway and poked his head inside. The first thing he saw was a reverse-osmosis water purification system—a highly sophisticated device for a pet hobbyist. He took a few steps into the room and saw the first of the caged snakes. He recognized a spitting cobra, a timber rattlesnake, and a Mojave rattler. "My brain was about to explode," he recalled. He paused to admire the quality of the cages—"some of the nicest, most pristine enclosures you'll ever see." It was just that they enclosed "the hottest snakes on the planet." As he made his way through the house, he could see a woman's body in the bedroom.

As events in Putnam Lake unfolded, the story began to conform to a common and predictable pattern. For every three-paragraph item in *Science Times* on snake locomotion or a 100-million-year-old fossilized specimen, it seems there are at least twice as many accounts of melodramatic real-life human–snake encounters: the escaped cobra that terrorized a neighborhood in Raleigh, North Carolina, or the "really poisonous" brown snake that interrupted a tennis tournament in Australia. They share with horror films an artful yet cynical cultivation of fascination (at a safe distance) and vicarious dread (a fear that nonetheless leaves a residue of disgust). And they pose extraordinary public safety challenges to community officials, as was happening in upstate New York.

Cooper retreated outside with the nervous deputy at his side and walked up to the assembled group of law enforcement officials. "Even with my penchant for exaggeration and embellishment," he announced, "this is *bad*." Then he issued a puzzling list of requests. "I need a state helicopter on standby," he told the group. "I need an ambulance. I need Jacobi Hospital in the Bronx to be alerted. I need you to evacuate the neighborhood. And," he added, almost as an afterthought, "I need some Post-its."

The helicopter was to air-evacuate any potential snakebite victims. Jacobi Hospital had a snakebite center. Clearing out the neighborhood was for public safety; at that point, it was still uncertain if any of the venomous snakes were loose in the house or had escaped. "Do you know what would happen if there were neighborhood kids playing in the grass with a black mamba loose?" Ross, the SPCA chief, told me later.

And the Post-its?

Cooper went back into the house, peered into each case, identified each species as best he could, and methodically affixed a Post-it on each snake enclosure—a smiling face for a harmless snake, a frowning face for a venomous or potentially dangerous snake. He couldn't identify all the snakes in that first pass, but there were a lot more frowning Post-its than smiling faces. According to an inventory Cooper later prepared for New York State investigators, there were 75 captive snakes in the four-room cottage, 57 of which were "the most dangerous poisonous snakes on the planet." Not just venomous, but deadly in a way that anyone would find heart-stoppingly alarming. The official inventory filed with New York State environmental officials (and this is only a partial list) noted 13 cobras, including 2 spitting cobras, an albino monocled cobra, and a pair of Sri Lankan cobras that were endangered species in their own country; 21 vipers, including a black-headed Brazilian lancehead viper, 2 eyelash vipers from Central and South America, a saw-scaled viper (responsible for countless snakebite deaths each year in Central Asia and the Middle East), and a Gaboon viper from Africa ("They have the longest fangs of any poisonous snake," Cooper said, "they like come out the other side..."); an asp ("like the one Cleopatra used to kill herself"); a death adder ("It's called that for a reason; everybody it bites *dies*"); a green mamba and a black mamba; a Costa Rican fer-de-lance; a copperhead; and 12 rattlesnakes. Investigators even found a mouse-breeding operation in the basement to keep the dozens of reptiles well fed. "You bring that many poisonous snakes in one place," said Ross, the SPCA officer, "it's just a time bomb."

The "snake guy" noticed a few other unusual things on that second pass. He saw that several framed photographs in the hallway leading back to the bedroom had been knocked askew, as if someone had staggered into them while passing by. And when he entered the bedroom, he found the woman lying atop a perfectly made bed, with crisp white sheets. Cooper noticed two sets of fang marks on her left wrist (the final police report later referred to seven fang incisions on the body). "She was not breathing," the official police report later noted, "and was cold to the touch." Cooper also observed that, unlike all the other snake containers in the living room, which had double locks, the locks were not entirely closed on one case; one lock lay on the floor. That case held the black mamba.

In the hierarchy of terror that venomous snakes inspire, the black mamba lords over virtually every other species. It is routinely described as "the most feared snake of the African continent." Its unique and evolutionarily honed venom, containing potent neurotoxins, attacks the nervous system instantly; roughly stated, toxins in the venom short-circuit the body's ability to transmit nerve impulses—to the brain, to the heart, to the autonomic nerves that control breathing (asphyxiation is the typical cause of death). In a faintly hyperbolic departure from the normally constrained language of the scientific literature, a recent review paper in *Nature Reviews Chemistry* about snake venoms described the species as "very aggressive when threatened, extremely fast, intelligent and has highly toxic and fast-acting venom." Of its quickness, Raymond Ditmars, the legendary curator of reptiles at the Bronx Zoo, once likened it to "the shaft of a traveling arrow."

"Black mambas will bite you for no reason," said Cooper, who occasionally slithers into hyperbole. "They're extremely, extremely poisonous."

In their initial passages through the house, investigators only had the boyfriend's word that all the snakes, including the black mamba, were still secured in their enclosures. It fell to Cooper to confirm that the reptile was still in its cage. He now believes the woman in the bedroom, later identified as 54-year-old Aleta Stacey, barely made it to the bed after being bitten by the black mamba. Those dislodged picture frames along the wall? Cooper speculated that they were inadvertently disturbed when Stacey, venom spreading rapidly through her body, began to lose consciousness as she careened toward the bedroom. "She's dying," he thought to himself, "and she knows it."

The scene of the "unattended death" remained in tumult for hours. Upward of three dozen law enforcement responders milled about outside the house, trying to figure out what to do with dozens of venomous snakes in a quiet neighborhood. An emergency helicopter waited on standby at a nearby ball field. Investigators from the New York State Department of Environmental Conservation had to be called in; the agency prohibits keeping exotic wildlife, such as venomous snakes, in private homes without a license. Herpetologists from the Bronx Zoo agreed to come the following day to remove and relocate the reptiles. (Jacobi Hospital had been alerted because, as the closest among all hospitals in the New York metropolitan area to the nearby Bronx Zoo, its snakebite center can administer antivenom treatments for a wide range of exotic snakes.) A van from the coroner's office arrived around 9:00 p.m. Because of the snakes, the coroner at first refused to set foot in the house.

Investigators initially considered the possibility of foul play in the death of Aleta Stacey, according to Cooper, but also included "possible suicide" in their reports, and the final New York State DEC report suggested she took her own life. The autopsy findings, according to that report, noted that a large quantity of drugs was found in her system, along with the fang marks. The tell, for Cooper, was her behavior. "You've just gotten bitten by this incredibly poisonous, dangerous snake," he said. "You run outside. You scream, 'Help, help, I've been bitten by a snake!' You call 911. You call your boyfriend. You do anything and everything possible to let someone know. You don't go lay down in your well-made bed and die."

In a subsequent, mordantly melodramatic cable TV documentary, friends of Stacey recalled her being despondent in the days leading up to the mamba episode. "She was off her meds and depressed," SPCA officer Ross said. "It looks like she unlocked the case with the black mamba, opened the lid, and stuck in her arm. She was struck twice. She closed the lid, laid down in the bed, and passed." "She basically committed suicide by snake," Cooper said.

This is more than a sad, lurid, gothic, decade-old story that confirms the lethality of the black mamba; that fact is well known, even among people who don't think (or care) a lot about snakes. "Black Mamba," after all, was the nickname adopted by one of the greatest basketball players in NBA history, the late Kobe Bryant, as testament to his quickness, deadly accuracy as a shooter, and killer instinct in the waning minutes of big games. "Black Mamba" was the code name for Uma Thurman's character in the film *Kill Bill*. Black Mamba has become such an ubiquitous and domesticated marketing term signifying a sort of racy danger that it is the brand name for a line of high-end (all vegetative) accessories like women's purses, as well as the name of an Italian rock group.

The tragic Putnam Lake story started small, first appearing in local upstate newspapers like the *Daily Freeman* of Kingston, New York. Soon after, it popped up in the *New York Post*. Local CBS News picked it up, as did ABC News, whose story aired in Los Angeles, Houston, and Chicago. Reuters and United Press International beamed out dispatches all over the world. The story appeared in the *London Daily Mail* and the *Independent*, on Australian television, in a South African daily, and on a website in Ghana. It even made its way onto a right-wing MAGA website called Before It's News. Animal Planet aired a docudrama about the episode, called *Fatal Attraction: Snake Secrets*, in September 2012.

The World Health Organization estimates that up to 138,000 people a year die of snakebites, mostly in impoverished developing countries, but to many in the developed world, those are distant, anonymous catastrophes. Closer to home, popular culture thrives on these creepy, terrifying, and idiosyncratic stories that poke that excitable "fear module" and rev up collective antipathy toward snakes; they reinforce a fear of death that is, actuarily speaking, irrational (less than half a dozen people on average in the US die of snakebite each year, mostly because they refuse medical attention). Yet each horror tale reaffirms the Otherness of snakes as dangerous and unlikable and legitimately terrifying organisms.

But snakes are impresarios of paradox. They force us to consider two other aspects of the Putnam Lake story. Illegality and hoarding aside, more and more hobbyists have easier access to more snakes (venomous and otherwise) than ever before. Snakes as pets have become incredibly popular; in the decades prior to the Putnam Lake incident, the number of pet reptile owners in the US alone, according to some estimates, doubled to 4 million people, although lizards account for many of those pets. The market in "morphs"—snakes bred in captivity to have unusual coloration, such as tessellated corn snakes and albino ball pythons—has exploded. Animal dealers imported nearly 300,000 Indian and Burmese pythons into the US over the three decades ending in 2011, according to government estimates, and John Virata, the editor of *Reptiles* magazine, told me that there are probably many more than the estimated 400,000 owners of the single most popular species, ball pythons (*Python regius*). When beloved Michigan "reptile influencer" Brian Barczyk died in 2024, his mourners included more than 5 million followers on YouTube and more than 7 million on TikTok.

Snake tales like the Putnam Lake story; like the possibly apocryphal tale of the cuckolded Russian snake-handler who livestreamed his fatal black mamba bite on YouTube; like the fearless snake-handler at the 2022 rattlesnake roundup in Freer, Texas, who suffered a fatal bite in the shoulder; like the male resident in Charles County, Maryland, who died in 2022 from a bite by one of the snakes he kept in his everyman suburban home; like Grace Wiley, onetime curator of reptiles at the Brookfield Zoo, who died by cobra bite; like Karl Schmidt, the legendary herpetologist at the Field Museum in Chicago, who died by the bite of a boomslang; like the well-known herpetologist Joe Slowinski, who succumbed to the bite of a many-banded krait on a field trip to remote northern Myanmar (a tragedy that inspired a book)—all these grisly stories with unhappy endings find a home in the popular imagination because they ride that razor-thin, morbid but irresistible line between dread and fascination. Referring to the case of a man in the Midwest who was bitten by one of the 100 snakes he kept at home, Chris Cooper told me, "As he was dying, he was on Facebook telling people how he got bit. Now 911 is not the first call you make."

One of the venomous snakes removed from the Putnam Lake house the following day was a pretty pit viper common to the American Southwest known as a black-tailed rattlesnake (*Crotalus molossus*). Blacktails don't boast of the same horror quotient as cobras and mambas, but they are stunningly beautiful creatures—usually with dark two-tone diamond-shaped patterns etched with Dürer-like crispness on a yellow-orange or tawny background, as precise and geometric a weaving as anything Nature (or the indigenous Apache, who have shared their habitat in the Southwest with the rattlesnakes for millennia) has created. The crisp skein of bright diamonds ends in the namesake black tail. This species is the protagonist of a very different snake tale, a story also a little scary that left not a trace in the daily news.

~

Harry Greene had what he calls his "gumption trap" moment about 30 years ago in the mountainous terrain of southeastern Arizona. The term comes from Robert Pirsig's bestselling 1974 book *Zen and the Art of Motorcycle Maintenance*. "That's when you have an incorrect starting assumption that keeps you from making progress," Greene told me. His incorrect assumption was that a rattlesnake implanted with a radio transmitter that he was tracking should have been down on the ground, beneath the branches of a juniper tree, right there in front of him. It was, he was about to discover, much closer.

Greene, a herpetologist who served on the faculties of the University of California at Berkeley and Cornell University for decades, is a name largely unknown to the general public. But his 1997 opus *Snakes: The Evolution of Mystery in Nature* resides on the bookshelf of virtually every living herpetologist and holds as well a cherished place in the libraries of hobbyists and of people who love to hunt and keep reptiles. It is at once a coffee-table book of exquisite photographs of rare serpents and a scholarly love letter to this most reviled class of animals.

In *Snakes* and in his ruminative 2013 memoir *Tracks and Shadows*, Greene lays legitimate claim to being the poet laureate of contemporary herpetology—a biologist with deep, hard-earned scientific knowledge about snakes along with a naturalist's sensibility in appreciating them and a writer's gift for describing the world where these creatures intersect with us, in all its wonder and tragedy and humbling complexity. In *Tracks and Shadows*, his field descriptions (forget the animals) settle into a blunt, wondrous lyricism. Of a field trip to the Mojave Desert, he wrote, "Everything is weathered. The air *tastes* clean. As afternoon shadows deepen, distant ranges turn flat lavender gray. Creatures that perish here dry quickly, minus the tariffs of scavengers, and drift off on the wind."

The son of a US Army Air Corps serviceman, Greene moved around often with his family, and one senses that nature became the real constant in his life. He has described himself as an "academic gone a bit feral," but in truth Greene spent a fair amount of time on the wild side before ever entering the academy, as a former ambulance driver, mortician's assistant, army medic, self-described mediocre student, and someone who has had to process death, in strangers and children as well as lovers, up close and personal. He also draws on a broad range of cultural nourishment, citing the lyrics of country singer Rita Hosking in one paper, snagging a blurb from novelist Jim Harrison, invoking the poetry of Robinson Jeffers in a recent essay, and talking snakes over dinner with Norman (*A River Runs Through It*) Maclean. Most of all, he is a fanatic evangelist for the worldwide cult of evidence-based knowledge known as science, not to mention an excellent field biologist.

It was in that last capacity, and perhaps cashing in on his roguish celebrity, that Greene called out his guild about a decade ago and criticized what most of the larger scientific community believed for many, many years: that snakes were "unworthy" of serious scientific study. Beginning in the 1980s, and especially in the last 10 or 15 years, Greene's argument that studying snakes could be revelatory has gone from minority pipe dream to eye-popping reality. His shadow falls on many chapters in this book: as scientist, mentor, raconteur, thinker, commentator, contrarian, inspirational figure, and a profanely exuberant lover of snakes.

At the time Greene started out in science in the 1970s, snakes were considered cold, asocial, uncaring, unthinking, uninteresting animals. And rattlesnakes? Nothing but malevolence with scales. But Greene earned his PhD at the University of Tennessee under the guidance of Gordon Burghardt, an early apostle of reptile ethology, so he was already groomed to appreciate behavior, even in supposedly asocial animals. And Greene ventured into field studies during the first burst of transformative technology: the implantable radio transmitter. "I was in Berkeley teaching field natural history and herpetology, in a place where there were rattlesnakes still on campus," he told me. "And telemetry was starting to just emerge, so all that came together. I mean, holy shit, I might get to actually watch rattlesnakes *do* something!"

He had initially started his fieldwork in Costa Rica studying bushmasters, the world's largest vipers (some reach eight feet in length). He was mainly interested in their feeding biology and defensive behavior. After this initial work, Greene gave a talk about bushmasters at a scientific meeting, and a stranger named David L. Hardy Sr. came up to him afterward. "I'm a physician," Hardy told him. "I study snakebite as a hobby. I've published some papers about it. And I'd like to help pay for your research and be your field assistant." Greene didn't exactly warm to the idea of research support, but he told Hardy, "I'll let you come along once and we'll see how it works out." It worked out quite well. "He was, like me, a military brat, and we both were crazy about snakes," Greene said. That was the beginning of a 20-year collaboration that revolutionized scientific thinking about snake cognition and behavior.

Hardy, an anesthesiologist with a prominent practice in Tucson, had a summer home in the Chiricahua Mountains in southeast Arizona. Greene visited for the first time in the summer of 1985 and fell in love with the place. "And here were these *magnificent*, absolutely mind-blowingly beautiful black-tailed rattlesnakes," Greene recalled. "And we could *find* them." That may seem like an obscure point, but it is one of the reasons that the bulk of scientific research on snakes has focused on mainly two groups of serpents: garter snakes and rattlesnakes. Why? Because of all snake species (at least in North America), they are the ones most likely to spend time above-ground and out in the open, where they can more easily be found and studied. The Arizona rattlesnakes also tolerated the new telemetry surgery "extremely well," Greene said, so the focus of his research over the next two decades shifted to black-tailed rattlesnakes.

In those early technological days, the radio transmitters were about the size and shape of half a human index finger, and they were inserted manually down the rattlesnakes' throats, without anesthesia. "We just smeared them up with mineral oil and take a pair of forceps and insert them into the gullet—carefully, of course," Greene explained. Once the device was past the rim of the mouth (and fangs), Greene and Hardy would use their fingers to palpate it down the snake's body until it lodged in the stomach. The transmitters would last between one and four weeks before the snakes would defecate them out. Later on, with a more elaborate technique, they surgically implanted radio transmitters about two-thirds of the way down the length of the snakes and then threaded a 12-inch wire just underneath the skin, turning the snake's entire body into a mobile, slithering antenna. "It really transformed snake biology, I think," Greene said. "That just changed our lives."

Beginning in 1988, Greene and Hardy began to track a population of black-tailed rattlesnakes that inhabited a small field site in southeasternmost Arizona, northwest of the town of Portal and, as Greene once wrote, "literally and figuratively on the road to Paradise." The two-mile site followed Silver Creek, flanked to the north and south by mountains. This remote terrain is very hospitable to snakes in general (more than 30 species inhabit the area), and to black-tailed rattlers in particular. Greene and Hardy thought this field site offered a unique opportunity to observe snake behavior in the wild. With the radio transmitters, the researchers could pinpoint the location of dozens of telemetered snakes, trace their movements over a 24-hour cycle, and begin to parse out the otherwise occult behavior of these secretive reptiles.

They used one snake to find others. "Number one was a male," Greene said. "We found a blacktail across the road. We caught him. We put a radio in him, we let him go. Ten days later, he's courting this female. She becomes number two. Three weeks later, another male has a try at her. He becomes number three. Eventually we had 50 blacktails with radios in them for periods of up to 12 years." The most profound impact of this technological trick was that for the first time in the wild, herpetologists could identify a radio-tagged snake as an *individual*, with all that individuality implies: behavior, temperament, personality, habit, predilection, idiosyncrasy, life cycles, life rhythms. They recorded meals, defecations, courtships, copulations, pregnancies, birth sites, hunting ranges, peregrinations, winter refuges.

Over nearly two decades, Hardy and Greene recorded more than 5,000 interactions with the rattlesnakes in this particular area. They numbered the snakes. They watched them court and give birth. They watched the snakes lie in ambush, waiting patiently for hours to strike their typical prey, white-throated woodrats and rock squirrels. And what they saw changed the way people thought about snakes. On one occasion, Greene and Hardy were observing a blacktail through their binoculars as it took up an ambush position outside a chipmunk's lair. They couldn't believe their eyes when they saw the snake use its head to tamp down a fern plant in order to have clearer sight lines to prey that might not appear for another week or so. Snakes weren't supposed to be able to see well, yet here they were, peering into the future.

And then there was the Sunday morning when Hardy called Greene in Berkeley, his voice trembling, reporting social behaviors previously unimaginable. "You're not going to believe what I just saw," Hardy said.

"Well, tell me!" Greene blurted.

"Well, Female 21..." he began.

Female 21 was one of the most closely observed (and beloved) blacktails in the study site, and at the time she was pregnant. "I've been going out to her gestation site every day," Hardy continued, "and sneaked up on her—and she's sitting with babies. She's not abandoning them! They're all coiled up on top of her. And when I got a little too close, they crawled in the burrow behind her, and she rattled and backed in after them." Hardly anyone had suggested rattlesnake maternalism—seemingly a naturalistic oxymoron if ever there was one—much less documented it before.

Greene's reaction? "Holy shit! What we expected was, she gives birth and she crawls away. At that point, she hasn't eaten in almost a year. But she doesn't crawl away." When Hardy went back the next day, "She's sitting there again. The babies are on top of her. And they're tongue-flicking each other." What they had observed, in more than a one-off context for the first time in rattlesnakes, was maternal care and postnatal protection. It didn't last long. As soon as the pups shed their skin for the first time, about 10 days after birth, they all dispersed. But it was the first hint at much more complex social behavior than anyone had imagined.

"Harry Greene was the first person to really start doing this—standing back and watching the snakes instead of just going and catching them," said Emily Taylor, a professor at California Polytechnic State University. "You stand back and watch them, because that's when you can really see some of these cool things." In the early 2000s, Melissa Amarello and Jeff Smith went on to show that rattlesnake moms use "babysitters" with their broods, and Taylor, a former student of Greene's at Berkeley, has popularized these findings by live-streaming rattlesnake dens. "They have these complicated social systems," she said, "with friends that they like to hang out with and other snakes that they like to avoid, and they live for decades, and the mothers take care of the babies, and they seem to show these cognitive processes that we wouldn't associate with them initially. It's only come to light recently because of Harry."

But back to Female 21 and the "gumption trap." As Pirsig had presciently written, if you are stuck with a false premise, "you can fail to see the real answer even when it's staring you right in the face..."

Greene was searching the ground beneath the juniper tree for Female 21 because he'd been getting confusing signals on his radio receiver. The snake should have been right there in front of him, on the ground in the leaf litter beneath the tree. But he didn't see anything resembling a snake. He bent down to scan the ground for telltale snake scales. Still nothing. Greene kept searching for a glimpse of skin and scales in the brush under the tree. Still no trace of his favorite snake.

"I stupidly thought—stupidly because I actually had no reason to think this, I had never seen a blacktail buried in the dust and the leaf litter under a tree. But I couldn't see her. So I was like, *Shit, where is she?* I crawled around under this low-hanging juniper tree several times, hoping to get a sign from this particular patch of litter. And all I was getting was confusing directionality. So I sat up. Now I'm sitting on the ground. I'm sitting among the overhanging branches of this juniper tree." That's when he suddenly remembered the "gumption trap" passage from *Zen and the Art of Motorcycle Maintenance*: having an incorrect assumption that keeps you from seeing what's staring you right in the face.

"As I'm sitting there thinking about Robert Pirsig, I have this vague sense that there's something really close to my left eye," Greene recalled. That's when he realized that Female 21 was not on the ground but perched on a branch of the juniper three feet off the ground. How close? "Inches. Inches from my face. She doesn't tongue-flick. She doesn't rattle. She doesn't pull her head back as if to strike. Doesn't move."

Greene started "quietly lowering myself" and backing away from the tree. "I think there's a pretty good chance I missed what would have been a horrible bite, because I had not taught Female 21 that I was a threat," he said. "She's the one we had 12 years of encounters with." Indeed, Greene and Hardy had caught her five times over that period, anesthetized her with gas in the field, and surgically implanted or replaced her radio transmitter (including once while she was gestating). Five times! And Female 21's reaction? "I don't think I ever even saw her rattle in 12 years," Greene told me.

Now no one would have been surprised if this encounter had had a different ending. In addition to the surgeries, the researchers had logged nearly 600 separate interactions with her in the wild. But there was another, less acknowledged aspect of Greene's early research that may have made a difference (and influenced the field, too). During the entire study, Greene and Hardy had conscientiously refrained from applying some of the age-old—and often brusque—techniques of snake handling. Unlike in the days of his Texas youth, Greene did not "pin" Female 21—mash her neck down with a metal hook and let her writhe until he got a tight manual grip behind her head. The main predators of snakes, such as raptors, specifically attacked the same location, at the back of the head. Thinking like a snake, Greene and Hardy convinced themselves that the blacktails would experience pinning as the attack of a predator, which would elicit the only reaction these animals knew: a defensive strike at the antagonist, fangs unfurled like avenging spears. During their many years of observation, Greene and Hardy handled the snakes with as much gentleness as care allowed and got to know these animals intimately as individuals. "There's no way to prove this," he told me, "but I think it's at least possible this could have ended a lot differently if I had five times mashed that snake's head against the ground with a snake hook and picked her up while she was thrashing—you know, essentially taught her that I was a threat. Maybe I didn't teach her anything, but at least I didn't teach her I was a threat."

Fate was not quite as kind to Greene's wife, Kelly Zamudio, a respected herpetologist on the faculty of the University of Texas–Austin. On a morning in 2002, while hunting lizards in Arizona, she accidentally stepped on a black-tailed rattlesnake and suffered a severe bite on her left shin. In a paper co-authored with Hardy, she described a 12-day hospital stay, multiple surgeries (including the removal of approximately two-thirds of the muscle around her tibia because of necrosis), and a dogged recovery that saw Zamudio running marathons within two years of the incapacitating envenomation.

Unlike the broadcast bombast of the internet, with its periodic and predictably titillating tales about venomous snakes, Greene and Hardy's long-term observational studies of the Chiricahua blacktails barely left a whisper in the scientific literature, much less in the mainstream media. They boiled down 15 years of field observations into a 26-page chapter in a 2002 book titled *Biology of the Vipers*, published by a relatively small academic press in Utah. No matter. The research profoundly changed scientific, and ultimately public, perception of snakes in general, rattlesnakes in particular. In his memoir, Greene said he and Hardy "especially relished familiarity" with individual snakes, who went about their business in the wild. "The blacktails often seemed aware of us," he noted, "but typically paused only briefly, rarely rattling, and resumed activity." In other words, these humans believed they had—could have—*relationships* with these snakes. I asked Greene if it was possible that Female 21 specifically recognized him.

"Maybe," he replied, a little discomfited by the implications of the question. "I mean, if you'd asked me that question [back then], I probably would have more decisively said, *Probably not*. But knowing what I know now that other people have discovered and that we discovered?" He paused to give a little laugh. "I would be careful about being too dogmatic about those things. I think it's entirely plausible." Entirely plausible, that is, that a non-captive pit viper in the wild recognized a specific human being and laid off biting him because she understood that she was not in danger.

Two snakes, two close encounters, two very different narratives. It will never be unwise to regard a venomous snake with care, distance, and respect, but Harry Greene has long believed that the public appreciation of serpents has been impeded, in part, by our "ignorance of their lives as individuals." As he and Hardy were among the first to show (and as many other scientists will tell you nowadays), snakes have temperaments, habits, individual talents, individual tics. And Greene believes that curiosity about their behavior rather than fear is essential to their preservation as a class of animals. As he once put it, "Natural historians transform curiosity into science and thereby help save species from extinction."

Harry Greene has seen too much of the real world to be a naive evangelist, but it's not hard to find a message in his sermon. By studying serpents with rigorous science, we not only learn to appreciate them, Other though they may be; we learn more about our own biology, about ways to improve our health, about how to titrate that bedeviling mix of fascination and fear, perhaps even how snakes might teach us to convert loathing into tolerance, respect, and benevolence—all of which are essential ingredients in the protection against extinction.

Snake Road
State Highway 254, El Dorado, Kansas

There is a rural road in Kansas that has a certain iconic notoriety among herpetologists and snake-lovers because it was the outdoor laboratory for one of the weirdest "experiments" ever to find a home (though just barely) in the scientific literature. It involved a black rubber snake, a short piece of garden hose, and dozens of unsuspecting but occasionally malevolent motorists.

In the summer of 1987, three researchers at Butler County Community College in El Dorado, Kansas, cooked up a field experiment to, in their words, "test if drivers would intentionally hit a snake" on the road. Over a six-week period in the summer and fall, the three researchers—William M. Langley, Hank W. Lipps, and John F. Theis—hid in the bushes alongside Kansas Highway 254 after they had placed either a black, serpentine-shaped rubber snake or a length of ordinary black rubber hose on a stretch of the two-lane rural highway northeast of Wichita. They strategically positioned these objects so that motorists who wanted to strike them had to deliberately deviate from their path.

The "data," minimal though it was, suggested a larger truth about society's antipathy toward snakes: Ordinary motorists went out of their way (literally) to try to express their dislike. Truck drivers, the researchers noted, seemed especially keen to squash the reptiles. Overall, when the rubber snake was positioned on the center dividing line, motorists consistently edged over to hit it (20 times out of 21 opportunities); when it was placed in the middle of the lane, they did the same thing (15 out of 20 times). At least along this patch of road running through the Kansas prairie, the human loathing of snakes translated into swerving, spur-of-the-moment serpenticide.

This anti-reptile road rage found secondary confirmation when the Kansas researchers returned to campus and questioned students at the community college on their feelings about snakes. "A survey of attitude of 364 college students toward hitting an animal crossing a road showed that both males and females chose to deliberately hit a snake more often than any other animal," the authors wrote. Their findings appeared in 1989 in the widely unread *Transactions of the Kansas Academy of Science*. (One herpetologist later told me that experiments of this sort had a hard time finding a home after peer reviewers learned that it was against the law in many places to deliberately place objects on a roadway that might impede or distract drivers.)

At one level, Langley et al. is a laudable—albeit amusingly modest—attempt to quantify hatred of snakes in a natural setting. In a larger sense, it is an almost ingenious way of exposing the irrational malevolence that snakes evoke in a large proportion of humankind. Is there any other creature in the animal kingdom that inspires motorists to swerve to kill rather than swerve to avoid? What is it about snakes that triggers such singular animus? For decades, that question has been pondered and parsed by scientists and academics, who have produced a truly titanic pile of studies that investigate snake fear and loathing at the psychological, neuroscientific, historical, cultural, and physiological levels. The general (and clinical) fear of snakes goes by the name ophidiophobia. You can find that fear and loathing everywhere, from the Bible to an MRI machine.

Several well-known snake scientists—Gordon Burghardt, James Murphy, David Chiszar, and Michael Hutchins—cited the Kansas highway experiment among dozens of studies in a detailed, richly documented 2009 essay that essentially asked: Why do so many people hate snakes, and what can be done to change attitudes? The questions are not abstract, because the answers predict the survival of snakes as individual animals, as endangered species, and as integral components of complex and fragile ecological webs.

First, outright fear or phobia. Where does it come from? Is it innate or is it acquired? Are we born with an intrinsic, almost genetic fear of snakes, or do we learn to loathe them? There has been a surprisingly huge amount of research trying to come up with answers to these questions. Israeli neuroscientists not long ago designed perhaps the ultimate claustrophobia nightmare experiment. In a study that might be subtitled "This is your brain on snakes," they placed people with confirmed snake phobias inside an MRI machine and then encouraged them to maneuver, with the push of a button, a live snake closer and closer to their heads as their brains were being imaged in real time. The ostensible aim of the MRI study was to observe "courage" at the neurological level as the subjects either overcame or succumbed to their fear—a fear the scientists quantified by measuring a standard physiological marker known as skin conductance (basically sweating). The findings revealed that people who were able to overcome their fear apparently did so by uncoupling the sheer physiological arousal of seeing a live snake (sweating, rapid heart rate) from the cognitive sense of fear lodged in the brain. The researchers even proposed several sites in the human brain (the anterior cingulate cortex in particular) that became active in individuals who managed to detach their excitability from their fear, while people who succumbed to their snake fear showed heightened activity in the brain structure known as the amygdala. It was as if there were two different neural roads leading to snake fear, and as neuroscientist Daniela Schiller put it, writing in *Scientific American*, the results "revealed an interesting dissociation between fear reactions, a sort of internal disagreement paving the way to courageous acts."

Did the brain light up out of fear of snakes or simply as a neural mechanism to *detect* snakes? And was this neural trace of ophidiophobia inborn or was it the product of learning? To this day, the answer remains muddled, but probably boils down to a little of both.

Just as the MRI scientists tried to dissociate physiological terror from cognitive fear, it's equally necessary to "dissociate" the fear of snakes from the loathing part. Where does that profound terror come from? There is limited but intriguing biological evidence for some sort of innate response, at least in primates, and some of the research goes back nearly two centuries. As Darwin described in *The Descent of Man*, his curiosity about the behavior of monkeys in the presence of snakes was inspired after he read descriptions of earlier experiments by the German zoologist and writer Alfred Brehm. Fascinated by Brehm's account of curious monkeys uncovering a basket containing snakes and recoiling in horror, Darwin took a stuffed snake to the monkey house of the London Zoo to see what kind of reaction it would trigger—"and the excitement thus caused was one of the most curious spectacles which I ever beheld." Impressed by the alarmed behavior and cries of danger in reaction to the stuffed snake, Darwin moved on to the experiment using a live snake in a paper bag, leading to his "satiating their horror" observation.

More recently, primatologists discovered that newborn vervet monkeys were able to verbalize a distinct, snake-specific alarm call in response to seeing a snake; in other words, they seem to have been born with the ability to recognize a snake as dangerous, before they'd ever seen one in the wild, as well as the ability to communicate that alarm to others. Moreover, vervet monkeys issue these alarm calls (known as chutter) only in response to *dangerous* snakes. Pythons, cobras, and mambas (all of which pose mortal dangers to monkeys) trigger the warning response; harmless snakes do not. As Burghardt and colleagues noted, "We can conclude that vervet monkeys have an innate capacity to recognize dangerous snakes and to behave appropriately in their presence, including the ability to warn conspecifics about the danger."

In 2001, Swedish psychologist Arne Öhman and Susan Mineka of Northwestern University showed that human adults were "prepared" to detect images of snakes on a computer screen more rapidly than neutral images like flowers, for example, supporting a hypothesis popularized by biologist E. O. Wilson that humans and other primates possess what he termed a "fear module" in the brain that becomes activated when visual cues (seeing a snake) are associated with danger. Subsequent research has shown that human toddlers as young as three years of age share some of these precocious snake-detection skills. This "preparedness hypothesis" remains controversial, but there is growing evidence that evolution may have privileged the ability to spot snakes in the environment among all other potential dangers. As Burghardt and his co-authors noted, "In Africa, where hominids evolved, venomous snakes are common and there are no simple rules for visually discriminating harmless from truly dangerous species. Thus, detecting and indiscriminately avoiding *all* snakes was probably favored by natural selection." These early psychological experiments popularized the notion of the "fear module" triggered by snakes. But it's a little more complicated than that pop characterization. Detection is not necessarily fear.

No one has explored this idea of natural selection "selecting" for the speedy detection of snakes more provocatively than Lynne A. Isbell, a primatologist at the University of California at Davis, who has proposed that the fear of snakes in deep evolutionary time, tens of millions of years ago, influenced the development of the brain—specifically the visual system—in early mammals, eventually including our ancestral primates. Her daring, deeply researched Snake Detection Theory, published in book form in 2009 as *The Fruit, the Tree and the Serpent*, blends neuroscientific, paleoanthropological, environmental, evolutionary, and biogeographical evidence to argue that perhaps 70 to 100 million years ago, mammals in Africa developed *two* visual pathways from the eyes to the brain—one that's very fast and subconscious and another that takes a little longer to reach the conscious visual part of the brain. She believes this early visual system evolved as a defense against early constricting snakes long before those ancestral mammals had to contend with their other two main predators, felids (lions and other large cats) and raptors. This two-part visual detection system, which was fine-tuned in early primates, allows rapid identification of something ("an object in the environment," as Isbell characterizes the perception), while the second proceeds to recognition ("Oh yes, that's a snake. And yes, that's deadly"). The difference in speed is only about 200 milliseconds, which sounds negligible. But as Isbell has pointed out,

"It doesn't take long for snakes to rear up and bite. They come really fast!" (A viper's strike, in fact, takes less than 500 milliseconds.)

What makes Isbell's theory so intriguing is her argument that this rapid, two-part visual detection system was modified and honed to a special degree in apes, Old World monkeys, and humans (so-called catarrhine primates), which evolved in Africa and Asia. The emergence of venomous snakes, which arose roughly 35 million years ago in Asia or Africa, greatly affected the evolution of anthropoid primates, ancestors of humans, because exceptional visual acuity was necessary for survival. By contrast, New World monkeys, which populate the landmasses of Central and South America, did not commonly confront venomous snakes until much later (roughly 18 million years ago) and thus have not developed as reliable a snake-detection system. In describing this long evolutionary process, Isbell essentially argued that snakes are the reason humans possess the most advanced visual system in the animal world.

As even Isbell has conceded, there's no way to prove the Snake Detection Theory, but it suggests hypotheses that can be experimentally tested. The majority of studies that have come out since her 2009 book, she's noted, "are consistent with or even supportive of the idea that snakes hold a privileged place in our visual systems." Less clear is whether that rapid, subconscious detection system represents, as some scientists have claimed, an innate "fear module" or not. Isbell's current thinking offers a slightly more nuanced interpretation: The fast subconscious detection of a snake, she believes, triggers an innate excitability or arousal—but not necessarily fear. "Fear is part of it," she told me, "but you don't have to feel the emotion of fear to detect snakes quickly."

So there is evidence of innate fear, or at least innate excitability, wired into the human brain, but not all fear is innate. It can be acquired, too. And the acquisition can be experiential, cultural, social, familial, even spiritual.

One of Harry Greene's most interesting later papers, co-authored with ethnographer Thomas N. Headland, documented the complicated ecological relationship between giant Asian serpents and the indigenous Agta Negritos people of the Philippines, pre-literary hunter-gatherers living in remote rain forest areas on the island of Luzon. Fully one-quarter of the 58 adult males canvassed by Headland had survived predatory attacks by reticulated pythons (*Malayopython reticulatus*), and there were six fatalities (four adults, two children) over four decades. What makes this ecological (not to say psychological) relationship particularly complex is that the humans eat pythons, pythons eat humans, and both compete for other food sources in the rain forest, such as deer, monkeys, and wild pigs. So the indigenous Agta Negritos experience the fear of being potential prey of snakes directly (victims of attack) but also as a communal, culturally transmitted fear. When six of your neighbors are killed by giant snakes, that's a pretty direct—and legitimate—form of learned fear.

Even if the human fear of snakes is innate and stretches back deep in evolutionary time, the most ancient cultures we know about also embraced snakes—as avatars of power, as embodiments of deities, as part of fundamental creation myths. As Isbell noted, snakes played a central role in the origin stories of early humans in the same places where she believes the primate visual system evolved its prodigious capacity, including in Africa, Southeast Asia, and Australia. One of the oldest human ritual sites known to science is located in the Tsodilo Hills, an isolated upland region of the Kalahari Desert in present-day Botswana, where in 2006 Norwegian archaeologist Sheila Coulson reported the discovery of a 6-by-18-foot rock resembling the head of a python in a remote cave. According to the creation myth of the indigenous San people native to that region, all of humankind derived from the python; the rock Coulson identified in Rhino Cave, with serpent-like eyes and mouth, bore hundreds of human-made indentations, and she reported that stone tools unearthed in the floor of the cave beneath the python-rock dated back roughly 70,000 years, making it among the oldest human ritual sites identified by archaeologists.

If those and similar findings hold up, ritual reverence for snakes was perhaps the baseline *cultural* emotion of modern humans. So how did loathing overtake it? The data in the Kansas highway study is about as soft as that rubber snake, but there are some weightier historical voices and some heftier numbers that substantiate the emergence of loathing. Carolus Linnaeus, the 18th-century taxonomist, famously slandered all of serpentdom when he wrote, "These foul and loathsome creatures are abhorrent because of their cold body, pale color, cartilaginous skeleton, filthy skin, fierce aspect, calculating eye, offensive smell, harsh voice, squalid habitation, and terrible venom." (Almost all those assertions are scientifically inaccurate, by the way, but in biology as in society, facts have no purchase against deeply ingrained biases.) For every Linnaeus, there is also a Catherine Cooper Hopley, who claimed the beauty and mystery of snakes belonged as much to poets and aesthetes as biologists and taxonomists. This yin–yang of judgment runs throughout both secular and religious literature, ranging from the censure of Linnaeus to the rapture of cults.

In 1965, the zoologist Desmond Morris (of *The Naked Ape* fame) and his wife, Ramona, co-authored a survey of this tortured historical relationship in a book called *Men and Snakes*. The book teems with breezy historical anecdotes chronicling the unique impact serpents have had on human culture, but its overall point can be boiled down to a single thought: "We are still reacting towards the snake as if we were a bunch of medieval peasants." In a more quantitative update on that sentiment, the Morrises conducted a vast survey of young visitors to the London Zoo in the 1960s in an attempt to measure public attitudes toward snakes. They asked children between the ages of 4 and 14 to indicate the animals they liked most and liked least at the zoo; after collecting 50,000 responses, they boiled down the replies to a random sample of 2,200 children and found that snakes ranked far and away as the most detested of animals. Nearly a third of all children (28 percent) indicated they disliked snakes (spiders, a distant second, came in at 10 percent).

Because snakes are so good at exposing human contradiction, it's worth remembering that analogous surveys have repeatedly shown that zoo visitors rate a visit to the reptile house as among their favorite destinations; like the monkeys observed by Darwin, we apparently enjoy satiating our horror. The overall physics of this relationship between humans and snakes seems to boil down to a kind of love–hate, attraction–repulsion, dread–infatuation oscillation that produces quantifiable effects. And the ability of snakes to catalyze these internalized tensions may hint at why, among all creatures, they manage to slither so effortlessly, and penetrate so deeply, into the human psyche.

A different form of acquired loathing—more structured, institutionalized, and dogmatic—emanates from the pulpit or bimah. Christian,

Jewish, and Muslim monotheism have all demonized snakes. It's not just that the Garden of Eden story, of paradise permanently lost, fingered the serpent as the guilty party; the guilty party received a sentence of eternal condemnation. "Before Adam and Eve sinned and ate from the Tree of Knowledge," opined online rabbi Dovid Rosenfeld, expressing a not-uncommon sentiment, "they had no inner desire to sin. Their 'evil inclination' was an external force, embodied in the Serpent...After Adam and Eve sinned, the desire for evil entered them and man became a mixture of good and evil."

If, as Rosenfeld suggested, the serpent planted the "seeds of decay within us," the entire human story of moral fallibility traces back to the snake. So great has the indignation of ancient sages been about this snake-based snookering of humankind's innate goodness that the Talmud, which generally forbids the killing of any animal on the Sabbath, does make an exception for snakes under certain circumstances. It is as if, to mix animalistic metaphors, snakes have become scapegoats, blamed for human failure, human weakness, human venality. Is it not possible that the human hatred of snakes might therefore be a displaced expression of self-loathing?

It is impossible to exaggerate the cultural impact of this religion-endorsed hatred of snakes. As religious scholar Elaine Pagels argued in *Adam, Eve, and the Serpent*, the early history of Christianity shows "how certain ideas—in particular, ideas concerning sexuality, moral freedom, and human value—took their definitive form during the first four centuries as interpretations of the Genesis creation stories, and how they have continued to affect our culture and everyone in it, Christian or not, ever since." Pagels was talking about intrinsic human values like equality, morality, and autonomy, but she could just as easily have been talking about serpents, and the idea that snakes were the root of all the problems.

And woe to those who contradicted the standard interpretations. The Ophites, a snake-loving gnostic cult in the early Christian Era, for example, radically recast the Garden of Eden narrative to suggest that God sent the serpent to Adam and Eve *wanting* them to eat from the Tree of Knowledge to show humans that "they contain true divine light if only they would look for it." As Burghardt and colleagues noted, the Ophites were "brutally suppressed" for their heresy. If art and culture also serve as forms of indirect instruction, the history of Western religious imagery from the 16th century on appears as one long, breathtakingly beautiful blockbuster show of museum propaganda retelling (and reinforcing) the Genesis story, including the message that snakes are cunning, crafty, and shrewd agents of Satan. In a cultural sense, they are apex moral predators. They expertly prey upon human vulnerability and spiritual frailty.

In ancient Greece, the cult of Asklepios (in Latin, Asclepius)—a demigod of healing long associated with serpents—predated the Hebrew Bible by centuries. But Emma and Ludwig Edelstein, the husband-and-wife scholarly team that attempted to track down every mention of Asclepius in the texts of antiquity, noted in the foreword to their massive 1945 collection of testimonies that "the worship of Asclepius, beyond its medical significance, came to play such a role in the religious life of later centuries that in the final stages of paganism, of all the genuinely Greek gods, Asclepius was judged the foremost antagonist of Christ." That stunning observation may help explain why Christianity—in its dogma, in its art, in its ruthless centuries-long suppression of pagan culture—justified a kind of spiritual extinction of the animal most associated with Christ's archrival.

Learning to hate snakes, however, need not come from the pulpit or scripture. From a purely practical and familial point of view, the herpetologists who do public outreach—visiting schools, letting children handle tame "ambassador snakes"—routinely describe anecdotal evidence of the way parents mentor their children in snake dislike. "It's not innate!" insisted Whit Gibbons, the well-known South Carolina naturalist and writer. He described visiting a group of students about 10 years old with their parents and passing around his ambassador snakes. "And I said, 'Okay, before we looked at all these snakes, how many of you were kind of scared of snakes or a little bit timid about them?' And most people were. And I said, 'How many of you now *like* snakes?' Nearly all of them put their hands up, except one. That was a little girl. She put her hand up, and her mother pulled her hand down. Does that tell you something about where our fear of snakes comes from?"

How to make sense of all this fear and loathing? Leave it to a poet to distill this complex psychological plumbing. In his poem "Snake," the British writer D. H. Lawrence deconstructed an impromptu encounter with a snake on a hot July day in Sicily, when he discovered the serpent drinking from "his" water well. Lawrence despised the creature; he hurled a piece of wood at it, chasing it away; and then he was overcome with empathy, shame, remorse…and understanding. He acknowledged that the serpent is a kind of king, "one of the lords of life," and that his all-too-common response to it was mere human pettiness. Some of the hotheaded poetic revulsion he expressed may have been innate, but as Lawrence made clear by the end of the poem, much of it was learned, as he "despised myself and the voices of my accursed education."

Ophidiophobia matters because it translates into loathing of snakes. Loathing makes violence against them acceptable. Snakes—real or rubber—get killed all the time. Loathing is not just a personal, emotional, "yuck factor" issue; it is a conservation issue. That is precisely the point upon which Gordon Burghardt and company focused their 2009 essay pleading for more snake benevolence: "The conservation of snakes is more difficult than for other vertebrate groups owing to the general bad reputation that snakes have in many regions of the world. They are loathed in ways that render rational discourse insufficient for their conservation." In lieu of "rational discourse," Burghardt and his colleagues ticked off a list of possible (although largely unsurprising) countermeasures to change public perception—more outreach, better zoo education programs, even a prototypical "Fear Zone" (tested at the Staten Island Zoo) to give people hands-on experiences with snakes to overcome their trepidations. But even back then, the clock was ticking.

"The most important question remains," they wrote 15 years ago, "can these approaches change the overall negative view of snakes held by humans quickly enough to make a difference in their protection and long-term survival? Sadly, we think the challenge is great and prognosis discouraging. But all is not lost…"

2

Snake Guys

~~~~~~

Herpeto-Sociology

O n a steamy, 90-degree day in the summer of 2022, I paid a visit to the World of Reptiles at the Bronx Zoo—a kind of obligatory pilgrimage for anyone interested not only in snakes, but also in the sociology and history of our *understanding* of snakes. The architecture of the building harks back to the days of pioneering zoological collections, with its Beaux Arts brick and terra-cotta facade, its Greek columns, and its frieze of reptilian heads (crocodilians, turtles, and lizards) poking out like cold-blooded gargoyles from the weather-worn cornices. Completed in 1899, it was among the first three animal houses built at the zoo and is now the oldest original structure. By look alone, the building endures as a temple to the glory days of serpents on public display at this legendary zoo.

When you step inside the darkened serpentarium, however, you're forced to confront the great paradox of serpent exposition. Zoos are the place where the vast majority of the public encounters living snakes, but those visitors don't see much. Two of the most electrifying traits of snakes are their graceful, quicksilver locomotion and their lightning-fast reflexes, especially when striking prey or provocateur. You see nothing of this virtuosic agility in zoo displays. Since time immemorial, snakes have lain in their enclosures like bumps on (or, in the case of the Bronx Zoo's king cobra on this warm day, *in*) the proverbial log—desultory, mostly inert, sedentary, unmoved to move. They are, in one sense, being true to who they are, because hours, days, sometimes weeks of serpent existence pass by in motionless ambush; many species restrict their activity to the hours of darkness; and almost all excel, as a Darwinian badge of honor, at their ability *not* to be seen. Still, for the most part when they are on display, they barely move. There was a famous scandal some years ago when the Houston Zoo replaced its specimen of a coral snake with a plastic model. "It was extremely realistic," Harry Greene told me, "and it was there for *years* before somebody went, 'That snake has not moved in three and a half years!'"

To be fair, the roster of serpents at the Bronx Zoo makes up one of the apex collections of reptiles in the zoological community—certainly in the US and probably throughout the world. And the stars are all there—the gorgeous Indian python (*Python molurus*), the murky and muscular green anaconda (*Eunectes murinus*), a mottled bushmaster (*Lachesis muta*), exotic and lesser-known rattlesnakes, the Asian cobra (*Naja naja*), a surprisingly chunky Ethiopian mountain viper (*Bitis parviocula*), vipers with horns, vipers with white lips, vipers with simply that malevolent triangle of skull housing slit eyes and heads tapered like the tip of a spear. There are also a few non-venomous entries, like a beautiful green tree python (*Morelia viridis*) from Australia and its South American doppelgänger, an emerald tree boa (*Corallus caninus*), posing a visual evolutionary question that most people rarely pause to ponder: How do animals separated by millions of years of evolution on continents divided by thousands of miles of ocean arrive at the same solution to the problem of life? The answer is genetics, but not the genetics of inheritance from a common ancestor. Rather, a more supple and improvisational genetics that relies on duplicating and tinkering with DNA over evolutionary time frames to converge on similar destinations from widely disparate points of departure.

Ever since zoological parks began the systematic display of snakes (London Zoo's pioneering reptile house opened in 1849), these emporiums of fear and wonder have emphasized the age-old metrics of Big, Deadly, and Rare. In his very first collection of essays, David Quammen (who yearned to be a herpetologist growing up before becoming one of our finest observers of the natural world) made the point that, when it comes to snakes, the human mind hungers for exaggeration. "Give us a huge snake," he wrote. "A monster, a serpent out at the far fringe of imaginability." We are obsessed with bigness.

For decades, the metric of reptile awe has been attached to the most predictable of superlatives—the longest (a reticulated python measuring 32.8 feet), the heaviest (a green anaconda, about 500 pounds), the longest fangs (the Gaboon viper of West Africa, with two-inch in-house hypodermics), the deadliest venom (the inland taipan of Australia). It was true in 1934 when Raymond Ditmars returned to the Bronx Zoo with "the deadliest and most treacherous New World reptile," a bushmaster caught in Trinidad, and it was true in the spring of 2022, when Ian Bartoszek and his colleagues at the Conservancy of Southwest Florida reported the capture of an 18-foot, 215-pound Burmese python outside Naples, Florida—the "most massive" python, a conservancy press release noted, ever caught in the state.

Zoo people know this. "The major challenge for the zoo herpe-tologist," one academic paper bluntly put it, "was to keep the viewer interested in an animal that was often immobile or difficult to see." But zoo people also know what pleases the crowds; press reports about that "deadly and treacherous" bushmaster captured by Dit-mars in 1934 produced an estimated 100,000-person bump in atten-dance at the Bronx Zoo. Awe and fear is the alternating current of sensation in the reptile house. And even the fear, for obvious rea-sons, is carefully curated and titrated. Every creature lies behind a thick pane of glass. Only in front of the huge, catastrophically neu-rotoxic king cobra (*Ophiophagus hannah*) at the Bronx Zoo is there a frisson of less-than-vicarious fear. The sign reads: DANGER. DO NOT TAP ON THE GLASS. WHAT WOULD YOU DO IF IT BROKE?

That warning hints at real, unstated danger—although the dan-ger is undoubtedly greater to the snake, banging its annoyed head against the barrier, than to the visitor. Still, in earlier days, when the Bronx Zoo put a spitting cobra (*Naja nigricollis*) on display, snake-handlers had to scrape venom off the glass every few days, so ferocious was the spray of deadly spittle from the antagonized ani-mal. Fast-forward to 2021, and a group of scientists (including Harry Greene) plumbed the hidden genetic history of spitting cobras, showing how three separate lineages of cobras, all of which arose independently in either Africa or Asia, evolved not only the same physiological equipment to expectorate venom but also the same behavioral trait of aiming that stream very precisely at the eyes of a potential predator. Convergent evolution affects not just coloration and camouflage but also anatomy and skill sets.

The Bronx Zoo is as good a place as any to ponder the pre-molecular, which is to say the "prehistoric" era of snake science and herpeto-sociology. "Snake studies," if you will, was a bracing mix of citizen science, Gold Rush–style specimen collection, gentleman naturalism, Barnum-esque public outreach, dismal animal husbandry, and unconscionable wildlife smuggling, with research conducted almost exclusively by men, beginning with Aristotle, who is credited with producing the first systematic scientific descriptions of these creatures. That is not to say there was no good science or good scientists in the 19th and 20th centuries. But it was an era of taxonomy without genetics, experiments without much statistical power (because there weren't enough snakes), compelling scientific questions without much interest from funding agencies, and an interrogation of nature without anyone other than white males asking the questions. It was the era of the Snake Guys, and all those threads extended then, as they still do today, into the world at large— into politics, societal bias, conservation, hobbyism.

Some of the species on display these days in the House of Reptiles—the timber rattlesnake, the bushmaster, the pythons, even the little tree frogs—make cameo appearances in this early history, too. And in history as in habitats, if you just follow the snake, it will lead you into all sorts of unexpected places.

～

Benjamin Franklin was a snake guy. In fact, one could argue that he is the quintessential snake guy of the entire American Experiment, not because he was a herpetologist or popularizer (although he clearly had more than superficial familiarity with reptilian behavior), but because he recognized in the temperament of a fiercely independent and self-protective serpent the perfect symbol for the nascent American identity. His tour-de-force rattlesnake metaphor on the eve of the Revolutionary War is well known. It wasn't, however, his first foray into literary allusions of a serpentine sort.

In the mid-18th century, England had the habit of exporting its criminal miscreants to North America, a policy the Crown cynically celebrated, as Walter Isaacson noted in *A Benjamin Franklin Reader*, "as a way to help the colonies grow." The colonists did not appreciate the help. On May 9, 1751, writing to the *Pennsylvania Gazette* under the pen name Americanus, Franklin dipped his rhetorical pen in an ink pot of serpent venom for the first time in a brief and witty letter called "Rattlesnakes for Felons." In a sly and indirect style of opinion writing that is as endangered these days as some rattlesnakes themselves, Franklin noted that venomous reptiles "we call rattle-snakes" populated the less inhabited precincts of the colonies; he likened the vipers to "felons-convict from the beginning of the world" and gently suggested that they shared a kindred outlaw spirit with the imported criminals. He went on to heap faux praise on Mother England for its magnanimous policy, in which "all the Newgates and dungeons in Britain are emptied into the colonies." Americans should be so grateful for British generosity, he continued, that he proposed a tongue-in-cheek exchange: sending North American rattlesnakes back to England as a token of appreciation. As Franklin noted (correctly), the snakes were "feeble, heavy, slow and easily taken" when they emerged from their dens in the spring. Thousands could be collected for a modest fee and sent as gifts to England.

"There I would propose to have them carefully distributed in St. James Park, in the spring-gardens and other places of pleasure about London; in the gardens of all the nobility and gentry throughout the nation; but particularly in the gardens of the *Prime Ministers*, the *Lords of trade*, and *Members of Parliament*; for to them we are *most particularly* obliged." (In the perverse and improbable ways that satire begets latter-day tragedy, Franklin's modest proposal was adapted for the purpose of real-life vengeance in 1978 when members of the

California cult Synanon placed a large, de-rattled rattlesnake in the mailbox of a lawyer suing the group, resulting in a near-fatal envenomation and hospitalization; Synanon's leader reportedly said, on audiotapes discovered during the ensuing police investigation, "Don't mess with us"—a latter-day sentiment, too, that has echoes of Revolutionary-era defiance.)

Young Ben Franklin's opinion of snakes in 1751 was not entirely positive, as is apparent in the concluding paragraph of the spicy dispatch from Americanus: "*Rattle-snakes* seem the most *suitable returns* for the *human serpents* sent us by our *mother* Country." And "human serpents" persists as an all-purpose political insult; as recently as the 2024 election season, for example, former vice president Mike Pence was likened to a snake by a wary voter. But in the two decades following the Americanus letter, images of a snake cropped up in many colonial publications, where it was understood to symbolize the need for unity. And Franklin's view, equating human miscreants with squamates, apparently evolved over subsequent decades, because by the time of the American Revolution, he made an astonishingly good case that the rattlesnake would have been a more fitting symbol of strength and independence for the United States than the bald eagle.

This second invocation of the snake as metaphor took shape in Philadelphia late in 1775. The colonists wanted to intercept British naval shipments of arms to their troops, and young men were being recruited in the city to form naval units to accomplish this task. Franklin happened to notice that the image of a timber rattlesnake had been painted on the drum of one of the newly recruited marines marching through town, along with the legend "Don't tread on me."

Soon thereafter, an anonymous contribution—signed "An American Guesser"—appeared in the pages of the *Pennsylvania Gazette*:

"The Rattlesnake as America's Symbol." Being the citizen-scientist he was, Franklin displayed a knowledge of reptile behavior in this much more famous essay, published on December 27, 1775. He understood that rattlesnakes (like all snakes) temperamentally prefer to avoid confrontation; their first line of defense is to slither off. But if perturbed or provoked, they will respond with a swift, painful, and potentially lethal response. Weaving that nugget of serpent behavior into political commentary, Franklin created one of the shrewdest metaphors in the history of American political rhetoric.

After noting that "the ancients considered the serpent as an emblem of wisdom, and in a certain attitude of endless duration," Franklin set out to enumerate all the distinguishing qualities of this particular animal. Familiar though the passage may be, it is worth pondering the political echo it still sends hurtling into recent times— something like the hum of the microwave background in our current political universe, echoing the Big Bang of America's creation. Here is what Franklin had to say about the timber rattlesnake:

> I recollected that her eye excelled in brightness, that of any other animal, and that she has no eye-lids. She may therefore be esteemed an emblem of vigilance. She never begins an attack, nor, when once engaged, ever surrenders: She is therefore an emblem of magnanimity and true courage. As if anxious to prevent all pretensions of quarrelling with her, the weapons with which nature has furnished her, she conceals in the roof of her mouth, so that, to those who are unacquainted with her, she appears to be a most defenseless animal; and even when those weapons are shown and extended for her defense, they appear weak and contemptible; but their wounds

however small, are decisive and fatal. Conscious of this, she never wounds till she has generously given notice, even to her enemy, and cautioned him against the danger of treading on her.

Was I wrong, Sir, in thinking this is a strong picture of the temper and conduct of America? The poison of her teeth is the necessary means of digesting her food, and at the same time is certain destruction to her enemies. This may be understood to intimate that those things which are destructive to our enemies, may be to us not only harmless, but absolutely necessary to our existence...

'Tis curious and amazing to observe how distinct and independent of each other the rattles of this animal are, and yet how firmly they are united together, so as never to be separated but by breaking them to pieces. One of those rattles singly, is incapable of producing sound, but the rising of thirteen together, is sufficient to alarm the boldest man living.

Brilliant both as metaphor and prophecy, Franklin's warning about "the danger of treading on her" echoed the by-then omnipresent image of the rattlesnake (including one engraved by Paul Revere) that for decades had appeared on numerous colonial broadsheets; specifying "thirteen" rattles, clearly referring to the 13 colonies, makes the political intent of the essay unmistakable. Almost simultaneous with the publication of Franklin's rattlesnake essay, a delegate to the Continental Congress named Christopher Gadsden created a flag that visually recapitulated Franklin's powerful text: It displayed the form of a coiled timber rattlesnake, fangs exposed and tongue aflicker, above the words DONT TREAD ON ME, upon a bright-yellow background. A native of South Carolina and a colonel in the Continental army, Gadsden presented the banner that same month to Esek Hopkins, the first commander in chief of the Continental navy, and it was actually used to signal naval maneuvers during the war. It has ever after been known as the Gadsden flag.

Franklin's rattlesnake metaphor—and Gadsden's flag imagery—has slithered through American history ever since. Nearly a dozen states have offered vanity license plates featuring the rattlesnake and DON'T TREAD ON ME legend. The right-wing Tea Party adopted the Gadsden flag as its banner in 2010. And that same yellow flag was waved by some participants in the January 6, 2021, assault on the US Capitol—a reminder that symbols can be powerful, but also perverted. The essence of the metaphor is that the snake never seeks confrontation or attacks its enemy. On that infamous day, it was as if Ben Franklin's rattlesnake had turned and bit the Republic itself in the ass. Meanwhile, in Gadsden's native South Carolina, the state's Department of Natural Resources now notes on its website that timber rattlesnakes "have been eliminated from much of the piedmont through the activities of people."

That is not the only instance of earlier herpetology sending a wrinkle into contemporary affairs. You might think that the widespread protests over the 2020 murder of George Floyd would be but a distant, muffled sound to the field-and-formaldehyde world of professional herpetology, but on June 18, 2020, the American Society of Ichthyologists and Herpetologists—one of the main professional organizations of scientists who study reptiles and amphibians—posted a brief, four-paragraph announcement on its website titled "Black Lives Matter." The note disclosed that the executive committee of ASIH had taken steps to change the name of the society's venerable journal *Copeia*, which had been in existence for more than a century. In defending the proposed name change, the journal's editor wrote that "a critical step is to recognize one of our own historic problems."

The name of the journal had no obvious connection to snakes or reptiles, but *Copeia* did pay homage to arguably the most prolific snake collector, species discoverer, and all-around taxonomist of the 19th century in the United States, a wealthy Philadelphia-born naturalist and paleontologist named Edward Drinker Cope. Cope left fingerprints all over the natural world, from the tiny tree frogs that bore his name to the massive bones of dinosaurs now in the collections of natural history museums. The problem, as the blog post circumspectly pointed out, was that Cope "held and published racist and misogynist views that our current membership finds abhorrent."

E. D. Cope, as even the contrite journal admitted, was a "titan" of herpetology. He was also a walking paradox—a pugnacious Quaker or, as one eulogist put it, a "militant paleontologist." Born in 1840, he published his first scientific paper at age 19 and described hundreds of new species during his career. With a broad forehead and prominent jaw, Cope was a prodigious, almost relentless field observer. Although he never finished the equivalent of high school, he published more than 1,300 scientific papers (some of which were criticized for the haste with which they were rushed into print), churned out 1,000-page monographs on vertebrate anatomy, and described 300 new species of reptiles, 300 new species of fish, and more than 1,000 fossils of vertebrate species.

During his long and often tumultuous career, he advocated an anatomically based view of evolution, highlighted the link between birds and dinosaurs, picked academic fights with relish, and turned 19th-century vertebrate taxonomy into a contact sport. He famously engaged in a rock 'em, sock 'em academic brawl with Othniel Charles Marsh, a scientific tiff known informally as the Bone Wars, over their fierce competition to unearth dinosaur fossils in the American West. At the time of his death in 1897, Cope was president-elect of the American Society for the Advancement of Science, forerunner of the country's most prestigious scientific society, the AAAS. He weighed in on everything from the taxonomy of mollusks to what he termed, in one essay, "the evolutionary significance of human character." It was in these late forays into philosophy and sociology that he got into latter-day trouble.

In a 1888 essay titled "The Relation of the Sexes to Government," for example, Cope argued that women lacked the physical and mental capacity of men and should not be allowed to vote because of their "inferior mental co-ordination, and greater emotional sensibility which interferes more or less with rational action." A Cope apologist, in a ham-handed attempt to clean up these messy assertions, later characterized Cope's view of women thusly: "They would vote through emotional suasion either with or against their husbands and lovers, he thought, and similarly in economics, although some women might be successful in business or professional undertakings, even they would be dependent upon capital produced by masculine effort."

Cope's views on race were even more extreme. The essays in "The Origin of the Fittest" are littered with references to "savages," along with Cope's assertion that the Negro and "Mongolian" races were clearly inferior to the Indo-European. But this racism reached a zenith in an 1890 essay in the journal *The Open Court*, where the anatomist argued that this inferiority was biological in origin and supported by scientific "evidence," specifically in the anatomical trait known as quadrumanous—a technical term indicating an animal's possession of opposable digits on both hands and feet, a character-istic of primates. "As every anatomist knows," Cope wrote, "quadru-manous features in all parts of the structure are far more frequently observed in the negro than in the white race." He went on to assert that Indo-European skulls were flexible enough to allow physical (and intellectual) growth of the brain throughout life, whereas the "sutures" of the Negro skull become fixed at an early age, "preclud-ing much further growth" of the brain.

Cope followed this tragic reasoning to an inevitably repulsive conclusion: the recommendation that all Negroes in the United States be sent back to Africa. "It appears to the writer," Cope, the president-elect of America's most prestigious scientific society, noted near the end of the essay, "that this is the only solution of a dan-ger that threatens, first, the purity of the race over the fairest part of our country; and second our political harmony, and perhaps even our national integrity." This project of mass racial deportation was "essential to our self-preservation."

There has always been a dark side to American excep-tionalism, and in Cope's case it is that you can be the greatest herpetologist-ichthyologist-paleontologist of the 19th century and also one of the most virulently racist scientists of your era at the same time. And this dark side of "genius" might well have never been acknowledged. As late as 1929, in a biographical memoir of Cope published by the National Academy of Sciences, Henry Fairfield Osborn wrote, "When we think more deeply of what really underlies human progress, we realize that only to a few men with the light of genius is it given to push the world's thought along, and that Edward Drinker Cope was one of these men."

Most mainstream 20th-century herpetologists remained oblivi-ous to, or untroubled by, Cope's problematic history, but the circles in which he traveled, and was admired, overlapped with the most prominent zoological institutions and scientific leaders of his era. Osborn, Cope's eulogizer-apologist, served as curator of vertebrate zoology at the American Museum of Natural History, was instrumental in launching (and staffing) the Bronx Zoo, and mentored a young snake enthusiast named Raymond Ditmars, who ultimately became the inaugural curator of herpetology at the zoo in 1899. In a sense, Cope's sociological shadow fell across the field of herpetology for the better part of a century—even more if you consider that his name graced the field's major academic journal right into the 21st century, where it collided with the national anguish over the murder of George Floyd.

It took someone outside the prevailing demographic of herpetology to set in motion the events that ultimately led the editors of *Copeia* to jettison Cope's affiliation with the journal's name. The decision to change the name happened very quickly, over roughly a three-day period shortly after Floyd's death, but that sudden change came after a 30-year gestation period. It began in the early 1990s, when Tamekia Wakefield, a black student at Rhodes College in Memphis, encountered some of Cope's racist writings in a non-science class. "Wow, that's messed up," she remembers thinking. Later, while taking a comparative anatomy course at Rhodes as part of her pre-med curriculum, Wakefield was surprised when her professor, Alan Jaslow, extolled Cope's scientific contributions during a lecture. Jaslow, a well-known herpetologist, had studied "Cope's tree frog" early in his research career and considered him "The Man" in 19th-century American science. Wakefield worked up the nerve to approach Jaslow after the class and inform him of Cope's racist writings. Like so many herpetologists, Jaslow had been unaware of this part of Cope's oeuvre. He immediately and permanently dropped mention of Cope from his lectures. "I think it was more significant for him than for me," Wakefield told me.

Shocked by Wakefield's revelation, Jaslow informally brought up the issue of a name change with at least one mentor who belonged to the American Society of Ichthyologists and Herpetologists in the 1990s. "The suggestion was seen as laughable," he noted later, and "not ever a thing that would happen." Still, the issue bugged him, and in June 2020, he had an opening to revive it again.

On June 12, the Diversity, Equity, Inclusion, and Belonging committee of the herpetological society, organized by mostly younger scientists, issued a statement condemning Floyd's death. At the time, the committee and a majority of the society's membership remained unaware of Cope's problematic views. Unaware, that is, until three days later, when Jaslow emailed to the committee a note petitioning the society to change the name of its journal. Jaslow recounted how, nearly three decades earlier, Wakefield "told me it was hurtful to hear my praise of Cope and that no doubt I was ignorant of his miss-use [sic] of science and authority, which I was."

Jaslow's email shocked the diversity committee. "Finding out about Cope, it was like there's only one decision to be made here, which is we have to change the name of the journal," said Rockwell Parker, a biologist at James Madison University and one of the organizers of the diversity committee, who wrote an account of the name change in the newly christened *Ichthyology & Herpetology*. "How do you tell a black person to join your society when an abject racist is the moniker of your flagship journal?" Although Parker said a portion of the membership resisted the change, there was a certain poetic—even naturalistic—justice to the name change. Just as snakes have thrived for 130 million years or so precisely because of their diversity, so, too, has recent snake science.

~

When the reptile house opened at the Bronx Zoo in 1899, two years after Cope's death, among the very first residents of the new building were timber rattlesnakes—the most symbolic of American animals. That is because a New York City teenager, Raymond Ditmars, had collected them personally during forays into the backwoods north of the city, years before he was named—at the urging of one of E. D. Cope's most fervent admirers—the first curator of reptiles at the zoo.

To people who know and love snakes, Ditmars is a self-taught, multi-talented demigod; to outsiders, the name barely rings a bell. He is easily the most influential (not to say beloved) herpetologist of the early 20th century. Just as Einstein spawned an entire generation of 20th-century physicists, Ditmars—through his animal husbandry, his research, his public persona, and his many books— spawned an entire generation of 20th-century herpetologists. Of all the many anecdotes about Ditmars, however, the one that sticks with me the most is a story from Dan Eatherley's comprehensive biography *Bushmaster*, in which he describes how Ditmars strapped a caged king cobra to the hood of his car and drove around in freezing weather for 15 miles to anesthetize the ailing animal before treating an eye infection. The story captures the enormous empathy Ditmars felt for snakes, but also the lack of knowledge and tools in his era to fully empower that compassion.

Ditmars became the most celebrated "snake man" of the early 20th century. *The New Yorker* profiled him—in 1928. Edward Steichen took his formal photograph for the National Portrait Gallery. His exploits were chronicled not only in the *Proceedings of the New York Zoological Society* but also in *Photoplay* magazine and the *New York Sun*. One of his early animal films premiered at the Brooklyn

Academy of Music. He tirelessly performed in what are now called public outreach sessions, but his lectures often offered a bit more frisson, like the time a cobra got loose in a public gathering. When he set off on a collecting expedition to South Carolina or Central America, tabloid newspaper reporters in New York crowded the pier or airport to document his departure. Long before Marlin Perkins and Steve Irwin appeared on a television screen, Ditmars pioneered some of the earliest nature cinema in the 20th century. His books—notably *Snakes of the World*—achieved mammoth sales. If he had been born 25 years later, he would have been a regular on the Johnny Carson show instead of Jim Fowler; if he had been born 50 years later, he would have had his own series on Animal Planet or the Discovery Channel.

Born in 1876 in Newark, New Jersey, his roots reached deeply back into the era of Edward Drinker Cope. The *L* of his middle initial stood for Lee, as in Robert E. Lee; his father served in the Confederate army during the Civil War and tried to groom his son for a military career. Raymond Ditmars's early childhood was closer to the end of the War Between the States than to the invention of the automobile or the airplane. Like Cope, he never attended college nor earned an advanced degree (he attended a military academy instead of high school); like Cope, he was a dedicated autodidact. But he had the "Madness"—that deep, inexplicable, almost uncontrollable love of snakes.

Similar tales of Ditmars's reptile madness are well known, not least because he recounted many himself in *Snakes of the World, Thrills of a Naturalist's Quest,* and *Strange Animals I Have Known*. As a teenager, after his family moved to Harlem and then the Bronx, he prowled the northernmost corners of Central Park in search of snakes (when there were still snakes to find there) and peppered snake-handlers with endless questions at the Central Park Menagerie (forerunner of the Central Park Zoo). He founded the Harlem Zoological Society when he was 16 and established a private herpetarium in the attic of the family home that would raise the blood pressure of any parent. The enclosures in the house on Bathgate Avenue held an eastern diamondback rattlesnake, cottonmouths, copperheads, timber rattlesnakes, fer-de-lances, and even a bushmaster—all venomous, all potentially deadly. As author Dan Eatherley noted in his biography of Ditmars, "He seemed to have an uncanny knack for handling venomous snakes." That was part of the Madness.

Perhaps the most surprising aspect of Ditmars's career path may help explain how he became the greatest 20th-century popularizer of snake culture: He was a journalist before he was a naturalist. His last job prior to taking on a curatorial role at the fledgling Bronx Zoo was as a court reporter for the *New York Times*. By day, he would chronicle trials and legal tussles in the downtown Court of General Sessions; by night, he would give public lectures about snakes, including the display of live specimens, which he stored in the city coroner's office while he was off reporting. He learned very early on how to read a room: In his first public talk, he recognized the excitement in his audience. "I could feel it," he later said. Even though he worked for little more than a year at the *Times*, he displayed a flair for storytelling, explanation, popular communication, and crowd-pleasing show-and-tell skills that infused his lifelong work as the public face of the serpent world. As his reputation grew during the early years of the 20th century, he assumed the role of a public celebrity.

There is a photograph of Ditmars from around 1913 that shows a dapper, dark-haired young man with an extravagant, broad-bar mustache, narrow face, and deep-set eyes, wearing a high-peaked white hat and a light suit with a short dark tie; as an older, established zoo curator, he was frequently photographed milking a deadly cobra or cottonmouth while wearing a three-piece suit and tie, with cuff links. When a boxful of imported cobras got loose in the basement of a Greenwich Village store, the local tabloids reported Ditmars's heroic capture of the venomous serpents. When he died in 1942, the *New York Times* obituary acknowledged that the newspaper itself had been the "accidental means" of launching him on his career in herpetology, referring to an assignment interviewing the director of the just-opened Bronx Zoo that resulted in a job offer. "Literally millions will note his passing with a sense of paternal loss," the obituary concluded. "As the champion of the lowliest, most detested, least understood forms of animal life he did more, single-handed, than any writer of his generation to convince a shuddering world that reptiles have their place in the scheme of nature."

Some eight decades later, Ditmars's impact on the public perception of snakes remains strong, but the legacy has become a bit more complicated. In 2015, Eatherley, a British filmmaker, published his richly detailed and unabashedly reverential biography of Ditmars (the full title is *Bushmaster: Raymond Ditmars and the Hunt for the World's Largest Viper*). The book reads like a project originally conceived as a two-pronged documentary film, pairing a biography of Ditmars with a quest by the author to find a bushmaster in the wild. But reading between the lines of Eatherley's narrative, you can't help but sense that the patron saint of 20th-century snake culture possessed at least a few qualities that have not aged well.

Eatherley describes episodes where Ditmars and colleagues used a bamboo stick to shove food down the throat of a python that refused to eat—a form of force-feeding widespread in the zoo world at that time, but now known to risk the fracture of vertebrae and ribs, essentially killing the animals. In a 1938 episode that would violate every ethical tenet of contemporary zoo practice, the zoo's curator of reptiles and amphibians was returning from a collecting trip in Trinidad with a dozen rare frogs known locally as mountain chickens (*Leptodactylus fallax*) when he decided to treat his fellow travelers to a meal of the animals (it "wasn't Ditmars's finest hour," even Eatherley concedes; the frogs are now endangered).

There were also incidents that, in retrospect, would have liability lawyers quaking. In an iron-horse prequel to *Snakes on a Plane*, Ditmars was transporting a bunch of rattlesnakes on a train trip down to Washington, DC, when the animals escaped their cage and, unbeknownst to other passengers, circulated underfoot in the railcar. While filming uncaged venomous snakes in his home studio in Scarsdale, a spitting cobra launched itself at his camera operator, who happened to be his wife, Clara. Ditmars claimed never to have been bitten by a venomous snake, but his steely discipline in snake handling created a lasting anatomical anomaly—his right hand was enormously enlarged, to the point of appearing swollen or injured, because of the firm grip with which he held venomous snakes behind the head.

In some of his pioneering nature films, Ditmars also crossed what would now be considered an ethical line. In a short film called *Killing the Killer*, he staged a fight to the death between a cobra and a mongoose. In *The Jungle Circus*, he posed frogs on dollhouse-size diving boards and glued the wings of a fly to a chair to create the illusion of a weight-lifting, bench-pressing insect. Some of these popularizations veered into stunts. As curator of mammals at the zoo, he sometimes dressed up orangutans in human clothes and sat them down at a table with eating utensils; at another event, staged for the press, he played Beethoven's Ninth Symphony to the zoo's king cobra while recruiting the writer Edgar Lee Masters to write a poem about it. As herpetologist William Lamar told Eatherley, "Ditmars's showmanship invariably led to some excessive or inaccurate characterizations."

Like so many celebrities, Ditmars discovered that public adulation and overexposure can cut both ways. It is hard for 21st-century imaginations to wrap around the extent to which Ditmars's collection expeditions in the 1920s and 1930s aroused so much public expectation and, on the rebound, so much public disappointment. When he arrived back in New York empty-handed from a 1932 trip to Panama in search of a bushmaster for the Bronx Zoo, for example, local newspapers from the *New York Times* to the *Brooklyn Eagle* greeted his return with headlines like "Ditmars Fails in Snake Hunt" and "Curator Returns Without Big Snake." Rarely has a naturalist's fieldwork been graded on such a tough curve.

Still, you rarely hear a discouraging word from contemporary snake experts about Ditmars. "With regard to what he actually *did*, I think it would be wrong to demonize Ditmars" is the way one respected herpetologist approached a conversation about the Ditmars legacy. There is little question that his books animated public interest in reptiles during an age when zoos were cultural temples of curated wildness. And for an autodidact who never attended college, Ditmars combined keen observational skills, genuine passion, and biological humility to become a top-notch naturalist.

For all his evident goodwill and expertise, Ditmars perpetuated a larger sociological legacy that permeated the society of snake-lovers for decades—a culture that Ditmars himself once likened to a "sort of freemasonry." Like that ancient fraternal organization, herpetology was a tight-knit, overwhelmingly male guild that honored tradition, celebrated toughness, and did not welcome women. In a tiny bit of ephemera that speaks volumes about male domination, professional condescension, and amateur acumen, a herpetological journal reproduced a letter from Grace Olive Wiley, a librarian in Minneapolis in the 1920s, to Ditmars in which she claimed to have successfully bred rattlesnakes at home. Ditmars responded with a curt, two-sentence letter dripping with disinterest—even though, as Ditmars conceded, it may have been the first successful example of captive rattlesnake breeding ever reported in the United States.

~

As popular as Ditmars and his contemporaries made snakes in the cultural imagination, what often gets overlooked is how substandard the care of animals was in the generic reptile house for decades. Roger Conant, one of the preeminent herpetologists of the 20th century and author of the most popular guide to reptiles and amphibians in North America, lamented the treatment of herps during a period that essentially coincided with the zenith of Ditmars's career. "Our efforts of half a century ago," Conant wrote in 1980, "seem crude in the extreme. In those days we often considered ourselves lucky if a majority of our animals lived a year or two, and what little captive breeding occurred was fortuitous, not planned." Massive loss of snake populations in zoos due to disease was not uncommon, and much of the herpetoculture, to put it bluntly, was ignorant. It was just a more respectable form of serpenticide, as some zoo officials now acknowledge.

Joseph Mendelson III, director of research at Zoo Atlanta, is fond of teasing audiences at public lectures by displaying an old photograph of snake-handlers at the London Zoo. The photograph shows six burly men standing outdoors, struggling to hold a gigantic python while the curator attempts to shove a rabbit down the snake's throat with a stick.

"What aren't you noticing in this picture?" Mendelson asks his audience. After a pause, he answers his own question.

"They're force-feeding a python," he says. "One, that tells you that the python wasn't eating on its own, which is the python saying, *You're not keeping me in proper circumstances*. And the other thing, which is really subtle, is that all the men are wearing huge heavy overcoats. It's *way* too cold for that snake. And by force-feeding it and putting food in its stomach, that's a death knell right there. It can't digest that rabbit. The rabbit is going to decompose, and the snake is going to die. They were just so fundamentally misguided on the biology of these animals that, in their attempts to help, they were actually making it a lot worse. It's just sheer ignorance." They did the same sort of force-feeding at the Bronx Zoo. Not only would Raymond Ditmars force-feed fasting snakes, he would invite local journalists in to document the event. On occasion, the zoo even invited members of the public to participate in these force-feedings—for a fee. If the snake died, they would go out and get another one. And that became another part of the problem.

"In the best zoos in the world in the 1930s, there were no breeding programs," Mendelson said. "When your tiger died, you ordered another one from the animal dealer, and it came straight from the wild. How is that not poaching? The animal just simply dies in a different place and under a different circumstance, but the end result is exactly the same. Zoos, as popularizers of wildlife, were exploitive consumers of wildlife. That was Ditmars's era, and surely with no second thoughts whatsoever, that's what he did. You put on a show for people."

The inability to breed snakes (or any other wild animal) in captivity created a parallel, decades-long problem. The market for exotic reptiles in zoos begat an underground of shadowy and then outright illegal animal trafficking. Jennie Erin Smith's 2011 book *Stolen World* vividly depicts the cynical commerce in reptiles and amphibians that resulted in prison sentences for the smugglers, shocking mortality of smuggled animals, and embarrassing resignations of several prominent zoo curators. Among the zoos exposed as dealing with reptile smugglers were the San Diego Zoo and the Knoxville Zoo. Illegal wildlife trade might have even indirectly led to the fugitive Burmese python that slithered into the Florida wilds sometime in the 1970s and forever changed the ecology of the Everglades.

Although the Bronx Zoo remains justifiably famous for its House of Reptiles, the names of other zoos crop up often when the revolution in the care and display of reptiles is discussed, beginning in the 1960s and 1970s, including Zoo Atlanta, the Dallas Zoo, and the Smithsonian Institution's National Zoological Park in Washington, DC. It is probably no accident that James B. Murphy worked at all three.

Murphy, now retired, became infected with what he coined the Madness at an early age. When his family lived in the Chicago suburb of Arlington Heights in the 1950s, Murphy's huge home collection of snakes was so notorious that the local police and fire departments sent circulars to their staffs warning against entering the Murphy residence, and the local utility company refused to allow its meter readers to set foot in the house. As a student at Xavier University in Cincinnati, Murphy kept boa constrictors and pythons in his dorm room, and his periodic pet feedings became so popular that the college newspaper published the schedule; hundreds of students would attend. Despite all that, Murphy's professors discouraged him from majoring in biology, so he ended up studying philosophy and English literature, which served him well in his later career as a zoo curator, reptile popularizer, historian of herpetology (and snake art), and self-described "bon vivant." He can tell you which country has the longest history of herpetoculture (Germany, going back 400 years) and which visual artist most accurately depicted snakes (either Albertus Seba or Ferdinando Sordelli). Most important, he is widely credited with popularizing innovations in husbandry that dragged zoo treatment of snakes and other reptiles into the era of modern science, with an impact that has spilled over (and benefited) the people who keep snakes as pets.

After working briefly as a herpetology keeper at Zoo Atlanta, Murphy moved to the Dallas Zoo in 1966, where he served as a keeper and curator for 30 years and championed a series of innovative techniques in the care of captive snakes. As improbable as it may seem, despite centuries of amateur and professional herpetoculture, reptile curators didn't even begin to understand how to create the optimal conditions for reproduction, not to say survival, of captive snakes until the 1970s and 1980s.

The clues came together piecemeal. Joe Laszlo of the San Antonio Zoo harangued his fellow herpetologists that captive snakes bred better with carefully orchestrated seasonal fluctuations in temperature, the kind that mirrored seasonal cycles in nature. Harry Greene, then at the University of California at Berkeley, noticed that a previously reclusive California mountain king snake (*Lampropeltis zonata*) he kept in the lab suddenly got more active, indeed almost frisky, when it was exposed to ultraviolet light; research soon showed that UVB was essential to healthy immune function in reptiles, and by the 1980s, ultraviolet lighting rigs became a fixture in hobbyist cages. Closer observation of snakes in the wild inspired zoo curators to more realistically simulate their natural environments, including temperature changes that mimicked circadian cycles and thermal gradients that provided warmer and cooler spots within an enclosure. And at the Dallas Zoo, Murphy and colleagues "discovered" that venomous snakes could easily be induced to crawl into plastic tubes, dramatically reducing the danger of handling them.

Murphy not only incorporated and popularized these ideas but also led pioneering efforts in captive breeding. The Dallas Zoo discovered that something as simple as creating larger enclosures enhanced courtship and reproduction. "We tried that with several different species, and it seemed to make a real difference," Murphy recalled. "It allows the time frame and the space for them to basically strut their stuff. We got much better at looking at behavior, and much better at looking at things like temperature." The only problem, he conceded, is that most reptile houses in most zoos had nowhere to grow and thus no way to increase the size of the enclosures for their snakes. "Nobody has been successful in the zoo business at building something twice as big or three times as big as they're currently building them," he said. "They should be including larger enclosures."

In 1996, Murphy joined the staff of the Smithsonian Institution as a research associate, where he worked with Dale Marcellini, then-curator of reptiles at the National Zoo, on an ambitious plan to reinvent reptile house displays at modern zoos. In 1988, Marcellini led a study that quantified the same sort of visitor ennui that I experienced at the Bronx Zoo decades later: Reptile house visitors spent only eight minutes total in the building, the research revealed, and a mere eight seconds on average looking at each display.

The study sent "shock waves throughout the profession," according to Murphy, and led to a large National Science Foundation–funded study at three institutions (Zoo Atlanta, the Dallas Zoo, and the National Zoo) to reimagine the reptile house. The curators came up with plans to increase interactive displays, and even tested the prototype of a hands-on-snakes visitor experience at the Staten Island Zoo. Despite the massive improvements in place by the 2000s, the philosopher in Murphy stepped back and posed an inconvenient question to his zoo profession colleagues. "Enrichment has become one of the newest trends in the zoo world," he wrote in 2016, "but I'm not certain how we measure whether our changes to an animal's living space and activities actually improves their lives or whether it just makes us feel better."

That kind of institutional introspection was not always part of herpetoculture's worldview, and that in turn might reflect the sociological bandwidth of the people who dominated the field for centuries. Throughout the long history of snake culture, there has been a pattern, and you don't need a PhD to detect it: Aristotle, Franklin, Cope, Osborn, Ditmars, Conant, Murphy, Laszlo, Greene, Marcellini, and so many others. All snake guys. It was as if snake expertise, like an unusual form of snake reproduction itself, was a kind of parthenogenetic process that produced only single-gendered offspring. And much like the specialized chemosensory world of the animals under study, there was a distinct musk hovering over the professional world of snake study: male, often macho (especially when it came to venomous species), fraternal, daring, sometimes reckless, trained to see certain things in the biology and natural history of snakes, but perhaps doomed to overlook others. As one professional herpetologist said of his own subculture, "There are different cliques in biology, and herpetology is more like a motorcycle gang. We're like a bunch of ruffians who drink and work hard and play hard."

The overwhelmingly male cast of herpetology was hardly unique in 20th-century science, but it perhaps persisted a little longer and privileged risk-taking a little more than did other branches of biology. Neither Murphy nor others could name a single female head curator of reptiles at a major American zoo in the 20th century, with the short-lived, ill-fated exception of Grace Olive Wiley's stint at the Brookfield Zoo. And that sensibility created a vibe that even a curator of reptiles like Dale Marcellini couldn't ignore. "Snake guys," he told me recently, "are really kind of weird."

Two relatively recent developments combined to revolutionize the way scientists study snakes. One was the advent of new tools like radio tracking and molecular biology, which changed the kinds of questions that scientists could ask. The other big change was *who* asked those questions.

———

In the summer of 2019, Emily Taylor was delivering the plenary lecture at a popular scientific conference in Rodeo, New Mexico, known as Biology of the Pit Vipers when she paused at the end and looked out at the audience. Taylor, who runs the Physiological Ecology of Reptiles Lab at California Polytechnic State University and specializes in rattlesnake physiology, had composed her talk as "basically an ode to pit vipers" (the formal title was Anatomy of a Pitviper: How These Iconic Animals Inspire Art, Fear, Respect and Knowledge). As she was wrapping up, she veered off topic to describe some still-unpublished research she and her students had compiled on the proportion of female authors in scientific studies of reptiles and amphibians. The good news was that the number was rising; the bad news was that the proportion was still very low. As she finished her talk, she paused, leaned into her microphone, and implored all the female scientists in the audience to come up onto the stage and join her. Chairs rustled throughout the auditorium.

The very first thing Taylor said to me, on the first of many times we have spoken, was: "Just to let you know, I get really excited when

I talk about snakes, so sometimes I talk too much." She was making up for lost time, because she had never even touched a snake until she was practically in her 20s.

Taylor grew up a military brat; her father served in the navy, and the family moved every couple of years, living in places like Italy and Hawaii. A self-described tomboy, she saw herself as ambitious, athletic, competitive, and occasionally impulsive as a young person. She wasn't much interested in nature, either, spending virtually all her free time through high school and college on the soccer field. When she entered the University of California at Berkeley in 1994, she majored in English literature while planning to become a doctor or veterinarian. But as part of her pre-med training, she looked for some "fun" biology credits, and someone urged her to take a famous Berkeley course called IB-104 (Natural History of the Vertebrates), taught by a guy named Harry Greene. On a class field trip, Greene handed her a California king snake (*Lampropeltis californiae*) that he found under a rock. Taylor's reaction was instantaneous. "Omigod," she thought, "this is the coolest thing I've ever seen in my entire life!"

Not long after, Greene introduced Taylor to rattlesnakes, which turned out to be the perfect serpentine doppelgänger to Taylor's brash, gung-ho personality. In trying to explain the attraction, Taylor struggled to find the right words. "They were almost kind of forbidden, you know? They were dangerous. And there weren't any women working with rattlesnakes, or very few. I had the kind of personality that made that attractive to me." She relished traipsing out in the field to find them. They were ambush predators, meaning they remained aboveground, coiled in a particular spot, waiting for prey to come along, and thus were easier to locate than a lot of other snakes. And, as Taylor would soon discover, they had very cool social behaviors. "That's why I fell in love with rattlesnakes."

Soon after, Taylor admitted, she became "one of the boys" in the very male world of rattlesnake researchers and decided to pursue a PhD in biology at Arizona State University.

When Taylor applied to graduate school in 1998, she recalled, "There was no option to study with a woman. I applied to four labs. All of them were headed by men." This was not an unusual state of affairs in academia in general but was especially true of herpetology. You could almost count the prominent female snake scientists of the previous century on one hand—allowing for several missing digits due to snakebite. One was Catherine Cooper Hopley, a self-taught English naturalist, whose 1882 book *Snakes: Curiosities and Wonders of Serpent Life* is an attentive and scientifically detailed celebration of the habits and behaviors of snakes. Helen Thompson Gaige helped make the University of Michigan a powerhouse of global herpetology with her studies on frogs (the university's museum of zoology now houses a staggering 70,000 snake specimens, believed to be the largest collection in the world) and her longtime stint as the editor of *Copeia*. Mimi Halpern did pioneering work on the chemosensory neuroanatomy of snakes in the 1970s at, of all places, an urban medical center, SUNY-Downstate in Brooklyn. And that list doesn't even include the checkered career of Grace Olive Wiley, who became curator of reptiles at Chicago's Brookfield Zoo in 1933, in part because the zoo apparently envied her private collection of 330 snakes (representing 115 species), and perhaps also her early success at captive breeding of rattlesnakes. Infamous for her belief that venomous snakes could be tamed and free-handled, she was fired after allowing 19 escapes of snakes over a two-year period.

Early in her career, you could barely tell the difference between Taylor and the traditional Snake Guys; she admitted as much herself. "I was very hotheaded back then, and I was very much one of those people who was 'one of the guys.'" That was about the only way to muscle into the rattlesnake science crowd. "I really wanted to work in that field," she continued. "And I think in a way my sort of confident, loud, assertive demeanor was helpful, because that tends to help women in male-dominant fields." But she ultimately outdid her mentor and isn't shy about saying it. "Let me tell you, I have tracked orders of magnitude more snakes than Harry Greene," she said. "I've spent hours and hours and hours of my entire adult life in the field with rattlesnakes. And I will 100 percent tell you, they absolutely have personalities. They are utter individuals."

In the style of Greene, Taylor lay back and just observed the snakes, but she did him one better. Inspired by another female rattlesnake enthusiast, Melissa Amarello, executive director of Advocates for Snake Preservation (ASP), who had placed time-lapse cameras on rattlesnakes in Arizona, Taylor and colleagues trained time-lapse cameras, and later video cameras, on rattlesnake dens in Colorado and Central California, then recruited 6,000 citizen-scientists to help analyze the data. In the latest iteration of the technology, Taylor can manipulate the camera of her current "Project RattleCam" channel from anywhere in the world and interact with observers in real time. In May 2024, she and colleagues began livestreaming from a second RattleCam trained on a high-altitude den in Colorado inhabited by about 2,000 rattlesnakes.

Despite the dearth of female mentors, Emily Taylor earned a PhD at Arizona State University in 2005, established herself as one of the world's foremost researchers on rattlesnake physiology, and has become a force of nature in contemporary herpetology—not just because of her science, and not just because by 2022 she had become president of the American Society of Ichthyologists and Herpetologists, but because of what might be called her sociological muckraking and her guerrilla approach to public outreach on behalf of snakes of all stripes. In addition to her citizen-science projects, she and Amarello organized a letter-writing campaign by school-children, pleading with residents of Sweetwater, Texas, to stop massacring serpents at their annual rattlesnake roundup.

But she never stopped being hotheaded about the lack of diversity in snake studies. As she prepared her plenary talk for the 2019 meeting, Taylor recalled an earlier Biology of the Snakes conference in 2017 at the same New Mexico venue. All of the invited speakers were male; of the other "contributing" speakers like Taylor, only seven were women out of more than 50 scientists. She stewed over that imbalance for two years, especially whenever she stared at a photo of the small female cohort from the 2017 meeting. In fact, when she returned to San Luis Obispo, she and her students at Cal Poly embarked on that systematic study of the number of women in herpetology. They tracked the authorship of papers dating back nearly half a century. Not that the data was surprising, but it put quantitative meat on the bones of a long-standing problem. In 1970, only about 1 in 10 authors of peer-reviewed papers about reptiles and amphibians were women; by the decade ending in 2019, the number had risen, but was still only about one in three, with much of the female authorship focused on amphibians, not snakes.

Which is why Taylor decided to pull her "stunt" at the 2019 Biology of the Pit Vipers meeting. After she finished her ode to pit vipers, she "asked all the women herpetologists in the audience— I mean, it was a pit viper conference, so most of them were snake biologists, but broadly speaking herpetologists—I asked them to all come up onstage," she said. An estimated 100 or so female scientists flowed to the front of the room; someone memorialized the moment with a photo. "So many of us!" Taylor later told me. "Not all were speakers, but many were. It was a powerful moment because everyone could see how quickly things are changing in the sciences, even in a male-dominated field."

Sheer numbers do not a qualitative change make, and by most accounts, the Snake Guy vibe of herpetology, both professional and amateur, is still alive and well in the field, on Facebook, among vendors at snake expos, and in the authorship of papers. But Emily Taylor's path through this field is more than a dutiful celebration of diversity and change. Like snakes themselves, the people studying them these days are not just more diverse than ever before; they are posing different scientific questions and using diverse new tools to discover new insights about these ancient animals. All those lowly and loathsome descriptors snakes have borne since their expulsion from the Garden—venomous, cold-blooded, unperceptive, asocial, limbless, and, most recently, invasive—have, upon closer scientific scrutiny, revealed remarkable qualities. Snakes are much more sophisticated, and snake science is much more revelatory, than almost anyone imagined a generation ago. And to use a word that crops up repeatedly in conversations with the people who study them, snakes are "cool."

In an interview that appeared in the *Washington Post* nearly a century ago, a well-known snake-lover named Ditmars made an interesting, prophetic prediction, pointing out that snake hunting was "far more interesting" than fishing, adding, "There is not only the danger and excitement of the chase but the captives remain living things to be observed and studied, and, believe me, the creatures are well worth all the attention scientists give them."

Those were the marching orders, as it were, for contemporary snake science. The speaker, by the way, was not Raymond Ditmars, but Clara Ditmars, wife of the quintessential Snake Guy.

Snake Road

Interstate 78, Hamburg, Pennsylvania

There is an interstate highway in eastern Pennsylvania that leads to one of the most famous emporiums in snake hobbyist culture: the modern reptile exposition. To be sure, snake expositions with buyers and sellers are not a new phenomenon. Reptile shows date back at least two centuries in Europe and nearly 130 years in the US, including a famous 1898 exposition at the Grand Central Palace in New York City. But reptile shows have become an omnipresent feature of American pet culture for at least three decades, a fixture on social media, and a market for surprisingly pricey custom-bred snakes.

At its inaugural event in 1988, the Hamburg (Pennsylvania) Reptile Expo mustered a mere 16 vendors; when I visited the October 2022 edition, more than 200 vendors crowded a huge fieldhouse, just off the main street in Hamburg and minutes from the interstate (as of 2024, the show had moved down the road to Morgantown, Pennsylvania). And they weren't just selling snakes. The live merchandise included turtles, lizards, frogs, and scorpions; the inanimate merchandise included all the sophisticated infrastructure of modern herpetoculture (terrariums, sure, but also frozen rodents, variable heating pads, organic tree-bark hideouts, coconut husk ground cover, and other environmental paraphernalia). The founders of the show were two local snake guys from Hamburg, Walt Loose and Troy Reppert. The person running the show now?

That would be the woman with the spiky purple hair collecting receipts at the door of what is widely considered the premier expo in the Northeast. Denise Readinger is the sister of David Hess, who bought the business from the original owners in the 1990s; she took over the operation following her brother's death in 2007. "When my brother first put on the expo," Denise told me, "the people who came were exactly what you'd expect—people with tattoos, all freaky looking. Now it's much more family-oriented, not just the freaky reptile crowd. People use snakes for emotional support animals as much as fluffy puppies." As if to punctuate that point several months later, Transportation Security Administration officials confiscated an "emotional-support" boa constrictor in the carry-on bag of a woman boarding a plane at Tampa International Airport in December 2022. It's not easy to hide a four-foot creature with 200 vertebrae and ribs from an X-ray machine.

The range of out-of-state license plates in the jammed parking lot—New York, Connecticut, Delaware, New Jersey, Maryland, Massachusetts, Vermont, New Hampshire, Virginia, and South Carolina—testifies to the far-flung appeal of these shows. Yes, you still see lots of tattoos, lots of untamed beards, lots of black T-shirts; Chris Cooper, the Putnam Lake snake guy, told me the typical herper demographic is "males with tattoos wearing a Black Sabbath / Metallica T-shirt—but they're also very smart." You're likely to see a little sideshow dissonance, too, like the guy selling F**K BIDEN baseball caps. But there are also lots of children, lots of families, lots of young people, lots of women. And you see a lot of exotic reptiles.

You can buy a captive-bred boa constrictor (up to $1,000) or a venomous Gaboon viper native to Africa (only $75)—Hamburg is famous in part for allowing vendors to sell venomous species. You can buy a rock python or a sidewinder rattlesnake, baby corn snakes or a black mamba. You can stock up on pre-packed, frozen meal kits of "medium"-size mice (50 count for $40) or rats to feed your pet snake, or mealworms or other insects to feed that $1,200 tokay gecko. The snakes, hundreds of them, lie in stacks of clear-plastic deli cups. The venomous species have red tape and clips around the rims of the containers; expo organizers advise that the venomous snakes are for the "advanced hobbyist."

This vast fieldhouse might seem like an unusual place to think about snake conservation, but there has been an argument at least since the 1990s fueled by the slogan *Conservation thru Commercialization*. The idea is that if you can breed rare species in captivity, for example, and sell the animals in expos or online, you don't have to poach them in the wild; if you can get the animals to procreate in your basement, you're ruining the market for smugglers and preserving populations in their native environment.

Before captive breeders learned the ropes, a rogue's gallery of wildlife smugglers raced into this marketplace vacuum. In her bravura job of behind-the-scenes reporting in *Stolen World*, Jennie Erin Smith captured this illegal subculture in all its gory amoral detail. One after another, she described episodes of traffickers smuggling baby pythons in cassette tape containers, vipers stuffed in sneakers, snakes concealed in clothing. She described how federal agents, setting up reptile sting operations, arrested a number of smugglers for violating the global CITES treaty, which outlaws the import of endangered foreign species. Several protagonists in her tale, including Henry Molt and Tom Crutchfield, received jail sentences for illegal wildlife trafficking—but not before they had extensive dealings not only with well-to-do private collectors, but also with some of the largest zoos in the world. Even trade in non-endangered species became incredibly sophisticated. Major US breeders of ball pythons, for example, pooled resources to provision the largest exporter of reptiles in West Africa with a digital camera, laptop computer, and satellite phone; when snake hunters in Africa caught a particularly unusual mutant, they would take a photo, email it to the stateside breeders from the field, and ship the reptile within an hour of capture in the wild.

The black market in exotic snakes was created in part by biological ignorance, because most zoos struggled in their captive breeding efforts. But self-taught snake nerds rushed into the breach. Beginning around the 1980s, private collectors arguably knew as much about reptile husbandry as zoo curators. More important, they began to learn how to get reptiles to reproduce in captivity, reducing the market forces that drove much (but not all) of the illegal reptile trade. In that respect, many of the vendors are the progeny, philosophically and sociologically, of Ditmars; he never attended college, and to hear some of the vendors tell it, they, too, didn't need to take a single biology course to gain empirical expertise in the art of breeding a beautiful snake.

As a result, the present-day reptile expo is an especially impressive Technicolor bazaar of DIY captive breeding, a vast inventory of limbless semi-GMOs, where patient matchmaking creates a kind of living, slithering jewelry. Breeders turn mutants and accidents of nature into money, creating albino snakes, snakes with pinkish or greenish tints, snakes with dazzling textile-like patterns. Self-taught captive breeders can be heard dropping serious genetic jargon like "codons" and "double-dominants" as they describe their efforts to create the unusual color patterns prized by collectors—"morphs," in the trade—by mating specific animals in their basements or garages. Many of them consult the website "Genetic Wizard," which guides amateur snake breeders in figuring out the type of crosses they need to create to produce the desired coloration patterns. Their skill in captive breeding clearly reflects this new era of citizen science and a democratization of expertise via the internet.

And high-end captive breeders have become rock-star entrepreneurs; as a recent *New Yorker* article documented, rare genetic mutants can carry eye-popping price tags, such as the "leucistic" ball python described by writer Rebecca Giggs that sold to a Belgian collector for $200,000. The products of all these genetic manipulations create an eerie echo of the 19th-century "pigeon fancy" in Europe, where self-taught breeders, including Darwin himself, created exotic, extravagantly peculiar, undeniably beautiful, and utterly unfit mutants through careful matings. That may turn out to be true of the snake morphs, too; herpetologists report hearing anecdotal accounts of cognitive deficits and other ailments in these hothouse genetic confections.

There is little question that amateur captive breeding has diminished the smuggling trade. "There really has been a sea change," Zoo Atlanta's Joe Mendelson said, crediting the role that home breeding by amateurs has played in significantly reducing the catching (and selling) of snakes in the wild. "And while I think that absolutely was a threat a few decades ago, for certain species in certain areas, the captive breeding industry has pretty much obviated that."

"Having the latest new wild-caught species in the reptile hobby is really no longer the case," John Virata, editor of *Reptiles* magazine, told me. "It is now those with the latest designer morph who get all the accolades." Commercialization has also modified behavior in the field. Many modern-day herpers, Mendelson said, have graduated to digital catch and release: they nab their snake in the wild, take a selfie, post it on social media, let the snake go, and then order a pet on the internet. One industry analysis estimates that 75 percent of reptiles on the pet market are now captive-bred.

The commercialization part of the equation is undeniable. The total value of the reptile pet business in the United States alone is $1.5 billion, according to a Gitnux Marketdata Report at the end of 2023, fueled by the approximately 4.5 million households that possess a pet reptile. But the conservation part is still a little murky, which adverts to some background tension that still exists between amateur snake enthusiasts and professional snake scientists.

There's a sharp divide, for example, over practices like picking up and handling venomous species with one's bare hands (free-handling), hoarding large numbers of snakes at home, and breeding some of the more exotic color patterns despite the possible genetic deficits that might come with such extreme breeding. In the age of the internet, Emily Taylor told me, there are Facebook groups showing snake enthusiasts (usually young men) free-handling venomous snakes—no plastic tubes for them. In that sense, Taylor is no longer one of the boys. "There's a lot of people out there—most of them are not scientists—who do free-handle their snakes," she said. "I'm in a Facebook group called Crotalus. It started out as a group of scientists and field herpers who are kind of responsible types, who would share field photography and stories with one another. And then it got basically taken over by these...frankly, just these people who are hoarding snakes. Who are going out and just catching snakes, and showing themselves free-handling the snakes. Basically, if someone is free-handling a venomous snake, any respectable snake keeper or herper or scientist is going to think that person is irresponsible. There's a solid line. Like, free-handling is it. You do not cross that line, ever."

For their part, captive breeders wonder aloud why their creations can't be used to repopulate stressed snake populations in the wild. At a reptile expo in New York State, a breeder from Connecticut displayed a beautiful array of California garter snakes in clear plastic deli cups—slender neonates with a mosaic of green and blue blotches accented by a striking red stripe. Garter snakes are thought to be ordinary and uninteresting, but the rare California species—and the even rarer, endangered San Francisco garter snake (*Thamnophis sirtalis tetrataenia*)—are considered among the most beautiful snakes in the world. Tom Romano argued that captive breeders could help conservation efforts by releasing their home-bred animals. But, he added, environmental regulators prohibit the release of endangered captive-bred species into the wild. "They won't let us do it," he said.

One of the reasons they won't let them do it is because of a controversy over what might be called genetic integrity. Can captive breeders produce the complete genome sequences of their home-made creations, and do those sequences more or less match species in the wild? If not, captive breeders might, in the interest of conservation of species, accidentally or inadvertently or indifferently flood the natural environment with human-made knockoffs that might pollute, biologically speaking, the natural gene pool. And captive-bred snakes might also introduce novel parasites and pathogens into wild populations. So the Conservation thru Commercialization slogan masks a much more complicated ecological debate.

You don't have to turn over any rocks, however, to discern the long-term impact of having a pet snake. Hayley Crowell got her first corn snake when she was nine years old; she's now finishing up her PhD in snake genetics at the University of Michigan (and, 25 years later, still has that same corn snake!). Sean Bush's grandfather gave him his first rattlesnake when Bush was five years old, growing up in Texas; he's just finished his term as president of the North American Society of Toxinology and has been the lead character in the cable TV show *Venom ER*. And Derrick Rossi, a co-founder of the biotech company Moderna, which produced one of the first lifesaving Covid vaccines, shared his childhood home with half a dozen snakes; each of his three daughters now has a corn snake morph in her bedroom. "I grew up with snakes," he told me. "I wanted to be a herpetologist. I have five snakes upstairs right now." They're nice anecdotes about serpent love, but they signify even more: how enthusiasm about snakes has merged with cutting-edge 21st-century science.

3

A Pandemonium
of Molecules

~~~~~~

Venom

As a businesswoman who prides herself on being genially efficient, Karin McElhatton knew she was in trouble one morning in 2019 when she sat in her car and started drooling uncontrollably. She'd had an inkling of trouble a minute or two before, when she had stepped briefly around the side of the car, felt a sharp pain, and looked down to see two ominous trickles of blood running down her right leg.

Sharp pain.

Fang marks.

Venomous snakebite.

Throughout the world, thousands of times daily, a variation on this rapid, three-step progression of awareness signals that an ordinary day has instantly become a nightmare.

After absorbing the initial shock, McElhatton paused to compose herself, considering where exactly she was and where she should seek help in her sparse arid corner of Southern California. She told herself she could drive to a nearby hospital, about 10 minutes away. But then her head started to droop, too, as she sped down Highway 126 toward Castaic, a town on the northern edge of Los Angeles County. That's when she knew she was *really* in trouble.

McElhatton has spent her entire professional life training—and loving—animals. You may not recognize the name, but you've surely seen her work. As co-founder and owner of Studio Animal Services, she has supplied animals for movies, TV shows, commercials, and ad campaigns for decades—the St. Bernard puppies in *Beethoven II*, Chevy Chase's golden retriever in *National Lampoon's Vacation*, Reese Witherspoon's canine sidekick Bruiser Woods from the *Legally Blond* movies, the Taco Bell Chihuahua, the AFLAC duck. On the company's sixty-acre ranch in Castaic, McElhatton and about a dozen staff trainers groom, as the website puts it, "dogs of every demeanor, multitalented cat teams, smart squirrels, dynamic ducks, and any other animal—large or small." Well, actually not *any* other animal. She usually doesn't do snakes—there's a separate Hollywood subculture of animal trainers devoted strictly to serpents. But she nonetheless had meaningful lifelong interactions with venomous reptiles. Early in her career, McElhatton shared a scene—shared her bare legs and feet, to be precise—with a live cobra around which she danced on a television show; that cameo role allowed her to get her first Screen Actors Guild union card.

More to the point, Castaic, about 40 minutes north of downtown Los Angeles, is surrounded by steep, dry canyons that make up prime rattlesnake country, and she was hardly oblivious to the danger venomous snakes represented. She had lost two of her studio dogs to snakebite. She kept half a dozen snake sticks on the property to wrangle and remove any that she found. And in the more distant (but relevant) past, one of her great-grandfathers back in Texas had died by snakebite.

Despite all that, she told me, "I try not to kill them."

One almost killed her.

Everything in McElhatton's life changed on that very hot Saturday morning in early April 2019, and like so many life-altering events, it grew out of a series of tiny and mundane deviations from routine that glided seamlessly into catastrophe. Normally she would drive down the canyon road in her own car; that morning she took her daughter's vehicle to get it washed. Normally she would take a turn at the bottom of San Martinez Grande Canyon Road and head toward the car wash; that morning, she decided to stop at the intersection where the canyon road met Highway 126 to check on the progress of a little roadside garden she had cultivated with native plants—a pretty little stand of poppies, California sage, brittlebush, river willow. Normally she would have hopped out of the car onto the roadway; that morning, driving an unfamiliar vehicle, she mistook the emergency brake for the hood release and had to walk around the vehicle to close the hood. As she did that, she wandered into a knee-high patch of foxtails by the side of the road. "I never saw it," she said. "I stepped into the dry grass and got nailed."

"Nailed" doesn't quite capture the immensity of the encounter. By stepping on a rattlesnake, McElhatton provoked a do-or-die defensive strike. The unseen snake was a southern Pacific rattlesnake (*Crotalus oreganus helleri*), according to local rattlesnake experts, a tawny brown pit viper that often blends imperceptibly into the water-starved vegetation of Southern California. The faint brown blotches that run down the backs of these snakes end in the rattle, which on this morning did not rattle—at least not until *after* the bite. And by stepping on the rattlesnake, McElhatton involuntarily surrendered to what Mexican venom expert Alejandro Alegón once aptly described as a "pandemonium of molecules."

"The pain was off the charts," she told me one September morning, describing the episode four years later. Half a dozen hummingbirds darted over a nearby garden fountain as McElhatton—with a broad, friendly face, an ebullient flow of silver hair, and an easygoing, self-deprecating manner—relived the experience all over again. "Within a minute, within 30 seconds, I felt a sensation in my lips, mouth, tongue, then the inside of my mouth. It was almost like a vibration." An odd metallic taste flooded her mouth, too. The "grotesque, agonizing vibration" spread from her face to her neck, then to her hands and arms, then through her entire body. She climbed back into her vehicle, thought about driving back up the canyon road to her ranch for help, but changed her mind. "I could tell I was losing function," she recalled, "and I started drooling into my lap." Instead, she decided to drive immediately to the nearest hospital, about six miles away. She never made it.

McElhatton had no way of knowing, but at least piecemeal scientific evidence suggests that she had experienced the worst possible scenario for a snakebite victim. Researchers at Loma Linda University Medical Center, about 100 miles east of her native garden, had compiled data showing that some California rattlesnakes typically inject a maximum amount of venom in a so-called defensive strike—not a bite intended to subdue prey, in other words, but one to confront a threat. The animals seem to view such an interaction, not unreasonably, as an existential threat, and respond by unleashing a high-pressure gush of venom; physicians who have treated snakebite also say that bites on the lower extremities tend to be more serious. "Defensive bites are more severe," one expert told me. "And if you step on a snake and it's afraid, like it's being defensive, that's going to be more severe. And lower extremity bites, just in and of themselves...they're also more severe." Not that it's any consolation, but a young Brazilian researcher, wearing special protective boots, "softly" stepped on South American vipers (*Bothrops jararaca*) some 40,000 times for a 2024 study showing that touching the snakes near the head provoked the strongest defensive strikes.

In the world of toxic mixology, there's nothing quite like the cocktail of potent bioactive ingredients that snakes cook up in their venom glands. Of the roughly 4,000 species of snakes in the world, some 600 species (15 percent) are venomous, and each species has perfected the proportion of ingredients in its toxic cocktail as an adaptation to its environment. Venom is "intrinsically ecological," according to the scientists who study it; its chemical composition reflects the kind of habitat, the kind of prey found in that habitat, age, gender, even the range of temperatures within a given habitat. Some rattlesnake venoms, for example, include a quick-acting, "knock-down" neurotoxin to prevent a mobile prey, such as kangaroo rats, from hopping too far away before the rest of the venom fully kicks in.

The venom of the Mojave rattlesnake (*Crotalus scutulatus*), common in the more desert-like environments of California and Arizona, is considered among the most toxic of all vipers, because it attacks both blood and nerves. By some estimates, a dose as small as 15 milligrams—roughly 0.003 teaspoon—can be lethal to an adult human. Yet researchers have detected dramatic differences in the composition of venom in Mojave rattlers that are practically next-door neighbors. Those found in Pima County, Arizona, which includes Tucson, for example, cook up a different cocktail of toxins than those a little farther east in Cochise County. "I can show you roads where, if you drive 20 miles from one end to the other, you'll get a 10-fold difference in the mouse lethality of Mojave rattlesnake venom," said Wolfgang Wüster, a venom expert at Bangor University in Wales. Yet some Mojave rattlesnakes lack "Mojave toxin," the component that creates devastating neurological effects, while other species of rattlesnakes, like the southern Pacific rattler that likely bit McElhatton, sometimes do have it. When you add in the vagaries of human biology, you can understand why Emily Taylor said, "Every bite is dynamic; there're no rules. Not only is every batch of venom different, but every person's *reaction* is different."

The rattlesnake that nailed McElhatton probably wasn't large—a four-year-old adult may be only 18 inches long—but size doesn't necessarily matter when it comes to a massive, or "bolus," injection of venom. By the time she looked down at her right calf and saw the first trickle of blood, the venom had already begun to wreak havoc. Popular accounts of snakebite often imply that the venom immediately hits the bloodstream, but snake toxins more typically inflict local damage, including swelling and pain, while spreading through the lymph in a slow but fluid ooze before reaching the bloodstream. Some components in the venom are so-called spreading factors—chemicals that facilitate a relentless diffusion through tissues until the venom reaches a vessel.

It is not uncommon for snake venoms to contain 100 or more discrete toxins. As McElhatton became instantly aware, some of the ingredients cause intense pain. Some are powerful enzymes that attack and dissolve the walls of capillaries and blood vessels; fluid began to leak out of her vasculature, and her leg began to swell. Another enzyme attacks and digests muscle; out in the wild, these "myotoxic" enzymes initiate the process of digestion before the snake even begins to ingest its prey, but in a snakebite victim, they digest tissue at the site of the bite and cause necrosis (death of the tissue), which is why fingers, feet, and sometimes entire limbs have to be amputated. Yet another set of enzymes attacks the pathway in blood that orchestrates the complicated process of clotting, so that blood does not coagulate properly; the two rivulets of blood running down her leg were harbingers of internal hemorrhaging that the snake's venom could induce. And the neurotoxic components of the snake's venom unleash a molecular wrecking crew around the synapses of the nervous system, blocking both the dispatch and the receipt of nerve impulses; this venom-induced paralysis may have affected everything from the tone of McElhatton's neck muscles (hence the drooping) to the automatic control of her salivary glands (hence the drooling). Most ominous, the neurotoxic components of some rattlesnake venoms block the autonomic nerve signals that control breathing, leading to asphyxiation. All these reactions—physiological vandalism of the most dire and sophisticated and evolutionarily honed sort—began to unfold within a minute or two of that initial flick of pain on her leg.

"My head was starting to drop," McElhatton recalled. Neuro-toxic venoms tend to degrade the nervous system in a top-to-bottom fashion, starting in the head and neck and moving down. As Mc-Elhatton began to feel the venom hijacking her body, she had the presence of mind to change plans. She knew there was a fire station about four miles away. "I didn't think I could make it to the hospital," she said, "so I went there." She turned left onto Highway 126, ran a red light at the first intersection, and gunned it down the highway until she reached the Commerce Street exit. It takes roughly 4 minutes and 30 seconds to reach Los Angeles County Fire Station 76 from the site of the bite if you catch (or run) all the lights. She barely made it. By the time she knocked on the closed door of the fire station, she was already showing the first signs of respiratory failure. Despite drooling profusely, she maintained a sense of humor, wondering to herself if anyone had a "St. Bernard towel" for all her slobber.

The first responders kept her calm until an ambulance rushed her to the hospital; she still remembers the "horrified" look on the face of a young EMT attendant as she struggled to breathe. Upon arrival at Henry Mayo Newhall Hospital in nearby Valencia, she was immediately intubated, placed on a ventilator, and whisked into the intensive care unit. She received multiple rounds of antivenom and remained hospitalized for ten days, three of them in the ICU. It took another two months to resume anything like normal activity. Four years later, she still felt some of the aftereffects.

McElhatton's description of her symptoms seemed sadly accurate to William Hayes, a venom expert at Loma Linda University, who has analyzed the venom of southern Pacific rattlesnakes in the Castaic area and is very familiar with its potent effects. In Laurence Klauber's opus *Rattlesnakes*, there is a victim's description of a Pacific rattlesnake bite that echoes, in harrowing detail, the same experience that McElhatton described. Judging by the symptoms, Hayes said, "It sounds like she got bit by a big snake. She may have had a big dose, could have had 50 milligrams or more dry mass of venom, and that's a lot." And the sensation of "vibration," Hayes speculated, might have been due to a component in the venom that causes involuntary muscle contractions, or rippling, just beneath the skin. "Southern Pacific rattlesnakes, there's a lot of myotoxins in their venom," he said.

When we spoke in 2023, Karin McElhatton was still recovering physically, but also psychologically. The episode altered her health, her lifestyle, even her relationship with the beautiful sere landscape surrounding her ranch. "It was a game-changer for me," she said.

"I used to bushwhack through the hills. I don't do that anymore." Rattlesnakes still pester her business and homestead. They turn up in the rock garden behind her home, on the path between her house and office, on the front porch of her house, even under one of the dog beds on the patio. Rattlesnake sightings on the ranch are so common that her staff keeps a ledger—date, size, location—taped to the refrigerator in the Studio Animal Services office (only four through the summer of 2023, but nearly two dozen in 2022).

Although Karin McElhatton's encounter with a Pacific rattlesnake was an intense and life-altering experience, it was also a demographic aberration. Each year, an estimated 138,000 people die of snakebite globally; less than 6 of those deaths occur, on average, in the United States, according to a 2022 article in the *New England Journal of Medicine*. An estimated 2.7 million people worldwide suffer "envenomations" (the actual injection of venom during a snakebite); roughly 8,000 envenomations occur in the US. Most of those global snakebites occur in poor rural areas of Southeast Asia, Africa, and Latin America, which is why researchers often describe snakebite as a "disease of poverty." In 2017, the World Health Organization prioritized it as a "neglected public health disease." Former United Nations secretary general Kofi Annan told a WHO conference in 2018, "Snakebite is the most important tropical disease you've never heard of."

But as in all things snaky, there's a niggling paradox: The fear of venomous snakes, and of their bites, is out of proportion to the real risk, at least in developed countries. Over the three decades ending in 2018, for example, less than four people in the US on average died of snakebite per year. By contrast, in recent years an average of 28 people died from being struck by lightning (Centers for Disease Control and Prevention); an average of 72 died of bee, wasp, and hornet bites (CDC); an estimated 3,000 died from food poisoning (CDC); an estimated 7,850 pedestrians were struck and killed by cars (Governors Highway Safety Association); more than 44,000 died from accidental falls (CDC); more than 45,000 died in motor vehicle accidents (CDC); more than 102,000 died from unintentional poisoning (CDC); and…well, you get the idea. No one thinks of climbing a ladder to change a lightbulb or crossing the street as a horrifyingly mortal risk, but the mere mention of rattlesnakes inspires a deep, primal fear.

"Almost no one in America dies from a snakebite, because we're so damned efficient and quick about resuscitation," said Sean Bush, who has treated hundreds of snakebites as an emergency room physician in both California and North Carolina. But he hastened to add, "A rattlesnake bite in Southern California is no joke." That cautionary note cut very close to home one day in June 2006 when Bush was working an emergency room shift at Loma Linda University Medical Center. Doctors there received an alert: A medevac helicopter was en route with a two-year-old snakebite victim. The patient turned out to be Bush's toddler son Jude, who'd been bitten on the thumb by a baby rattlesnake while playing in the backyard of the family's Yucaipa home. (He received antivenom treatment, recovered quickly, and, Bush told me, entered college in the fall of 2022.)

As to venomous snakes, they are for the most part minding their own business and looking for their next meal or next mate; they certainly don't stalk people, typically become aggressive only when threatened, and, unlike many other dangerous animals, usually give you a warning, at least in the case of rattlesnakes. And although it's rarely mentioned, the evolution of venom means that snakes as a class of animal don't need to rely on brute force or physical size to prevail against predators and prey alike. Go at them if you will with shovel or pitchfork, but the vast majority of snakebite victims suffer the consequences by moving toward, rather than away from, a venomous snake, even a dead one (there are multiple cases of people being envenomated by a freshly killed or decapitated snake).

"Provoked bites," the authors of the *New England Journal of Medicine* article noted, "are more likely to involve males and the upper extremities." Also alcohol.

~~

During the Middle Kingdom in ancient Egypt, roughly 2,600 years ago, the "priests of Serquet"—physician-magicians who specialized in the treatment of venomous bites or stings by snakes and scorpions—would accompany field expeditions of miners, who journeyed to the southeastern corner of the Sinai Peninsula to extract copper, turquoise, and other precious minerals to bejewel the reigning pharaohs. The priests possessed a kind of medical field guide that listed various snakes, described the symptoms of their bites, and prescribed treatments, which often amounted to vain incantations and ineffective herbal concoctions. We know this from an ancient Egyptian text known as the Brooklyn Snakebite Papyrus, which has long been in the collection of the Brooklyn Museum but only recently been translated into English. Egyptologists believe the papyrus was based on an even earlier text, testifying to the fact that the search for a snakebite treatment is arguably one of the oldest quests in the annals of human health—older than Hippocratic medicine, older than the Greek or Roman alphabet, perhaps as old as writing itself. And for nearly three millennia, that text was essentially a testament to futility. Among the most frequently recommended remedies for venomous snakebites, according to the priests of Serquet, was onions.

It took another 2,500 years or so for science to come up with something better for snakebite than onions, incantations, whiskey, or amputation. In 1895, Albert Calmette, a French-born physician who had firsthand experience treating venomous snakebites while working in Indochina, invented the first crude antivenom treatment for cobra bites. The basic idea involved repeatedly injecting tiny amounts of snake venom into large animals (primarily horses or sheep, but also donkeys and camels) to stimulate an immune response; researchers then collected and purified the resulting antibodies. Generations of snake-handlers—Raymond Ditmars, Bill Haast at the Miami Serpentarium, researchers at the Instituto Butantan in Brazil, and elsewhere—milked venom from the world's most dangerous snakes to create antivenoms, but it was (and, to a certain extent, still is) almost an artisanal undertaking.

That's because snake venoms vary so widely from one species to the next that antivenom for rattlesnakes, for example, would be largely ineffective against cobra venom. Though lifesaving, antivenom treatments are far from perfect. They require refrigeration and intravenous administration, they can trigger anaphylactic shock, and they are exceedingly expensive, especially in the impoverished areas of the world where snakebite is prevalent. "They are probably the most expensive drug in rural hospitals," one venom expert told me. They're not cheap in the developed world, either; a single vial can cost anywhere from $8,000 to $14,000. But perhaps the most lethal limitation is that most venomous snakebites occur far from a medical facility equipped to administer antivenom (more than 75 percent of fatal snakebites claim their victims before they can reach a hospital). That gap, in miles and minutes, is a valley of death and disfigurement. Bridging that gap has challenged biomedical researchers ever since Calmette distilled the first antivenom.

Two and a half millennia after the priests of Serquet accompanied ancient Egyptian miners to the Sinai, another field expedition—this time to collect rare animals and plants in Southeast Asia—inspired, in a tragically indirect way, potentially the most radical new treatment for snakebite victims. In September 2001, a joint expedition of naturalists from the California Academy of Sciences, Asian scientists, and the forestry service of Myanmar (formerly Burma) set out on an ill-fated trip to collect rare specimens of reptiles, amphibians, mammals, and plants in the northernmost (and most inaccessible) reaches of the country.

The expedition—chronicled in Jamie James's 2008 book *The Snake Charmer*—seemed doomed from the start. The Burmese hosts guaranteed logistical support that never materialized. Incessant rain battered the scientists in the field, forcing them to slog for days on end along muddy, barely traversable paths. Leeches and insects besieged them. Most significant, local collaborators had promised but failed to produce a medical professional to accompany the group. After more than a week in the field, the expedition took a tragic turn on the morning of September 11 when the group's leader, herpetologist Joe Slowinski, stuck his hand in a bag to pull out what he thought was a harmless snake and got nipped on the middle finger of his right hand. Although they weren't his last words, his immediate response—"That's a fucking krait!"—was essentially a self-pronounced death sentence.

The venom of the rattlesnake that bit Karin McElhatton had obvious neurotoxic effects, but the venom of a many-banded krait (*Bungarus multicinctus*)—even the small, slender, juvenile specimen, less than 10 inches long, that clamped onto Slowinski's finger—is more deadly still. Krait venom attacks the junction between nerves and muscles; in the worst cases, it leads to a rapid and progressive head-down paralysis, often beginning with drooping eyelids and ending in respiratory arrest. Krait bites are often painless, so victims are unaware that they've been bitten; the snakes are even known to enter homes at night in search of prey in their native Southeast Asian habitats, sometimes biting humans in their sleep. They cause widespread fatalities and are considered one of the "Big Four" venomous snakes in South and Southeast Asia.

None of that fazed Slowinski, a renowned and much-beloved snake-handler, whose blend of fearlessness and boyish exuberance was legend in the herpetological community. The previous evening, after a boisterous dinner in the tiny village of Rat Baw and a night of drinking, he had thrown his arms in the air and declared, "I am the king of snakes! I can catch snakes better than anybody!" Hours later, he lay on a sleeping bag in the village's lone schoolhouse, helpless to stop the effect of the snake's venom.

Inch by inch, organ by organ, Slowinski began to lose the use of his eyelids and limbs, his ability to speak, finally his autonomic ability to draw a breath. The krait's venom blocked the nerves that moved toes, fingers, mouth, the muscles that breathe. Scientist colleagues on the expedition took turns administering mouth-to-mouth resuscitation for a staggering 12 hours. Despite heroic efforts to keep Slowinski alive until medical help arrived, he died on September 12, two months shy of 39 years old.

Slowinski's death shocked the herpetology community, but snakebite—with possibly lethal consequences—has always been an occupational hazard, captured by the phrase *not a matter of if, but when.* Karl Schmidt, the legendary director of reptiles at the Field Museum of Natural History in Chicago, died from the bite of a boomslang (*Dispholidus typus*), a rear-fanged snake that rarely claims victims; a scientist to the bitter end, Schmidt refused medical treatment and carefully wrote down details of the envenomation until he died. Robert Mertens, an esteemed curator at a museum in Frankfurt, Germany, died from the bite of an African twig snake (genus *Thelotornis*), which he had kept as a pet. William "Marty" Martin, widely considered the leading expert on timber rattlesnakes (*Crotalus horridus*) in the United States, died in 2022 from the bite of a captive timber in his West Virginia home. And Grace Olive Wiley succumbed to a cobra bite in 1948.

In the *not a matter of if but when* world of people who handle venomous snakes on a routine basis, reactions to Slowinski's death varied. Many felt stunned that such an experienced and respected researcher had succumbed. Others, aware of Slowinski's reputation for impulsive behavior, almost saw the Burmese tragedy as a death foretold. "Tales of his reckless snake-handling were legendary," James wrote. "Yet that, too, was within the expected range of behavior for a field herpetologist. A love of danger was acceptable, even tacitly admired—as long as your luck held." And then it didn't.

In the wake of the Slowinski tragedy, the California Academy of Sciences began to look for a physician-consultant who could advise the organization on future expedition medicine. That search led to an emergency room physician at the hospital of the University of California–San Francisco named Matthew Lewin. He had served as the expedition physician for scientists from the American Museum of Natural History on five field trips to the Gobi Desert. He was local. He liked snakes. For all those reasons, Lewin considered himself "a natural fit" for the Cal Academy post, and the Cal Academy thought so, too—in 2010, he became director of its Center for Exploration and Travel Health. And through a series of medical coincidences, that consultancy ultimately got him thinking about an unconventional approach to snakebite treatment: something you could take into the field (either a pill or a nasal spray), something cheap (perhaps a repurposed drug), something that didn't require a cold chain (that is, refrigeration up to the moment of use).

Matt Lewin has had a colorful and peripatetic scientific journey and doesn't go out of his way to conceal bits of eccentricity. When I first interviewed him on Zoom, he walked around with a parrot named Oscar perched on his shoulder and happily showed off the makeshift garage laboratory where he undertook his initial pharmacological experiments. Born and raised in Corte Madera, California, he chased snakes as a kid in Marin County but steered clear of the west side of Highway 101, where rattlesnakes were known to roam. He earned a joint MD/PhD degree at M. D. Anderson Cancer Center in Houston, Texas, where he specialized in the neuroscience of pain. After returning to the Bay Area, he worked as an emergency room doctor at the UCSF hospital and was later the medical director for an ambulance company in San Francisco. In that last job, he had to contend with a shortage of the anti-anxiety drug diazepam, which was used to treat seizures, tremors, muscle spasms, and alcohol withdrawal in the ER setting. That crisis led to a realization that doctors could administer diazepam as a nasal spray, which got him thinking. Could you create a nasal spray against snakebite, too? He toyed with the idea of repurposing an old anti-paralysis drug, neostigmine, as a nasal spray for emergency use in cases of snakebite. It was an already approved drug, a cheap generic, and could be administered on the spot by non-experts, much the way the drug Narcan is now available for accidental opioid overdoses.

In 2011, Lewin's noodling about snakebite treatments became more than an academic exercise when Cal Academy asked him to accompany an expedition to the Philippines. Venomous snakebites suddenly became more than an abstract medical concern. "I was contemplating the issue of what I would do if we encountered a venomous snake," Lewin told me. "At that point, I was only aware of what had happened to Joe Slowinski. But of course I did a literature search, and wow, there's nothing out there on this." That was when he realized that the Slowinski tragedy, writ large, was in fact a "humongous global health issue."

Upon his return from the Philippines (no bites), these ideas coalesced into an ambition that even Lewin admitted was crazy: the search for a cheap, already approved small-molecule drug that could be taken into the field as part of a first-aid kit, could be administered by non-doctors, didn't require refrigeration, and somehow offered broad protection against snakebites, despite the huge variation in venoms, that lasted long enough for people to get to a hospital. Mission impossible? "My initial response was just to dismiss it as a silly idea," Lewin recalled. "We've got antivenom. Somebody would have thought of this…I had a nice clinical career, I'm publishing at a decent clip, submitting a variety of papers, and exploring things I'm interested in. The idea just kept bugging me."

The idea inched closer to reality the following year when Lewin attended a party at the Marin County home of rock musician Jerry Harrison, formerly of the Modern Lovers and Talking Heads. Lewin was "keeping to myself in the kitchen" when Harrison walked in and said, "Does anyone have any crazy ideas lying fallow?" Like many an obsessed amateur inventor, Lewin couldn't help but blurt out his great idea to revolutionize snakebite treatment. Harrison, who in addition to producing records invested in healthcare initiatives, quickly set him straight. "First of all," Lewin remembered him saying, "nobody is going to take you seriously because it's not a business. Even if it's a great idea, somebody has to make the drug." Harrison mentioned that patent protection might be helpful, too. Then he said, "Call me on Monday."

Lewin didn't even have a candidate drug, but within days he had a patent lawyer and a partner. Harrison was intrigued by the idea— so much so that he agreed to co-found a company with Lewin called Ophirex, Inc., whose headquarters were basically Lewin's Corte Madera garage. And next, Lewin volunteered for a perilous experiment: allowing himself to be medically paralyzed, as if bitten by a cobra, to test whether a nasal spray form of neostigmine could blunt the effects of snake venom.

It wasn't as crazy as it sounds. Neostigmine had been used since the 1970s as a treatment for snakebite in India, but only as an intravenous medicine administered in a hospital. It had been used even longer for the treatment of myasthenia gravis, a neuromuscular disorder (in fact, researchers had used snake venom in the 1970s to figure out the exact neurological deficit that caused the disease). No one, however, had dared test whether a squirt of neostigmine up the nose could reverse the neurotoxic effects of snakebite until Lewin, in a wild, n = 1 experiment he arranged in a carefully controlled hospital setting, allowed himself to be paralyzed to test the drug. Doctors at UCSF gave him a constant infusion of mivacurium, a curare-like drug that imitated the neurotoxic paralysis induced by cobra venom.

Roughly two hours into the experiment, Lewin began to show the telltale signs of neuromuscular deterioration. His vision began to fail. He had trouble swallowing. He couldn't lift his head. He began to experience difficulty breathing. Only at that point, 115 minutes after the infusion began, did doctors give him a squirt of neostigmine up the nose. The drug kicked in rapidly and reversed Lewin's paralysis. An account of this unusual experiment, published in

Clinical Case Reports in 2013, described the volunteer subject only as a "healthy, 45-year-old male."

The study didn't fully answer the question about a portable snakebite medicine, but it certainly answered any questions about Lewin's commitment to his crazy idea. He just needed to find a better drug.

～～～

Even prior to his adventure in voluntary paralysis, Lewin knew—because of his background in neuroscience—that scientists had used snake venom in the 1970s to unravel the neural mechanism that caused myasthenia gravis. But most people are generally unaware of the dramatic ways snake venoms have contributed to human well-being—not just in terms of basic research, but in the contents of millions of pill bottles in medicine cabinets all over the planet.

For those who think the only good snake is a dead snake, perhaps they're still alive to harbor such ungenerous thoughts precisely *because* of a snake. Specifically, because of the South American pit viper (*Bothrops jararaca*). Most people have never seen these dangerous vipers, and probably would never choose to do so, but they have arguably had an impact on human health nearly comparable to that of the development of vaccines and antibiotics—largely because of a small army of scientists, beginning in South America, who took biochemistry classes from this unique reptile. Anyone who has taken the drug captopril, or enalapril, or lisinopril (my hand is raised here), or quinapril, or ramipril, or trandolapril, or moexipril, basically any of the dozen or so next-gen drugs known as ACE inhibitors—and there are tens of millions out there—has *B. jararaca* and its venom to thank.

This fantastic pharmacological story has its origins more than half a century ago in Brazil, where scientists pursuing a mistaken hypothesis stumbled upon one of the most important molecular pathways in the annals of cardiovascular medicine. Doctors knew that people bitten by *B. jararaca* suffered a precipitous drop in blood pressure and often went into shock. Their hunch was that plummeting blood pressure could be due to histamine, a signaling molecule of the immune system. That hunch led Brazilian scientist Mauricio Rocha e Silva and colleagues at the University of São Paulo to do a series of experiments in which they injected dogs with the viper's venom in an effort to figure out how its toxins so dramatically lowered blood pressure. High blood pressure (hypertension) affects nearly 1.3 billion people worldwide, including nearly 50 percent of adults (120 million) in the US; it is, according to the World Health Organization, one of the leading causes of premature death due to heart failure, kidney failure, and other, often "invisible" illnesses. Histamine turned out to be a red herring, but in 1949, Rocha e Silva reported the isolation of a small but remarkably potent component produced by animals (including humans) as a reaction to the snake's venom; this peptide, called bradykinin, instantly causes the muscles in arteries and veins to relax, which explained how jararaca venom caused blood pressure to crash.

But further laboratory experiments threw a curveball at the researchers: When they injected purified bradykinin into animals, the molecule blinked in and out of existence immediately, after a single pass through the pulmonary circulation. In some experiments, it disappeared in 17 seconds, and seemingly lost its power to send blood pressure plummeting. Brazilian scientists could not replicate in the lab what happened to snakebite victims. The snakes apparently knew something that the scientists didn't, because *their* venom not only triggered the release of bradykinin in snakebite victims, but also somehow protected it from being destroyed in seconds. It took more than a decade to figure out that animals (including humans) possessed a counterpunching molecule. This counterpunch took the form of an enzyme (ultimately known as angiotensin converting enzyme, or ACE) that, within seconds, demolished bradykinin in the bloodstream, canceling its power to lower blood pressure.

Clearly, there was something in the snake venom that prevented bradykinin from vanishing after a snakebite. That paradox puzzled Sérgio Ferreira, a would-be psychiatrist who switched to pharmacology and joined Rocha e Silva's lab in 1961. Ferreira turned his attention back to the snake and its venom in a search for clues. Over the course of four years, including many five-hour trips from his lab in Riberao Preto to the Butantan Institute in São Paulo where the snakes were milked, Ferreira eventually discovered nine distinct snake peptides—short chains of amino acids—that in a sense acted as molecular bodyguards preventing bradykinin from being degraded by the ACE-related enzymes; indeed, the small bodyguard molecules Ferreira discovered in snake venom actually enhanced the power of bradykinin to drive blood pressure lower and lower. The research showed that these snake-derived compounds made it "much more potent when you have both things," Ferreira told an interviewer in 2008. "So the snake was very, very clever."

Ferreira reported the discovery of these so-called bradykinin potentiating factors (or BPFs) in 1965, and he brought a small packet of the freeze-dried venom extract—"the world's only sample of BPF," according to one history—when he moved to England for a postdoctoral fellowship. He had intended to continue the research at Oxford University, but changed plans at the last minute when his wife accepted a research position at the London School of Economics, which is how he ended up in the laboratory of John Vane at the Royal College of Surgeons. Vane, whom *Nature* later eulogized as "one of the greatest pharmacologists of the twentieth century," worked with Ferreira's freeze-dried extracts to show how the bodyguard peptides in snake venom blocked the enzyme (ACE) that normally cleared bradykinin from the bloodstream—hence ACE inhibitor.

But they still had a problem. It took enormous effort to isolate these potent venom ingredients to do further experiments. In an attempt to lure a pharmaceutical company into synthesizing one of the BPFs, Vane passed a preprint of a 1968 paper to two scientists at E. R. Squibb and Sons (forerunner of Bristol Myers Squibb)—"so we did not have to milk 300 snakes a week!" recalled Yeshwant Bakhle, who worked alongside Ferreira in London. After initial enthusiasm, according to Bakhle, Squibb actually canceled the project at one point, but the two medicinal chemists tasked with creating a synthetic version of one of Ferreira's tiny peptides persisted despite the internal decision. The result was a drug to lower blood pressure that mimicked the action of the snake-derived peptide and could be formulated as a pill. That drug, captopril, became Squibb's first billion-dollar pharmaceutical.

The Food and Drug Administration approved captopril in 1981. Not only was it the flagship ACE inhibitor drug, spawning nearly a dozen progeny drugs that lower blood pressure, but the discovery also initiated an epochal transformation in cardiac medicine, specifically what is known as the renin-angiotensin system, which according to Bakhle "now underlies the successful treatment of almost 50% of the patients in cardiovascular medicine, with serious possibilities of extension to diabetes, Alzheimer's disease and cancer." And it is a reminder that "innovation" in drug development rarely follows a neat, logical narrative. As Bakhle later put it, "The progress from snake venom to the most widely used cardiovascular medicine is full of non-science like coincidence, chance and personal belief."

More than 40 years later, after tens of billions of dollars in revenue, ACE inhibitors are still one of the most lucrative classes of drug in pharmaceutical history. Vane received a Nobel Prize in 1982 (primarily for his research on the biology of pain); Ferreira, who returned to Brazil in 1975, was named a foreign member of the National Academy of Sciences 25 years later. Needless to say, the vipers do not receive a royalty.

Captopril was the first venom-derived drug, but it's not an only child. As researchers at the University of Porto in Portugal pointed out in a recent review article, snake venoms have provided "a rich playground for medicinal chemists." The very qualities that make snakebite deadly—disrupting clotting on the one hand, causing blood to coagulate on the other—have been repurposed into drugs that are now used to dissolve blood clots, block hemorrhages, and treat pain. In the US, the FDA approved tirofiban, derived from a toxin made by the saw-scaled viper (*Echis carinatus*), in 1998 to treat acute coronary syndrome, and that same year approved eptifibatide, based on a component of a species of pygmy rattlesnake (*Sistrurus miliarius*), for its anti-clotting activity; several other snake-derived drugs have been approved for medical use in India and China. "It is becoming evident that the ancients were right," Ana Oliveira and colleagues argued in their 2022 *Nature Reviews Chemistry* article, "as the venom of this splendid animal is an extraordinary library of bioactive compounds that has great medicinal potential."

If that were all that venomous snakes contributed to human welfare, it would be reason enough to slip them a Nobel Prize, too. But the reality is that snake venoms have proven to be exceptionally beneficial to basic research as well, advancing progress in neuroscience, genetics, and evolutionary biology. Sylvie Diochot and Eric Lingueglossa of the French research institute CNRS in Valbonne, France, have used the venom of black mambas to tease apart mechanisms of pain—specifically how an ingredient in the snake's venom they dubbed "mambalgin" disrupts the flow of ions at nerve junctions, which is essential to the transmission of nerve messages that specifically signal pain. They, too, had hoped to create a new drug for pain relief based on mambalgins, but investors gave up before the drug could advance to clinical trials.

In the field of genetics, Stephen Mackessy of Northern Colorado University, Todd Castoe of the University of Texas–Arlington, and their colleagues have studied the venom of rattlesnakes in the American West to reveal the sophisticated mechanisms by which snakes manipulate regulatory genes in their venom glands to adapt rapidly to both environmental and ecological changes. In one of the more mischievous puns in the snake science literature, their analysis of the venom of the Plains rattlesnake (*Crotalus viridis*) bore the title "Snakes on a plain."

And Nicholas Casewell, who directs the Centre for Snakebite Research & Interventions at the Liverpool School of Tropical Medicine, recently led the team that used the venom of three different species of cobras, living in three disparate locales, to show how the snakes independently evolved the ability to spit their venom to incapacitate predators. This example of "convergent evolution" exhibited by snakes, which merited a cover story in *Science*, reveals again that evolution can cook up the same useful solution to a meaningful life problem in animals separated by thousands of miles of habitat and millions of years of evolutionary history.

In 2023, this cutting-edge science circled all the way back to the

Brooklyn Snakebite Papyrus. Elysha McBride, Isabelle Winder, and Wolfgang Wüster of Bangor University in Wales, recently took a stab at identifying some of the vaguely described snakes in the ancient Egyptian document. In a shrewd and creative use of a technique called niche modeling, the scientists created an ecological approximation of the region roughly 6,000 years ago, merged that with historical records delineating the extent of Egyptian territory in ancient times, and assessed the possible "paleodistribution" of numerous venomous snakes that could conceivably have inhabited the area six millennia ago. In posing the question "What Bit the Ancient Egyptians?" the Bangor team reached a remarkable answer. They identified 10 snake species—including the Palestine viper (*Daboia palaestinae*), the puff adder (*Bitis arietans*), and the black mamba—none of which are found in present-day Egypt.

⌢

So humans have been unhappily acquainted with deadly snakes for thousands of years. But only recently have we understood what venom is, and what makes a snake venomous. And the answer is a bit more complicated than you might think.

It begins with this foundational paradox: Virtually *all* snakes are venomous, not that you'd notice. Even garter snakes, hognose snakes (*Heterodon platirhinos*), and ring-necked snakes (*Diadophis punctatus*), among the most routinely handled serpents on the planet, and even that dainty ribbon snake I twirled triumphantly in my hands as a kid—they all possess venoms surprisingly similar to the deadliest serpents in the world. "You know a beautiful little snake that's real common in the US called a ring-necked snake?" Joe Mendelson of Zoo Atlanta asked me once. "Those things are *shockingly* venomous. And nobody talks about it because they don't bite ever, ever, ever,

ever! I would give one to my niece and let her hold it and play with it without giving it a second thought."

"They're technically venomous," agreed Nick Casewell, the Liverpool venom expert. "So you can pick up a garter snake or a hognose snake or whatever and it bites you. You might get a little bit of swelling and a bit of pain, and that's because it's venomous. It's injecting you with a bit of venom. It's not going to put you in hospital or kill you or anything. But they are technically venomous snakes. They're just not medically important."

Why not? The difference lies not so much in biochemistry as in anatomy, according to Casewell, who initially set out to study parasitic tropical diseases in Africa and realized the medical importance of snakebite in that setting, jumped fields, and has become one of the leading international authorities on snake venoms. With front-facing hypodermic-like fangs, a capacious venom gland, and, perhaps most important, a sheath of powerful muscles to compress that gland like a piston, deadly snakes inject a huge dose of toxins, and do it instantaneously. "They can inject just a real bolus, a large amount of venom, very, very quickly," Casewell said. "So rapidly, in fact, it makes a huge difference. You would get a fraction, a hundred-thousandth of that, if a garter snake bit you. So the anatomical component is probably the most important." The vast majority of "technically venomous" snakes essentially have to bite and chew, which makes them harmless. The ones that can deliver a bolus of venom are what Casewell and other scientists describe as "medically important venomous snakes." They not only kill tens of thousands of humans each year but leave another 400,000 or so with permanent disabilities, both physical and psychological. And the psychological damage undoubtedly extends to an additional 2.7 million or so people who experience a "dry bite"—the snake bites the person but, for unknown reasons, doesn't inject any venom.

There's a well-worn cliché that snake venom is simply an exotic form of saliva, which like many clichés has a nugget of truth, but it doesn't begin to tell the whole story. Casewell, who has published extensively on the genes that encode snake toxins, points out that the breathtaking variation in the cocktails of venom is the result of backroom genetic tinkering over millions of years with what biologists refer to as housekeeping genes. These genes don't encode potent psychoactive molecules like dopamine or powerful growth signals like insulin; instead they encode nuts-and-bolts bit players that keep an organism biologically humming. One such gene, with almost comic echoes of the Garden of Eden story, is called ADAM28; it encodes an enzyme that normally governs interactions between cells and plays a role in such activities as fertilization, muscle development, and neurogenesis.

Snakes have essentially modified these genes to create potent toxins. "Most of them," Casewell said, "are related to preexisting housekeeping genes that have evolved and been weaponized, if you like." Like other venomous animals, snakes have turned their venom glands into an in-house genetic engineering laboratory, where they create copies of these ordinary housekeeping genes, tweak the copied genes with serial mutations, and then allow natural selection to pluck just the right combination of toxic compounds to maximize either the neutralization of prey or the deterrence of predators. Whatever the ratio of ingredients of various toxins, they wreak four main types of havoc: breaking down cells and tissues (known as cytotoxins), breaking down muscle to thwart movement and to get a jump start on digestion (myotoxins), disrupting the ability of blood to behave properly, either clotting too much or too little (hemotoxins), or short-circuiting the function of nerve impulses that affect motor control and breathing (neurotoxins).

The mixology metaphor is especially appropriate because, although snakes as a rule have a fairly limited liquor cabinet to draw on (less than a dozen families of proteins go into venoms), they can produce thousands of different "small batch" venoms by mixing these ingredients in so many ways. A common—and particularly nasty—component is an enzyme known as phospholipase A2 (PLA2 for short), which can eat through blood cells and disable the body's blood-clotting mechanisms like a wrecking ball; in the venom of some species, however, that very same enzyme instead demolishes the junction where nerve cells tell muscles what to do. Matthew Lewin characterized the versatile destructiveness of PLA2 in especially vivid terms. "How does this enzyme have so many tricks?" he said. "You can imagine that the business end of the enzyme is munching away—munch, munch munch, munch, munch. It's like a lawn mower shredding whatever comes in its path. But it's got a zip code. It's got something on it that's telling it either to stay in the blood and destroy the red blood cells, which is how you get hemolysis [breakdown of blood], or it's got a zip code that's saying, Go to the neurons and stop the firing."

Another class of enzymes, known as metalloproteinases (called that because their chemical structures incorporate an ion of the metal zinc), attack molecules crucial to blood clotting. Yet another class of compounds, known as three-finger toxins, figure prominently in the venom of snakes like cobras and mambas; they are especially neurotoxic, capable of disrupting nerve cells before, during, or after nerve signals have been sent. In a 2022 paper, a team of scientists noted—without apparent irony—that the three-finger toxin in black mambas had "an elongated middle finger."

So each snake species brews up a unique mix of these elements. But snakes take it a step further. Like medicinal chemists, they tinker with these elements to create variations through their reptilian riff on genetic engineering. If each snake's gene for metalloproteinase, for example, may be likened to a template, mutations can create 20 or 25 subtle variations on the basic molecular blueprint—ever-so-slight biochemical variations on the basic enzyme. These subtle variants (known as isoforms) can function as distinct but complementary components of venom. When a southern Pacific rattlesnake injects its PLA2 component during a bite, it's not just one enzyme but perhaps half a dozen variations derived from the basic blueprint, each capable of creating a discrete slice of physiological mayhem. Researchers have even shown that sibling snakes sharing the same parents and same birth den can cook up differing venom mixes, depending on whether they're male or female.

In nature, of course, nothing is static. When a predator like a southern Pacific rattlesnake tweaks its venom, its prey like a California ground squirrel will often develop some immunological resistance to the venom. Since at least the 1980s, biologists have marveled at how these squirrels have developed resistance to rattlesnake venom, to the point that the rodents would almost taunt and trash-talk their adversaries. In the age-old evolutionary back-and-forth between predator and prey, the snakes responded by reshuffling the genes in their glandular factories to produce more effective venoms; the squirrels responded by adapting to that change; and so on, ad molecularly infinitum. This evolutionary dialogue is often described as an arms race between predator and prey.

Matt Holding of the University of Michigan believes a different way of thinking about this evolutionary back-and-forth is less like a battlefield, more like a card table. Each animal, rattlesnake and squirrel alike, holds an array of genetic cards—variations of genes known as alleles—that can be called upon by natural selection and played when they help the predator capture food or help the prey blunt the effects of venom. It's not a ceaselessly escalating battle of ever-more-potent venoms, according to Holding. It's more like playing the best card in an animal's genetic hand to trump what the adversary is doing, which makes the encounter less about brute-force power and more about molecular diversity and versatility. "The card game analogy is good for two reasons," Holding said, "because of the value of the card and also the extent to which it fits a particular situation." Evolution may deal an animal the jack of hearts, for example, but gene duplication "can give you two jack of hearts, or both the jack of hearts and the jack of spades." And as in bridge, even a low card like the two of diamonds can win a hand under the right circumstances.

All this molecular nuance means not only that snake venoms are biochemically complex, but also that they are a medically moving target for doctors trying to treat snakebites. Victims might or might not survive depending on the composition of the venom and, crucially, the kind of medicine on hand. As a sobering example, Casewell cited the current situation in India, habitat of the Russell's viper (*Daboia russelii*), which he called "arguably the most important medical snake in the world, certainly in India." Snakebite is a huge public health problem there; a recent analysis by researchers in India suggested that snakebites cause nearly 60,000 deaths a year. Many instances of Russell's viper bites occur in northern India, but as Casewell pointed out, "Its venom composition changes quite considerably over the Indian subcontinent. Antivenom is currently manufactured using venom from the southeast of the country, and that antivenom has been shown, in animals models at least, to not work in the north of India because of differences in venom composition."

That is why it is so daunting to create a new kind of snakebite treatment. To be sure, experts in antibody engineering have been working on cutting-edge approaches to antivenom treatments, creating the same type of antibody drugs that have become a familiar (and indeed bestselling) form of therapeutic against cancers, arthritis, psoriasis, and other medical conditions. These drugs, known as monoclonal antibodies, can target a broad range of snake venoms, and Andreas Laustsen of the Technical University of Denmark has been a leader in developing these promising therapies. The new "molecular" antivenoms, however, share several shortcomings common to other monoclonal antibodies: They require IV administration (and thus treatment in a medical facility), need to be refrigerated, and have been, to date, extremely expensive, all of which complicate their potential utility in the poor rural settings where the majority of fatal snakebites take place, far from a hospital.

~

Eventually, convinced that a pill would offer a better form of emergency in-the-field medicine than a nose spray, Matthew Lewin scoured medical journals, looking for small molecule compounds that might blunt the powerful enzymes in many snake venoms. In his spare time, he began to order various snake venoms from vendors in the US and tested potential antivenom drugs in his Corte Madera garage. Lewin assembled a panel of 28 snake venoms from six continents, everything from *Bothrops jararaca*, the "captopril viper," and Mojave rattlesnake to four cobra species and numerous Asian vipers, and then tested them against a panel of FDA-approved compounds or generic or abandoned drugs in the hope that one of them might show some sort of activity. "All this work was done at the garage level," he told me, "because I was just doing it between clinical shifts. So the first few years of this was all sort of tinkering at home."

That tinkering eventually led to a surprising small-molecule candidate—a dud of a drug that had failed multiple clinical trials, but nonetheless appeared potent and promising against snake venom. It was called varespladib.

Lewin still likes to share the photograph from the November 2014 assay that revealed varespladib's promise as a snakebite medicine. He had assembled a panel of venoms, mixed each one with varespladib, and then added a dye, which would turn yellow if any venom remained in the test tube. If the drug neutralized the venom, however, the fluid would remain colorless. It was a clean sweep—a row of crystal-clear test tubes. Those results not only showed promise, but revealed a surprising twist as well: Varespladib appeared to neutralize the venom of very different snakes from Africa, Southeast Asia, and North America. The drug showed this broad activity because it had originally been designed to target and disarm that "munch-munch-munch" enzyme (phospholipase 2). In fact, the whole reason the drug had been developed in the first place was that PLA2, in a different form, was known to be a bad actor in human inflammatory diseases like rheumatoid arthritis and sepsis. And even though PLA2 was a relatively scarce component in the venom of species like cobras, kraits, and mambas, varespladib still seemed to have an effect.

The drug had been jointly developed in the early 2000s by Eli Lilly & Co. and Shionogi & Company, Ltd., in Japan, tested in multiple clinical trials, abandoned, acquired by another pharma company, Anthera, and abandoned yet again, until it arrived in Lewin's Marin County garage on the third hop. And as Lewin was delighted to discover, the drug had already been extensively tested in humans in no less than 29 clinical trials by the three pharmaceutical companies. On the one hand, those results produced an unrelievedly dismal record of failure against half a dozen diseases; to Lewin and his garage-band pharmacological colleagues, however, the trials also produced a treasure trove of data showing the drug was safe for human use. Indeed, Lilly and the two other companies had amassed a huge amount of pre-clinical data on safety, pharmacokinetics, and bioactivity that, if the drug showed promise, probably saved tens of millions of research dollars.

With its promising test-tube results, Ophirex finally moved out of the garage. In short order, the company conducted animal studies of varespladib to see if it worked in something more complicated than a test tube. In 2018, they reported that the drug rescued mice from the venom of the Papuan taipan (*Oxyuranus scutellatus*), one of the deadliest snakes on the planet; all the control animals died. That same year, they reported that varespladib reversed symptoms and protected pigs from the highly neurotoxic venom of the eastern coral snake (*Micrurus fulvius*); all the control animals died. A decade after Lewin's one-off voluntary paralysis experiment, more than two dozen pre-clinical studies, published by Ophirex and other scientific groups, all told the same basic story: Varespladib blunted the effect of many different snake venoms.

Small biotech start-ups make a splash in the business world not only with the amount of cash they raise, but also with the quality of the company they keep. With this initial tongue-flick of promise, Lewin became something of a pied piper of repurposed pharma, recruiting prominent collaborators and attracting the attention of significant funders. He met over dinner with several Lilly representatives when they were in San Francisco for a scientific meeting; they agreed to license their pre-clinical data to Ophirex in 2017, and Lilly's chief medical officer, Timothy Garnett, joined Ophirex's board of directors upon his retirement. Lewin intrigued representatives of the US Department of Defense, which has considerable interest in snakebite treatments for troops stationed in Africa, Asia, and the Middle East; the US military has given Ophirex $13.8 million in funding, and Rebecca Carter, who met Lewin when she was at the Defense Advanced Research Projects Agency (DARPA), later joined Ophirex as its chief development officer. He even coaxed the University of Arizona's Leslie Boyer out of retirement to join the crew. "He solved the one insurmountable problem that I'd been fighting my whole life, which was how to treat patients before they got to a hospital. I've been waiting three decades for something that could be administered in the field or in an ambulance." The company started gearing up for a clinical trial, and pretty soon, Jerry Harrison was giving talks at scientific meetings titled "Snake Drugs and Rock and Roll."

Then in 2021, Derrick Rossi, co-founder of Moderna and an avid snake-lover, gave an interview in which he not only discussed the messenger RNA technology behind the company's successful Covid vaccines but also suggested that another "wild" application of mRNA technology might be snakebite treatment. Rossi, who grew up with a nine-foot reticulated python in the family home, had never heard of Ophirex, but people at Ophierex soon heard about the interview, and within months he became an investor and joined its board. And he immediately understood the appeal of varespladib: "Quite frankly, it's a better idea than mRNA," he told me, "because mRNA still requires a cold chain."

The Food and Drug Administration seemed impressed, too, granting fast-track status to the drug, signifying its potential medical importance. The potential market was sizable—not just the 2.7

million people worldwide who suffered an envenomation each year, but another 2.7 million people who experienced "dry bites" and would reasonably want to pop a pill instead of waiting to find out if venom was coursing through their body. Not to mention dogs, farm animals, and other veterinary applications. And in 2021, nearly a decade after Lewin lay paralyzed in a San Francisco hospital room, Ophirex finally took the leap. The company launched a double-blind, placebo-controlled Phase II clinical trial of varespladib in snakebite victims in 16 sites, half in North America and half in India. All the signs were promising.

In the summer of 2024, before Ophirex had published any detailed results of varespladib in humans, the first hints began to emerge. One of the doctors who participated in the study presented preliminary results of the trial at a scientific meeting. On the basis of those sketchy details, and several background interviews with people familiar with the data, the company's initial Phase II trial could be summed up in a single word: snakebit. It wasn't just that the trial had a fitful rollout during the Covid pandemic, beset by unprecedented logistical problems. By its very nature, the design of the clinical test had so many variables that the outcome, as one source put it, was destined to be "messy." Among those variables: many different snakes, many different kinds of venom, and many different bodily locations of the bites, along with some very unlucky randomization of patients (five people in the trial, for example, suffered cobra bites; all five were randomly assigned to receive the placebo instead of varespladib). In 15 cases of snakebite, doctors didn't even know which species had bitten the victim.

The ultimate confounder, however, may have been ethics; all snakebite victims enrolled in the trial had to be treated with antivenom as well as the blinded pills. So when one of the participants, Malcolm Chandler, who suffered a copperhead bite in the backyard of his Durham, North Carolina, home, gave an interview to a local newspaper in 2022, he suggested the treatment had worked. "It couldn't have gone any better," he told the *Raleigh News & Observer*. "You hear horror stories about people who've gotten bitten, and I was afraid that was me." Instead, Chandler said, a month after being hospitalized, he was back to playing golf with his granddaughter. The problem was that, because of the way the study was designed, neither Chandler nor his doctors at Duke Health nor Ophirex knew if he had received varespladib or a placebo; Chandler did, however, receive nine vials of antivenom (reportedly at $11,000 per vial).

Finally—and this would normally doom any potential drug approval—Ophirex's clinical trial failed to meet its primary endpoint, worked out during preliminary discussions with the FDA. Lewin and his colleagues had designed the study to show that varespladib would demonstrate beneficial effects within six to nine hours after humans received the drug, just as their research had shown in mice. In what might be considered the revenge of the traditional model organisms, the mouse experiments turned out to be misleading. As Chandler's doctor, Charles Gerardo, told attendees at the biannual "Venom Week" meeting in June 2024, human patients did not respond as rapidly as mice, and the clinical trial failed to achieve its primary target. "While the study did not meet its primary endpoint in this Phase 2 study, a potentially promising signal of benefit was seen in the group of patients receiving treatment within five hours of the bite," Timothy Platts-Mills, chief medical officer, said in a statement.

At that point, it looked like that might be the end of Lewin's odyssey—great intentions, heroic persistence, no cigar. But there is a potential loophole—an ethically defensive one—in FDA regulations called the Animal Rule. It allows the agency to consider a drug for approval based on animal experiments alone when it would be unethical to conduct a human trial (to test the efficacy of a drug against the anthrax toxin, for example, you wouldn't want to expose human subjects to anthrax first). Ophirex representatives made that argument in discussions with the FDA in the fall of 2023, heartened by secondary data showing that snakebite victims who received varespladib within 5 hours had quicker and more complete recoveries, whether 7 or 28 days after being bitten, than did people who received antivenom and a placebo.

The FDA doesn't discuss pending drug considerations; without describing details, Lewin confirmed that the FDA had requested additional animal experiments, and he expressed hopes that the company would be in position to offer its novel treatment, either as a pill or as an intravenous injection (currently being tested), by the start of the 2027 snakebite season.

Until all the scientific and regulatory issues get sorted out, it won't be clear if Lewin's crazy idea turned out to be a good one, but experts in the community of venom researchers have been closely following—and cheering—Lewin's efforts to reinvent snakebite treatment. "He's definitely been a pioneer," Nicholas Casewell said, "no doubt about it. He deserves a lot of credit for this, no matter what happens next."

Even if it's shown to be effective, the Ophirex drug will never be a universal snakebite remedy. It only targets one of half a dozen classes of venom components, and in many snake species, PLA2 enzymes form a fraction of the overall venom. Casewell, whose Liverpool lab has also been developing novel snakebite therapies, expressed both optimism and a word of caution. "Small molecules offer some really exciting promise for certain types of toxins, particularly the enzymes," he said. "But, for example, to date there is no small molecule that has been identified as being broadly effective against three-finger-toxin neurotoxins. But there is a lot of antibody research in that space that is quite promising. So maybe we need antibody for one toxin type, maybe we need a drug for another toxin type." His current thinking: "We can't have a one-size-fits-all, so we're going to need combination therapies."

Casewell's group in England has been trying to repurpose two other already developed pharmaceuticals to see if they, too, might be effective against snakebite. One of the drugs, DMPS, is a "chelating" agent—a compound that attaches to and removes toxic metal from the bloodstream of people who have been exposed to heavy metal contamination, such as mercury poisoning; the other drug, marimastat, has been tested as a cancer therapeutic that targets the matrix of tissue surrounding tumors. As it turns out, both drugs neutralize "serine metalloproteinases," the metal-based enzymes that are a key component in many snake venoms. The Liverpool group has shown that these drugs block the activity of snake enzymes in animal experiments, and Casewell is on track to initiate a clinical trial in the Year of the Snake, 2025, to test the idea in humans. The Liverpool group also foresees possible combination therapies, building on a 2020 study in mice that tested a combination of one of the chelating drugs (marimastat) with the Ophirex drug (varespladib) against snake venom. "We did see a really big effect," Casewell said, "when we combined the two drugs." And Casewell has collaborated on research headed by David Baker of the University of Washington describing the creation of "de novo" never-seen-in-nature proteins that protected mice from the venom of cobra neurotoxin—a report that came out several months before Baker was awarded a Nobel Prize in October 2024.

For the time being, antivenom remains the best treatment for venomous snakebites. It saved Karin McElhatton's life and is the major reason for the small number of snakebite-related fatalities in the developed world. As the fitful progress of these early drug trials suggests, the process of discovering new treatments for "the most important neglected tropical disease you've never heard of"—with its staggering human toll of mortality and morbidity—is a painstaking, tentative, and unpredictable roll of the biomedical dice. The prospect of mitigating the devastating effects of snakebite appears more promising than it has since the invention of antivenom 125 years ago, but at present remains as elusive as in the days of the Brooklyn Papyrus. Lewin and Casewell have been heartened by the encouraging data shown by varespladib, DMPS, and marimastat. If the pre-clinical promise translates into regulatory approval, future treatments for the world's 2.7 million annual snakebite victims may well be cheaper than antivenom, available as a pill instead of an IV— and almost certainly better than onions.

Snake Road

Eastern Parkway, Brooklyn, New York

There is a road—a grand tree-lined 19th-century boulevard, to be precise, designed by none other than Frederick Law Olmsted—that qualifies as a "snake road" even though it probably hasn't seen a snake in a century. Eastern Parkway passes the main entrance of the Brooklyn Museum, which houses one of the world's most celebrated collections of Egyptian art. Statuary and panels in the museum's third-floor Egyptian galleries attest to the deep spiritual relationship between the pharaohs and the animals with which they shared their dynastic hours, years, millennia, and, in the case of the mummified animals (who had their own cemeteries), eternity. Serpents were considered to be animals with souls; indeed, as far back as the 1930s, museum curators used X-rays to detect the telltale skeletons of mummified snakes entombed in miniature coffins. The ancient Egyptians were arguably the first human collective not only to ponder our complex relationship with snakes—but to put it down in writing.

At a practical level, the so-called Pyramid Texts—inscriptions carved on the inner walls of small pyramids in Saqqara, about 20 miles south of the more famous pyramids of Giza and dating back to perhaps 2300 BCE—memorialize many snake-centric "utterances." These were exhortations designed to rid homes and agricultural fields of serpents. At the symbolic level, serpent deities played central roles in Egyptian cosmology. Ra, the sun god, fought a daily battle with the serpent Mehen during his circadian passage through the underworld. The evil god Apep, archenemy of the sun god, was associated with the cobra. The beneficent goddess Meretseger—"The One Who Loves Silence," her serpentine folds represented in a sinusoidal sculpture in one of the museum's galleries—prevented snakebite and protected against snakes. Embedded in these contradictory mythical powers is one of the earliest manifestations of the tension that snakes triggered in civilized societies. These texts and objects, surviving in some cases from more than 4,000 years ago, attest to the ancient realization that the human–snake relationship, even then, was complicated. "The Pyramid Texts are probably the first texts that could be described as accounts of the human-serpent relationship," said Edmund Meltzer, the esteemed translator of several Egyptian medical papyri.

But there is an even more important reason that the Brooklyn Museum lies on a snake road. Among its most treasured artifacts is one that has never been on public display.

On a rainy afternoon in March, Yekaterina Barbash, curator of Egyptian, classical, and ancient Near Eastern art, led me to a small, fluorescent-lit room in a fourth-floor study area of the museum, where she opened a wide shallow file drawer and delicately transferred a portion of the "Brooklyn Snakebite Papyrus" to a broad worktable. The section she brought out—at least 26 centuries old, brownish green, faint striations of plant fibers barely visible, like botanically lined paper, beneath the cursive script—lay sandwiched between two plates of protective glass. Its thickets of cursive symbols, mostly in black ink with "chapter headings" in red, are indecipherable to all but a sliver of Egyptologists who can "transliterate" the handwritten hieratic text back into hieroglyphs and thence into a modern language.

Remarkably, there were no English translations of this document until the last couple years.

The Snakebite Papyrus was first and foremost a medical handbook. It provided a list of dangerous snakes, the symptoms of their bites, and potential remedies (all natural, almost all fruitless) to offer the bitten. "Incidentally," Barbash said, pausing to indicate a particular hieratic sign on the page, a black vertical squiggle with an offshoot rising up to the right, "one thing I wanted to mention is that in Egyptian, the word for 'venom' and the word for 'semen' are homophones"— that is, words that sound the same but have different meanings. And since vowels are generally not written, she added, signs at the end of words known as determinatives help clarify the intended meaning. "This sign?" she continued, pointing out the determinative. "This is a cursive form of the hieroglyph for 'phallus.' It's quite recognizable. It's one of the determinatives of 'venom.'" Long before Jung and Freud, apparently, humans had made the connection.

Snakes were an intrinsic feature of life in ancient Egypt, not just in the arid deserts and riverine ecologies of North Africa, but in the very script of its foundational texts. The Pyramid Texts, the Book of the Dead, the surviving papyri—all were written in a hieroglyphic alphabet in which many letters are represented by snake figures, including several that are obviously cobras and others that are horned vipers. Almost all of those ancient texts describe snake deities. "Like many cultures," Barbash has written, "Egyptians saw serpents as creatures of the earth that embody primeval, chthonic qualities, involved in the process of creation."

The Snakebite Papyrus ended up in this fourth-floor workspace as an indirect result of one of the most infamous political scandals of the

19th century. Charles Edwin Wilbour, a journalist, Egyptologist, and political operative, ran a printing company in New York City that operated as a front for a kickback scheme organized by Tammany Hall; the city of New York contracted with the plant for all its printing jobs, and the company siphoned vast sums of money back to the city hall politicians. With the fall of Boss Tweed and his scandal-ridden political machine in the early 1870s, Wilbour's company became implicated in the ensuing scandals and lawsuits; he fully expected, according to one account, to be indicted. Instead he devoted his post-Tammany time (and, one presumes, Tammany-generated wealth) to becoming the foremost American Egyptologist of the 19th century. He traveled extensively in the Middle East, purchased a boat to travel up the Nile River, and purchased many ancient Egyptian artifacts, including papyri, during these visits. Long after his death in 1896, Wilbour's heirs donated a number of these papyri to the Brooklyn Museum in 1916. But another large cache of papyri was discovered in 1948 in a trunk that had been stored in the Paris hotel where he often stayed. One of those, stashed in an old biscuit box, was the Snakebite Papyrus. The toll of all that non-curated storage became evident when Barbash showed me a box of fragments—hundreds of cracker-size scraps of papyrus, some of which belong to the Snakebite Papyrus, piled up like jigsaw pieces to a puzzle that may never be solved.

For many years, the lone translation was in French by an Egyptologist named Serge Sauneron, who camped out in the Brooklyn Museum in the 1960s (Barbash pointed out some of his handwritten annotations along the edge of one of the glass panels). Sauneron's translation was published posthumously in 1989, following his death in a car accident. English-language versions have emerged only in the last several years. In any translation, the Brooklyn Papyrus is a grim testament to medical futility.

The first part lists 37 types of snakes (and one lizard, a chameleon), along with a description of the symptoms of their bites and a prediction of the fate of victims (the first page of part 1—and thus the first 13 snakes—is missing). A typical entry reads, "If it bites a man, he dies immediately" or "If he vomits, he will die." Those snippets come from *Snake Identification in the Ancient Egyptian Brooklyn Medical Papyrus*, a new English-language translation published in 2024 by Meltzer and Gonzalo Sanchez, a retired neurosurgeon and Egyptologist. Sanchez and Meltzer, who collaborated on an earlier translation of an important medical papyrus, enlisted the help of several prominent international experts on venomous snakes as co-authors—Nicholas Casewell, Wolfgang Wüster, and Gordon W. Schuett—to suggest the possible identity of snakes minimally described in the papyrus as, for example, "black entirely."

The second part of the Snakebite Papyrus, which Sanchez and Meltzer did not translate, lists recipes for potential cures, and it is here that the human desperation in the face of venomous snakebite, perhaps for the first time in print, emerges. Present-day physicians like Sanchez have good reason to view this compendium of Egyptian medicinal advice as a catalog of helplessness. Wendy Golding translated the entire text of the papyrus as her doctoral dissertation at the University of South Africa in 2020, and this futility is clear from the almost comic reliance on onions as a snakebite treatment—onions ground up and mixed with beer, onions mixed with wine, onions mixed with herbs and spices. Many of these onion treatments were intended to be emetics, although there is virtually no modern evidence to suggest that vomiting blunts the effects of snakebite. These early "pharmacological" interventions supposedly only worked when accompanied by specific incantations. In Golding's translation, one such exhortation reads: "Hail to you, Onion. Hail to you, tooth of the god! Hail to you, original/important tooth of Osiris! Hail to you, guardian who protects all the gods, by means of your name, that of Onion! May you enter into the stomach of X, son of Y. Ward off all the venom which is in there, by means of your name, that of Onion!"

Among other ingredients listed in various aspirational antidotes was everything from cumin and honey to pelican scat, hippopotamus oil, deer blood, sweet clover, a plant called baboon's hair, and bits of copper. Of these various remedies, Sanchez told me, "The treatments are essentially useless." Whatever their merit, the advice was clearly influential and heeded: The demand for one coveted herbal ingredient, silphium, became so great in ancient Egypt that the plant was stalked and harvested unto extinction. And Golding believes those seemingly helpless incantations may still have had medical utility in terms of calming snakebite victims.

The Brooklyn Papyrus confirms that snakebite commonly caused death in ancient Egypt, yet serpents also elicited reverence. Barbash reminded me of the central importance of snakes in early Egyptian cosmology, pointing out that the creator god Atum, from whom everything else in the world derived, was depicted as a snake. And, she said, the image of the goddess Wadjet, depicted as a rearing cobra from the Predynastic Period, which is to say at least 5,000 and perhaps 6,400 years ago, was repurposed as a kind of royal hood ornament known as the *uraeus* in the crown of kings—a hooded cobra poised to spit fire at enemies of the pharaohs.

Experts believe the Snakebite Papyrus dates back at least to the middle of the sixth century BCE, but Barbash doesn't rule out a date as much as a century earlier. And that's just a tentative time stamp on a much older Egyptian obsession with snakes. Scholars generally accept that the Snakebite Papyrus is based on an earlier version. How much earlier? No one knows, but the "Pyramid Texts" date back to perhaps 2300 BCE. And the recent Sanchez-Meltzer translation suggests that the snakes listed in the Brooklyn Papyrus may reflect ecological and herpetological conditions that existed as far back as 3500 BCE. In other words, Egyptians probably began to wrestle with the vexing duality of awe and dread of snakes at least five millennia ago—a tension that scribes immortalized, perhaps for the first time, in the form of the written word.

How did ancient Egyptians contend with that ambiguity? "Let me start," Barbash replied when I posed that question, "by saying they use the same word for 'reverence' and 'fear.'" She didn't need to say another word.

4

Dreams of Healing

~~~~~~~

Metabolism

The story of the Greek demigod Asklepios is the genesis of the universal symbol of the medical arts: the staff around which an entwined serpent nimbly rises. The rod is the logo of the American Medical Association. It signifies medical societies and pharmacies throughout the world. It graced the T-shirt of the young physician assistant who registered me for a Covid PCR test not long ago. It is one of those visual symbols that is so ubiquitous, so commonplace, so subtle an example of archaic branding that it almost ceases to rise to the level of consciousness. The intersection of the Snake and the Healing Arts goes so far back in time that it disappears beyond the horizon of the historical record and melts into myth. What is true and what is merely fanciful storytelling? It's virtually impossible to tell, but the question leads to one of the most astounding chapters in the lore of serpents as agents of healing in the history of humankind.

The myth part goes something like this: Apollo—son of Zeus,

god of disease and healing, source of sun and light—was the father of Asklepios. His maternal line is a little murkier. Some ancient sources claim that Asklepios's mother was Coronis, a mortal seduced by Apollo; others say Apollo alone sired his son (an act of parthenogenesis not unknown in the serpent world). In one story, Coronis cheated on Apollo, who condemned her to death without knowing that she was pregnant with Asklepios; as she succumbed to flames on a funerary pyre, he extracted his infant son from her womb. As a demigod, Asklepios began to learn the healing arts—in one version, Apollo sent his son to be schooled by the centaur Chiron, renowned for his extraordinary understanding of herbal cures and treatment of snakebites; in another version, Asklepios learned the secrets of healing directly from the serpents themselves, befriending a snake that licked his ears and whispered secret medical skills. In yet another variation on the theme, Asklepios watched a snake revive a dead serpent with the use of a magic herb—an herb that Asklepios then used to revive dead mortals. Asklepios put this snake-imparted skill into practice to resuscitate Glaucas, the young son of the king of Crete, who had fallen into a cistern of honey and suffocated. Multiple stories in the ancient literature describe roughly half a dozen instances when Asklepios brought the dead back to life.

In the petty vainglorious world of Greco-Roman deities, however, the god of the underworld, Hades, complained to Zeus that Asklepios was hurting business by reviving people instead of letting them die. Zeus, whose godly powers apparently extended to omniscient demography, worried that Asklepios's skill at immortalization, if it persisted and spread, would overpopulate the planet and doom the world. So he hurled one of his thunderbolts and smote Asklepios—nipping this immortalization movement in the bud—and, with him, earthly knowledge of his magical skills. Apollo, appalled, successfully negotiated with Zeus for the return of his son as a deity, as long as he renounced the habit of bringing back the dead. For his pains, Asklepios is still heralded as the original Greek healer.

If literary references were the same as job references, Asklepios would have the most impressive medical résumé in all antiquity. In *The Iliad*, Homer described him as "that unfailing healer." The Greek poet Pindar praised "the craftsman of mild remedies…whose hero's hands warded from weary bodies all disease." In a long passage in *The Metamorphoses*, Ovid described the journey of Asklepios, in the form of a snake, from Epidaurus to Rome, where the god "put an end to grief and pestilence; and as he came—the bringer of good health—so he remained." Playwrights such as Aristophanes in *Ploutos* and Sophocles in *Philoctetes* acknowledged his skill as a healer; historians such as Strabo, Pausanius, and Plutarch purported to document his miraculous cures.

If we accept current thinking that the "crystallized" version of *The Iliad* (the first print version) dates back to around 750 BCE, and the oral version perhaps two centuries earlier, the aura of Asklepios as the primordial healer of humankind shines across three millennia of history. And for the last 2,500 years, in the very first lines of the Hippocratic Oath, attributed since antiquity to the namesake physician who himself is said to have been a familial descendant of Asklepios, doctors have pledged fealty to Asklepios (right after Apollo) in the oath that binds, to this day, the professional conduct of physicians.

What does all this have to do with snakes, either now or in the distant past? What often gets obscured by the mists of mythology is that the popular belief in Asklepios as a healer left a vast brick-and-mortar residue of sanctuaries sprinkled throughout the ancient world—sanctuaries where snakes apparently played a central role in the birth of medicine, the creation of centers for healing (a kind of proto-hospital, if you will), and perhaps even the invention of holistic medicine. Within two centuries of the publication of *The Iliad*, roughly in the sixth century BCE, a cult devoted to Asklepios took root in ancient Greece; it gained prominence in the fifth and fourth centuries BCE and persisted well into Judeo-Christian times. And as classicists and archaeologists have documented over the last two decades, this cult of healing erected sanctuaries, hundreds of them, throughout the ancient world and practiced a mind-bogglingly wild form of healing that employed "sacred snakes" as an essential part of its therapeutic rituals.

A rural hillside outside the Greek city of Epidaurus in the Peloponnese, for example, became the site of one of the most elaborate sanctuaries dedicated to Asklepian healing—indeed, one of the most elaborate architectural complexes discovered from the ancient world. Part temple, part spa, part sanitarium, part psychotherapy institute, part arts center, and perhaps even part reptile house, this sprawling complex, known as an Asklepeion, became the mother ship of the cult and a magnet for the ailing and the disturbed. In a serene setting amid forests and streams, carefully sited on a hill surrounded by nature, patients would walk the grounds and relax during the day and then retire to a massive dormitory-like inner sanctuary (the *abaton*) at night for what was known as the incubation or temple sleep. Arrayed on couches or on the floor, these pilgrims partook, by some accounts, of an archaic form of psychedelic to promote dreams. "Snakes were encountered at every step," according to Carl Kerenyi in his biography of Asklepios. And in the darkness, free-ranging non-venomous snakes may even have been allowed to slither unrestrained throughout the gymnasium-like dormitory while the patients slept (to this day, one of the most common species of rat snake in Europe is known as the Asclepian snake). In the morning, patients would recount their dreams to the attendant physician-priests, who interpreted the dreams and translated them into a strategy for healing. "In the case of Asklepios, more than symbolic, the snake is really a sort of incarnation of the god," said archaeologist Melina Melfi of Oxford University, who has conducted extensive excavations at Epidaurus. "The snake appears instead of the god, in the milieu of the god, for healing."

The complex at Epidaurus was not a one-off, cultish shrine. A German archaeologist has cataloged an astonishing 903 sanctuaries dedicated to, or associated with, Asklepios throughout the Mediterranean in ancient times, from Kos to Athens, from Rome in the west to Pergamum in Asia Minor, as far north as Germany, as far south as North Africa. "The cult became so popular that, although Asklepios never had a monopoly on divine healing—the Greeks and Romans worshipped many gods and heroes as healers—he far surpassed all others in the duration and geographic range of his appeal," wrote Bronwen Wickkiser, a classics scholar at Hunter College who has authored several books about Asklepios.

Many of those sanctuaries, including the one in Rome described by Ovid, required a kind of sacred certification in the form of a live snake brought from an existing sanctuary. In *The Metamorphoses*, Asklepios *is* a snake. And these were not storefront chapels: The sanctuary at Epidaurus featured a gymnasium (which hosted athletic competitions described by Pindar), an enormous theater (which still presents classic Greek plays in the summer), a dormitory 60 meters long, a thermal spring, a bath, a spa, and a curious, circular building called the Thymele, with a labyrinth in the basement that some scholars believed housed the "sacred snakes."

The Asklepian sanctuary in ancient Athens occupied a spot on the southern slope of the Acropolis—"the oldest and most sacred civic space in Athens," according to Wickkiser, just below the Parthenon and immediately adjacent to the most important artistic venue in the Greek world, the Theater of Dionysus. Wickkiser argued that the cult based in Epidaurus held such geographic, and therefore strategic, value for Athens during the Peloponnesian War in the late fifth century BCE that it was important to the balance of Athenian power. The classicist Robin Mitchell-Boyask has suggested that the Asklepian concept of healing initiated the exalting and regenerative qualities of the arts, noting that in many Greek cities, including Athens and Epidaurus, the temples to Asklepios were often sited in close, non-random proximity to the great Greek theaters of the day. Greek medical historians have suggested that the elaborate facilities at the sanctuaries created a "total healing environment," marking the birth of holistic medicine. And David B. Morris, a biocultural historian at the University of Virginia, went so far as to argue that the cult of Asklepios flirted with the birth of psychotherapy, with its emphasis on the inner life of the individual (including the dream life) and its non-Hippocratic embrace of the erotic.

A crucial aspect of the cult's popularity is that its practitioners welcomed patients with chronic diseases that physicians of the Hippocratic school were reluctant to treat. And in an era when doctors were primarily itinerant, the Asklepian sanctuary represented a fixed site where patients from far and wide convened and stayed in one locale, a forerunner of the modern hospital. Pilgrims who visited these sanctuaries left hundreds of votive tablets as a kind of hand-carved blurb—what scholar Lynn LiDonnici described as a "fourth-century promotional campaign"—attesting to their cures;

snakes slither through many of these testimonials as agents of healing.

One tablet described how a woman unable to conceive came to Epidaurus to be impregnated by the god, according to Nikasiboula of Messene, "slept in the sanctuary in order to be favored with progeny and beheld a dream. She dreamed that the god had come to her, followed by a snake with which she copulated. And within a year she gave birth to two boys." Another votive described a painful, debilitating abscess that racked the toe of an elderly man until a snake licked the spot, curing the affliction. And in one of the most amusing testimonial-anecdotes, a young mute girl was running around in one of the sanctuaries when she spotted a snake. "Terrified, she cried out for her mother and father," according to a later account, "then she went away, cured."

To these anonymous anecdotes we may also add the playwright Sophocles, who reportedly subscribed to the Asklepian school of medicine and may even have undergone the snake therapy in his home. And finally there are Socrates's enigmatic last words—"Krito, we owe a cock to Asklepios; pay it and do not neglect it"—an apparent reference to the traditional expectation of patients to sacrifice a rooster at dawn on the altar of an Asklepian sanctuary.

The cult of Asklepios—with its reputation of healing and its essential association with snakes—permeated all levels of Greco-Roman civilization for nearly a millennium. "It would take a new cult based on the worship of another divine healer, Jesus, to ultimately close down Asklepios's sanctuaries," Wickkiser has written, "and then only by mandate rather than by obliterating popular interest in him."

Given the 2,500-year run of increasingly sophisticated and rational Hippocratic medicine, it has been easy to dismiss the mystical, pseudo-therapeutic, magical interventions of Asklepios as more myth than medicine, more irrational longing than plausible treatment. The Irish classical scholar E. R. Dodds may have reflected this disdain when he described Asklepios as a "medical reptile." And mere use of the word *cult*, given the contemporary context of the term, only cements the idea of belief over evidence, credulity over rigor, the magical over the rational. The preposterous notion of immortalization and regeneration in the origin story of Asklepian medicine was reason alone not to take it seriously as medical science.

In museums throughout the world, there are statues of Asklepios, bearded and virile and indeed godlike, serenely gripping his serpent-entwined staff, often with the snake seeming to impart secrets of healing to him. For centuries, that whispered knowledge was considered serpentine blarney.

And then came genomics.

～～

In October 1998, Stephen M. Secor and Jared Diamond published an extremely unusual and extremely influential essay in the journal *Nature* that proposed an extremely provocative idea. Instead of studying seemingly convenient laboratory mammals like rats and mice—"with their lifestyle of frequent small meals" and their barely detectable fluctuations in physiological processes—scientists should turn their attention to animals that had much more extreme metabolic behavior. Studying extreme biology might provide unexpected clues about how vertebrate animals (birds, mammals, us) regulated fundamental life processes such as food intake, digestion, energy conversion, blood chemistry, and growth. The animal with the most extreme metabolic behavior, they suggested, was the Burmese python (*Python molurus bivittatus*).

For several previous years, Secor and Diamond had published studies in less visible journals on python metabolism. Long before intermittent fasting had become a fad, Secor, a postdoctoral fellow at the University of California at Los Angeles with a long-standing love of snakes, and Diamond, a professor of physiology at UCLA with a broad, big-picture perspective on biology who had just published his bestselling 1997 book *Guns, Germs, and Steel*, had been searching for a new, non-traditional lab animal with such extreme dietary behavior that research on them might unearth unexpected scientific (and perhaps medical) surprises. Among the animals they considered: lions, wolves, deep-sea fish, and 18 different species of lizards, turtles, frogs, and snakes. They finally settled on pythons.

The dietary habits of pythons intrigued them the most. The snakes ate huge, infrequent meals, and could go months without eating at all. They could ingest prey 1.6 times their own total body mass, which they likened to a 62-kilogram (137-pound) human eating a 100-kilogram (220-pound) meal "in one gulp." Such huge meals triggered extreme responses in digestion—huge increases in stomach acids, huge spikes in triglycerides in the blood (which turned plasma from clear to milky white), huge changes in the size of internal organs like the heart, intestine, kidney, liver, pancreas, and lungs, and, perhaps most remarkable, huge increases in metabolic energy demands to deal with all the other changes. In order to digest their gargantuan meals, pythons had to muster energy expenditures so extreme that the only comparable example Secor and Diamond could find in the natural world was a racehorse at full gallop. All these changes occurred within a matter of days, and then just as suddenly disappeared, subsiding into a quiescent phase of fasting. How did they do it?

In the popular imagination, cold-bloodedness has always seemed like an inferior kind of lifestyle. Warm-blooded animals like birds and mammals maintained their own internal temperatures to keep life processes humming along, albeit by keeping the metabolic furnace constantly stoked. Cold-blooded animals like snakes (known technically as ectotherms) had to regulate their internal temperature by relying on their environment, basking in the sun when they needed to warm up, shifting to shade or water when they needed to cool down, and usually retreating to caves or underground burrows when they needed to escape fatal summer heat or life-threatening freezes (snakes technically don't hibernate but rather "brumate" during the winter). From the perspective of warm-blooded chauvinists, they couldn't do it on their own; from the perspective of others, you might say they were ecological savants, creatures that developed an exquisitely sensitive understanding of their thermal environment as if their lives depended on it, which they did.

And Secor and Diamond were particularly intrigued by the energy requirements of cold-blooded versus warm-blooded animals. As they wrote in their *Nature* article, "Someone driving a car in normal traffic finds it cheapest to keep the car's engine running while stopping briefly at traffic lights, but turns off the engine, thus saving fuel consumption, and restarts it after stopping at a railroad while waiting for a long train to pass." Warm-blooded animals had to keep their engine constantly running. Snakes, like other cold-blooded animals, spent most of their time at the railroad crossing, turning the engine on and off as needed. It was almost like a temporary, reversible death.

Gasoline as an energy metaphor seemed to be much on Secor and Diamond's minds. They likened the python's metabolic response to a huge meal as "pay before you pump"—meaning the snakes had to turn on and regulate a massive biological machinery to process their meals before they could derive energy from them. How did a seemingly primitive animal deal with such a severe metabolic challenge? Secor and Diamond didn't offer any particular answers, but they ended their essay with several compelling reasons why researchers should consider using juvenile pythons as experimental animals. As vertebrates, their biochemistry more closely resembled that of mammals than did that of fruit flies or squid, which were common lab animals. They were "far cheaper, easier and cleaner to house, maintain and feed" than rodents (laboratory snakes ate at most two meals—and produced only two "semi-solid defecations/urinations"—per month). Juvenile pythons were docile and "less likely to bite than are rats." And, they added, "They do not arouse the controversy associated with medical research on similarly sized mammals." Because many people did not like and in fact loathed snakes, animal welfare was unlikely to be a public concern.

There wasn't a rush by scientists to pet stores to stock up on pythons for research, and although Secor and Diamond posed some fascinating biomedical questions, scientists still lacked the technology to take a crack at answering them. Nonetheless, the Secor and Diamond paper resonated with a lot of basic researchers. David Julius at the University of California–San Francisco, a future Nobel laureate, remembered reading it as he contemplated a series of experiments using snakes. David Pollock, a researcher at the University of Colorado Medical School, was influenced by it.

Leslie Leinwand, an expert on cardiac biology, went further. With the *Nature* article "absolutely the impetus," she began to experiment with pythons in her University of Colorado laboratory several years later, even recruiting Secor as a collaborator and at times housing up to 50 snakes in her lab. Her group published an article in *Science* in 2011 that got a lot of attention in both the scientific and the lay press, reporting that pythons created a unique group of fatty acids in their blood after a meal that seemed to dramatically increase the muscle mass of the heart without pathological consequences. Science writers were smitten with the idea. "And the day may come," Lawrence Altman wrote in the *New York Times*, "when doctors literally prescribe snake oil for heart disease." When I spoke with Leinwand in 2013, she made a point about how difficult the research was. "If I had realized how hard it is to work on them, I might have rethought the whole thing," she said. "These animals—you have to be careful. To a python, we are food." Her initial experiments were so promising that she even co-founded a biotech company called Hiberna Corp. to investigate python-related compounds as potential pharmaceuticals for heart conditions.

What was still missing was the ability to track what was going on at the molecular level in pythons in real time as they ate and digested one of those extreme meals. Genes could lead to molecules; molecules could lead to pathways; pathways might explain the astounding biological feats these serpents performed. That kind of experiment seemed far afield from the normal lanes of biomedical research. But Pollock, also at the University of Colorado, had a reputation for taking on weird projects. And he had just the person in his lab to work on pythons: an avid snake enthusiast and technology nerd steeped in all the latest genomics technologies named Todd Castoe.

~~~

In October 2011, exactly 13 years after Diamond and Secor's call to arms, a handful of scientists (barely 20, minuscule by the standards of most scientific meetings) gathered for three days to talk snakes and molecules in Vail, Colorado. It was the first-ever formal international gathering on snake genomics and, like so much of snake science, it probably seemed at the time like a biological afterthought, coming more than two decades after the launch of the Human Genome Project. The meeting was organized primarily by Castoe, a postdoctoral fellow at the University of Colorado School of Medicine at the time, and Secor, by then a professor at the University of Alabama at Tuscaloosa.

If you were polite, you might say this was fringe science; if you were blunter, like some of Castoe's colleagues back at the med school in Aurora, Colorado, you would tell him he was crazy to think genes could teach him anything about how a snake metabolically processes its dinner—it all happened much too quickly for genes to make a difference. Castoe and Secor and a few other colleagues were scientists at medical schools—*medical* schools—who thought it might be useful to read the entire genetic script (or genome) of snakes, not just pythons but also similar projects under way on the garter snake and king cobra. The Vail contingent was undeterred. "Applications of these first three snake genomes will include a better understanding of extreme physiological adaptation in snakes," they wrote in the meeting summary, "with applications to the human condition..."

Castoe had never seemed to care what others thought. He grew up in Poughkeepsie, New York, about 90 minutes north of New York City by train. He had a boyhood love of snakes and lizards ("nothing crazy, about average"), a geek's love of gadgetry and tech, and an exuberant, colloquial way of expressing his scientific passions and opinions. Those affections don't quite explain how he ended up at what was basically a small forestry college, SUNY-Syracuse. But during his PhD research at the University of Central Florida, he became fascinated by some genetic oddities in the metabolic machinery of snakes, specifically their mitochondria. By the time he headed west to the University of Colorado, he had found a kindred spirit in David Pollock, spending what would turn out to be a five-and-a-half-year postdoctoral fellowship in a lab that gleefully pursued non-traditional science. Castoe was the right person in the right place at the right time. As he recalled, Pollock was "really interested in non-traditional model systems, really interested in extremes of adaptation, really interested in molecular genetics."

At the same time, genomics—deciphering the entire genetic text of an organism by recording its sequence of DNA—was taking off, and a new wave of high-tech machines known as next-generation DNA sequencers began to hit the market. There was cutting-edge molecular method to this seeming snake madness. Castoe had a gift for coaxing brand-new next-gen sequencing machines from the biotech companies that manufactured them. "I was kind of a 'biostitute' for the early sequencing companies," he admitted. "They used to sponsor a lot of the python work, with free reagents, and I gave talks for them all over the place." It was, he recalled, "just like the Wild West at that time."

Castoe and Pollock started using those machines to compare genes involved in metabolism between snakes and other animals. "And we began to see some *really* weird stuff happening," Castoe recalled. They were specifically looking at the genes controlling mitochondria, tiny organ-like islands inside each cell that control energy. Practically since the birth of science writing, mitochondria have been described as "the cell's power plants" because of the way they hold the metabolic genes and biochemical instructions telling cells how to use oxygen to convert nutrients into the energy that sustains all living organisms, including plants. "Basically, the short version [of what we found] is, mitochondrial genes—the reason that you breathe oxygen, literally—had changes in them in snakes that are not observed in any other vertebrate, or any other animal, for that matter," Castoe said. "What that means, at the simplest level, is, 'Hey, snake proteins work fundamentally different than other animal metabolic proteins.' What the heck is going on?"

As Castoe and Pollock pondered what the heck was going on, they ran across Secor's research with Jared Diamond and realized that there was even more evidence than they had realized to support the idea that snakes had evolved an out-of-the-box form of metabolism, in part because they were cold-blooded and in part because of the long and unpredictable intervals between meals. So they reached out to Secor to see if he was interested in collaborating.

Secor was more than eager to dive in. He, too, had grown up in upstate New York and, unknown to Castoe until many years later, attended the same small forestry school, SUNY-Syracuse, a few years earlier. As a herpetologist, Secor had done research on sidewinder rattlesnakes at Arizona State University, where he got a "lifetime get-out-of-jail-free card" (Castoe's take) when he survived a dry bite to the groin by one of his rattlesnakes—a snake he'd imprudently been carrying in a cloth bag cinched to his belt loop. As an accomplished physiologist, he recognized the metabolic inventiveness of snakes, especially pythons, and brought that to the attention of Diamond when he was both a graduate student and postdoc at UCLA. For years, Secor had been looking for a kindred herpetological spirit who had chops on the genetics side, so when Castoe and Pollock approached him, he leapt at the chance. "It was a perfect match," Castoe said. "We worked with Stephen for many years, co-developing this system." After pausing a moment, he added, "Well, that's actually giving us too much credit. I would say that he developed the system, he developed the arguments for the system and informed science to the fact that it [the python metabolic system] is quite remarkable." Given Secor's early role in shaping this overall research, it's doubly tragic and sad that he is incapable of describing it from his perspective after suffering a debilitating stroke and officially retiring in 2021.

But he was hale and healthy at the first snake genome meeting in 2011, which was when the scientific forces began to align to tackle the python genome. No one had completed a snake genome yet, although several were on the way, and by the time the first DNA sequences came out a couple years later, it was clear that snakes possessed a kind of metabolic and biochemical wizardry that, to use another favorite Castoe phrase, "blew our minds."

Over the next two years, Castoe, Secor, and a large group of collaborators—39 in all, twice as many as attended the original meeting in Vail—scoured python DNA for clues as to how the snake could pull off the trick of regenerating its organs seemingly at will, pumping them up when needed and then whittling them down to size when digestion was accomplished. They did this with the knowledge that, at the genetic level, snakes seemed to pull off this trick with essentially the same basic repertoire of genes as other vertebrates. "Yeah, snakes kind of *look* different," Castoe told me. "But we've looked at a lot of other vertebrates in the past, and it's remarkable how much of the genome seems pretty highly conserved." That was another way of saying that, as extreme as python metabolism was, the basic genetic blueprint running the metabolic show was pretty similar, suggesting that what the researchers discovered in snakes might have relevance to other organisms, including humans.

When Castoe and company started doing their first experiments, looking to see which genes became active, or "expressed," 24 hours after the snakes broke their fast, colleagues in the medical school were withering in their skepticism about the python experiments. "Everybody in the medical school said, 'You're an idiot.'"

They believed that it would be impossible to see palpable physical manifestations, known as the phenotype, in snakes mere hours after a gene (or genes) became active. "Stop looking," they told him. There would never be enough time for the genes to churn out enough new proteins to wreak wholesale changes in the organs of the animal. After a theatrical pause, Castoe added, "Well, they were wrong."

The first evidence emerged in 2013, with the publication of the Burmese python genome in the *Proceedings of National Academy of Sciences*. Castoe and a huge army of collaborators collected python DNA before the snakes ate a meal, and then re-collected their DNA 1, 4, and 10 days after eating. With the new technology, you could see if a gene was turned on (and therefore ordering the creation of a new protein) or remained turned off. The researchers observed a "rapid and massive" activation of roughly 2,000 genes that leapt into action within 24 hours of the snake eating a meal. These genes weren't just opening the spigots on some digestive juices; the wholesale genetic activation seemed to orchestrate changes in both the size and the function of multiple organs.

Indeed, the researchers identified modules within the overall genetic architecture of the snakes—"suites of radical adaptation," they called them—that allowed the animals to evolve quickly in the face of environmental change or challenge. These fast-response genetic modules influenced the entire process of eating and metabolism, from the evolution of a "kinetic skull" that allowed snakes to eat prey larger than their head to the rapid creation of digestive proteins to an unprecedented ability to "remodel" their own organs after feeding. And many of the genes that drove these changes seemed "intriguingly" similar, the researchers wrote, to human genes known to be associated with metabolism, development, and pathology.

That initial report generated a lot of interest in the press; the occasional oddball snake story was always worth a short feature in the popular media. But it's in the afterlife of a major paper that a lot of the nitty-gritty details, painstakingly worked out in subsequent experiments, begin to coalesce into a bigger, often more profound picture. In 2012, Castoe set up his own independent lab at the University of Texas–Arlington, and roughly every two years thereafter, he has added a new chapter to the python metabolism story—more precise, more molecular, and more surprising.

In the initial follow-up work, Castoe and his colleagues started tracking changes, genetic and physiological, that occurred within a narrower time window. In 2015, they reported on experiments where they collected python DNA, for example, at three hours after a meal, six hours after a meal, and so on, with follow-ups at 12, 24, and 72 hours. These experiments involved frenzied dissections in Secor's Tuscaloosa lab as the team separated out each organ at each time point, weighed the wet mass, and instantly froze the tissues for later analysis. (Secor, according to colleagues, even preserved some of the python vertebrae to fashion jewelry in his spare time.)

There is a common misconception that DNA is DNA; the same instruction manual is in every cell, and the text doesn't change. But one of the techniques of modern genomics allows scientists to look at different tissues and see which parts of the text are being read (and used), almost the way a well-thumbed page in a book reveals a key passage. In the case of genes, this technique reveals which genes have become activated (or silenced) at different time points. In Castoe's telling, they saw "humongous" shifts in gene expression within three hours of a meal. After the long train of fasting had passed, turning the python's engine back on required a massive metabolic mechanism kicking into gear much sooner than anyone thought possible— in snakes or any other vertebrates—and involved nearly a quarter of the entire genome. "We have gone down and measured that to three hours after eating," he said, "and we still see thousands of genes that we can significantly detect are at different concentrations. Which means they were turned on the *second* that thing started happening [that is, the python began to eat]. Totally wild. And I think it also reset our expectations. We stopped listening to people's expectations about how fast certain processes could affect phenotype." In other words, how quickly gene activation translated into changes you could actually *see*.

And then things got *really* weird.

~~~

It would be an exaggeration to say that pythons invented intermittent fasting maybe 70 million years ago—but not by much. And they weren't alone as innovators. Pythons—so-called Old World snakes that evolved in Africa and Asia—developed a completely unique system to process food and manage energy. But so did boa constrictors, so-called New World snakes that followed a different evolutionary path in a different hemisphere, but still came up with the same metabolic tricks. That became clear by 2017, when Castoe's team showed that, in order to handle extremes of both gluttony and fasting, several lineages of snake evolved metabolic hardware and genetic software not seen in other vertebrate species. "You only see this, or see it arise multiple separate times, in snakes," said Castoe. "Without question, three very distinct lineages do this, and there's stuff in between them that does not do this. All of them eat *really* infrequently." Which left Castoe and colleagues scratching their heads over a very colloquial scientific question: Why do snakes do stuff so differently?

Whenever you talk about metabolism, you're basically talking about energy—how an organism obtains, processes, stores, and utilizes the energy necessary to stay alive. And biological energy can be thought of in economic terms, like income and expenditure. Warm-blooded animals—birds, mammals, us—spend a lot of energetic capital keeping the warm-blooded furnace stoked, so they have to eat frequent meals; there's always an energy bill to pay soon, so they're always hustling to make the metabolic rent. That type of system requires a high metabolic rate, a big budget to maintain the machinery that functions day in and out and, not coincidentally, causes a fair amount of stress at the cellular level that comes with burning up a lot of oxygen to continually manufacture that energy. And this is the first place where snakes in general, pythons in particular, do things differently.

Experiments conducted by Stephen Secor in the early 1990s had shown that pythons "have one of the lowest basal metabolic rates of any vertebrate ever," according to Castoe. "What that means is that a fasting python uses less energy per gram of tissue than any friggin' vertebrate you've ever seen. Somehow their idle is so low that it's really unique." How unique? A return to the Diamond & Secor Railroad crossing helps to put this in perspective. "Something unique about the python allows [them] to turn their motor down so low that they probably get humongous energy savings from that," Castoe said. "If they can keep it that low for eleven months of the year, and only throttle it up on short notice to eat a rat and shut it back down, it's like the trade-off of turning off your car at a train crossing but not at a stop sign. For organisms that eat once a year, most of the year they're stopped at a train crossing. Therefore, to keep their engines idling for the eleven months they're sitting there is stupid."

What do pythons and other snakes do with their energy savings?

They reinvest it and diversify. They produce more eggs. They reproduce annually instead of biannually. They *proliferate*. "Energy is constantly a commodity in snakes," Castoe continued. "For a frequently feeding organism, building all this stuff from scratch and throwing it all away when you're done, and then rebuilding it again tomorrow, is extremely energetically expensive. Because there's a tremendous amount of waste in throwing it all away." Another way of thinking about it is that warm-blooded animals are big spenders, energetically speaking, while snakes are penny-pinchers. Those different spending habits writ large (predation, resource depletion, garbage) have ecological consequences, too.

All this became clearer when the genome scientists took a closer look at some of the 2,000 genes that leapt into action when a python stretched its jaws to ingest a meal. The most obvious changes were immediately visible, with changes in the size of internal organs that you could see with the naked eye. Within three hours of a meal, massive changes began to take shape in the organs and chemistry of the animals, and it wasn't simply a matter of that anthropomorphic bulge in the belly so beloved of *New Yorker* cartoonists. Organs changed size, shape, and function.

The first thing to register change was the heart. It suddenly got a lot bigger. In humans, an enlarged heart is usually a pathological condition known as cardiac hypertrophy, a sign that the heart is not pumping vigorously enough; in pythons after a meal, an enlarged heart seemed to be not only healthy and natural but actually *necessary*. Shortly after pythons began breaking down their food, the fats and triglycerides in the snake plasma became so thick that Castoe likened the blood to "whipping cream—40 times the lethal dose of triglycerides and fatty acids that would kill a human." The python heart needed to get bigger, he continued, "to pump this sludge through the body." Once that task was complete, the snake seemingly canceled the enlargement part of the program, and the heart shrank back to normal size. Over the course of a lifetime (and pythons can live for decades), these snakes could regenerate and degenerate crucial organs at will, gearing up when digesting a meal and standing down to idle at the railroad crossing.

There were dramatic changes in the stomach. In a fasting snake, the pH of the stomach is 7—the same as water. "It's not even producing stomach acid," Castoe said. "It's not doing anything. Its physiology is off. It's not functional. It doesn't ramp down. It's *off*." In other words, no energy is being expended to do the expensive metabolic work of churning out digestive enzymes and all the other molecular bit players that, say, humans use to break down, convert, and utilize their three square meals a day. And when the snakes began to digest a new meal, which again could be a whopping one and a half times their total body mass, they turned the engine back on, producing a gush of acid that broke down muscle, bone, cartilage, tendons, alligator hide, whatever came down the gullet. That process began in the first hours of eating and lasted for about six days.

There were dramatic changes in the intestines, where the body absorbs broken-down nutrients. In a fasting snake, the intestine is a tiny, flaccid little tube. "When it's fasting, it's really hard to find!" Castoe said. After the snake swallows a meal, that tiny tube transformed into an immense, fluffy organ that looked like a rolled-up shag carpet. The "shag" part was an intestinal structure known as the villi, which increased five-fold in size to absorb the nutrients. "You don't need a microscope," he said. "You don't need anything. It's that obvious." There were similar changes in the liver and kidney—a spurt of growth followed by a kind of deliberate wasting.

Some of these changes had been observed and known for a long time. And even humans typically regenerate some tissues; there's a regular turnover of cells in the intestines and skin, for example, and people can regenerate parts of the liver. "Not *anywhere near* the scale that we're talking about in snakes," Castoe said. But it was more than just the scale of the changes; it was the totality of the process. Snakes knew how to throw this whole metabolic machinery into a reverse gear. They could flip an ON switch and regenerate tissues in key organs, and then flip another genetic OFF switch (multiple switches, really) and precisely cull away the new tissue, using their genetics as a molecular scalpel. These changes—dramatic growth and shrinkage, rapid cell proliferation and targeted cell death—were driven at the cellular level by well-known molecular processes that build and eliminate cells.

Despite all this seeming progress, Castoe faced a crisis known all too well by snake scientists: funding. The Texas group banged the can in 2015, submitting several ambitious grant proposals to study all aspects of organ regeneration in snakes, but the National Institutes of Health turned them down. "Almost out of desperation," Castoe said, his lab threw together what he called a "totally boring" grant proposal to the National Science Foundation for a project that focused solely on the changes in the python intestines. The proposal got funded. The results were anything but boring.

Proliferation is usually a dirty word in biology. It's the defining feature of cancer, where cells multiply uncontrollably until they form masses and tumors. And programmed cell death (a process known as apoptosis) is one trick many organisms, including humans, have evolved to try to stay one step ahead of an incipient cancer, by detecting proliferating cells and eliminating them. Pythons and other snakes had developed a work-around, a way to manipulate this life-or-death yin–yang in a profoundly novel way. In a creature that celebrates and illuminates dualities, this biochemical paradox may be the most impressive (albeit microscopic) feat of all.

"Not only do we see a scale of regeneration that's remarkable—it's absolutely unparalleled in any other vertebrate than pythons," Castoe said. "We see it reversed! So it's almost like you're undergoing cell proliferation, somewhat analogous to cancer, new developmental growth when you feed. Eight to 10 days later, those cells are undergoing very careful programmed cell death in many cases, and being purposefully killed and culled back out..." It was cellular addition and subtraction, on-demand regeneration followed by precision wasting, exquisitely orchestrated, again and again and again. As Castoe put it, "Pythons are really damn good at very carefully removing those recently proliferative cells." It's almost like the snakes create a short-lived form of cancer to meet the extraordinary demands of digesting a huge meal, then "cure" the cancer by cutting it out once digestion is complete.

In *The Giant Snakes*, Clifford Pope's famous 1973 book on pythons and other huge serpents, he devoted one sentence to metabolism and noted, "An animal able to take in 400 times its daily energy requirement at one ordinary meal must have interesting digestive processes." It took half a century to understand just how interesting. Most of the "data" Pope described about the digestive process of giant snakes came from experiments using X-rays. The images were cool, showing the bones of prey animals slowly dissolving and disappearing in sequential X-rays of the gut. But there was nothing about molecules. And Castoe's snakes seemed to break a few molecular rules, too. When mammals or birds ate a meal, it was like briefly turning on the metabolic furnace; when pythons ate a meal, they ignited a metabolic conflagration so big that it threatened to burn down the house. The animal consumed enormous amounts of oxygen to build up the organs and process the meal (this was the gal-loping racehorse), and that in turn inflicted an enormous amount of metabolic stress on the cells. Most animal cells, like furnaces, have a cutoff switch when things get too hot; pythons did not. Castoe and Secor kept adding bits and pieces to this increasingly detailed met-abolic story, but it all amounted to a puzzle. How did pythons use a set of genes, common to all vertebrates, to do something no other vertebrate could do—essentially let the metabolic fire rage, and enlarge all these critical organs, without shutting down or burning up? "We're missing something," they told themselves. "What's the missing link?"

They haven't completely solved the conundrum, but they've found that pythons and other snakes have evolved an almost bizarro-world, alternative form of biochemical processing that may ultimately affect our understanding of diabetes, cancer, and the abil-ity to regenerate organs and tissues. In a postmodern update of the Asklepios myth, perfectly adapted to our genomic age, the snakes have begun to whisper molecular secrets of potential healing.

~~

When humans have thought about snakes, they've been hung up on conventional, physical metrics—how many feet long, how much they weigh, what the symptoms of a venomous bite might be—as if those might be the most distinguishing traits of the animals. And even in the early part of the genomics era, the scientists focused on how those 2,000 or so genes correlated with changes they could see with the naked eye. They were still focusing on the big, but some-what misleading, picture. The most stunning differences between python metabolism and warm-blooded metabolism takes place in the invisible realm of molecules and pathways. Somehow snakes devised a way to run through a molecular stop sign obeyed by all other organisms.

That became clear by 2022, when a wonky scientific word—*non-canonical*—began to creep into the descriptions of the python researchers. When it came to basic metabolic biology, they began to think, snakes broke one of the cardinal rules of living things. Anyone who has dived into the small print of intermittent fasting has probably encountered a mumbo jumbo of jargon involving terms like insulin-like growth factors (IGFs), mammalian target of rapamycin (mTOR), and insulin itself. These are all molecules crucial to pathways that promote growth. It turns out that pythons speak that same mumbo jumbo—but use those terms to say profoundly different things.

When Castoe and crew dove deeper into the "boring" question of what happened in the intestines when pythons broke their long fasts, they stumbled on some of these same growth pathways. Within hours of feeding, there was that huge gush of nutrients in the snakes and therefore a huge burst of cellular growth, as you might expect, when the animals turned on the machinery to regenerate a bigger heart to pump the sludge and a bigger intestine to digest meals as large as alligators, feral pigs, and white-tailed deer, as pythons do in Florida. In most mammals, like us, however, the rapid growth of cells flashes a red light, because it potentially indicates not only the kind of wild, unchecked cellular growth typical of cancer, but also inflammatory conditions like Crohn's disease or irritable bowel disease. This kind of rapid proliferation, along with the conflagration of basic metabolism, normally triggers a lot of stress in the system, and too much stress in turn usually activates a pathway to a suicide button inside the cells, apoptosis, to halt the rapid growth. In most organisms, that is the "canonical" drama at the level of cells and molecular pathways. If you see something (a wild proliferation of cells), say something ("Stop!") and do something (hit the kill switch).

That wasn't happening in the pythons. Aundrea Westfall led the effort in Castoe's lab to drill down and look at what the snakes were doing at the molecular level, in the pathways that controlled growth on the one hand and responded to stress on the other. "What we see early on is that major growth pathways, like insulin and mTOR, are like one massive output," Castoe said. "That signaling was through the roof in the regenerating species." But they also noticed something very peculiar. Several pathways designed to detect cellular stress and shut it down went into overdrive, too. One of these stress-response pathways, driven by a signaling molecule called NRF-2, pricked their attention. Although there was relatively little about it in the scientific literature, NRF-2 was a kind of master gene, or transcription factor, that sat at the center of a complex regulatory network whose primary function was to manage stress—oxidative stress (from the metabolic conflagration), proliferative stress (from cancer-like growth), even chemical and carcinogenic stressors. Moreover, it had been turning up in research on humans who achieve super longevity. "That was weird," Castoe recalled thinking. The other pathway, which had only recently been discovered, was something called the unfolded protein response. "It was a pathway that was all over the place in our snake data," Castoe said.

These stress-response pathways allowed pythons to tolerate extreme growth signals—including signals for regeneration—that would be fatal in almost any other organism. "Python blood, its plasma, is really full of triglycerides and fats and stuff early on, more so than any vertebrate would normally live through," Castoe said. "We also know that insulin signaling is like 40-fold what would be lethal in a human being."

And it wasn't just in pythons. The boa constrictors, too, were intermittent feeders, and they ran through the same metabolic stop signs. By 2022, exactly a decade after the initial python genome paper, Castoe's lab reported findings on the metabolism of boas that received much less attention than the original genome paper but described a much more striking observation. Researchers claimed to have identified a shocking physiological trick evolved by these snakes to override normal biological guardrails that prevented unbridled growth—a trick that allowed snakes to regenerate organs, to overcome too much insulin ("insulin shock"), to avoid the usual, or "canonical," consequence of insensitivity to insulin, which in humans is the defining characteristic of type 2 diabetes.

"All this nutrient uptake fires up insulin signaling and mTOR, and that leads to increased and increased and increased levels of growth," Castoe said. "Cellular stress goes through the roof. High cellular stress immediately leads to cell death as a protection that's broad in all multicellular animals for protecting against cancer. What that means is, by default, this excessively high growth signaling kills cells in their tracks. There's a reason that other vertebrates don't have this. We'd all be dead if we had this level of insulin and pro-growth signaling."

So what do the snakes have that other vertebrates don't? Almost a metabolic magic trick. At the same time that the system is screaming, "Too much growth, shut us down," the parallel stress-response system jumps in and whispers, "No worries, we got this." "It's like literally switching two train tracks," Castoe said, hopping aboard the metabolic railroad again. "It basically switches from the pathway that [says] this much growth signaling leads to cell death, switches it over to the northbound track, which is, 'Nope, we're okay, cellular stress is not a problem, keep on trucking.'" By overriding programmed cell death in this extremely stressful post-feeding period, the snakes acquire the ability to regenerate organs. Again and again and again, after every intermittent meal.

The snakes had one more trick up their sleeves—a trick seemingly and surprisingly shared, among all animals on the planet, with a select group of humans who have undergone a very particular form of surgery.

～

Around 2020, the Castoe lab acquired a shiny new technological toy. It was a genomics machine that conferred unprecedented biological power—it allowed them to identify and segregate distinct classes of cells within a tissue (such as the intestine) and then reveal what genes were active in each subset of cells. In short, they could discern what specific subsets of cells were doing—what proteins they were making, what signals they were sending, what other groups of cells they were talking to—*at any time point during digestion.* This was another way of pushing past the myopic focus on "bigness." Organs were not simply a mass of identical cells; they were a mosaic of different cells doing different things at different times, and parsing their activities could reveal a kind of microscopic hierarchy. Now they had a tool to create an organizational flow chart revealing the executive genes that controlled python metabolism.

Using this new technology, they've identified certain populations of cells marbled throughout the lining of the intestines, tracked their activity through different post-prandial time points, and identified a subset of intestinal cells known as enterocytes that wheel into action about six hours after the snakes eat and seem to orchestrate many of these transformative events, from the regeneration of organs to the negotiation of tension between growth and stress. These cells are known as Best4+ cells, named for a molecular landmark on their surface known as bestrophin-4, and they were not even known to science until 2013. Castoe's lab has shown that these cells act like a kind of command-and-control center for extreme metabolism, leaping into genetic action about six hours after a meal, sending a flurry of signals to other cells, and not only triggering the regeneration of intestinal tissue but also resetting the entire metabolic process, including how insulin and other growth signals are interpreted by the organism. And there are two very general punch lines to this highly technical discovery.

One, snakes and humans possess Best4+ cells; laboratory mice don't. "Almost everything we know about the human intestine is based on the mouse model," Castoe said. "Everything we *think* we know. And it actually suggests that there is potentially regenerative capacity that we've *totally* missed in the human intestines, mostly because we have not been using a model that makes any sense with regard to the role of Best4 cells." Castoe believes the Best4 cells instruct other cells to respond to the gush of insulin signaling in a radical way.

In other vertebrates, cells typically become fatigued by the relentless prompts to grow and eventually tune out the message; they become insulin-insensitive, which is precisely the situation that defines type 2 diabetes in humans. "Every other vertebrate turns the volume way down as soon as it starts," Castoe said. Snakes don't. They let the insulin song play loud and long, like a heavy metal anthem, because they have a molecular way to prevent insulin insensitivity. And as part of this process of regenerating organs, the Best4 cells in python intestines appear to reawaken potent molecular pathways, dormant since embryogenesis, that control growth and wound healing. The python experiments lend weight to an important rethink already under way among biomedical researchers: that the gut plays a previously underappreciated role in controlling insulin pathways and metabolic processing.

Second, this striking constellation of phenomena—tissue regeneration, reawakened embryonic growth and wound-healing pathways, and a total metabolic reset—has been seen in only one other scenario: humans who have undergone gastric bypass surgery. For at least 15 years, doctors have observed a surprising and welcome side effect in patients who have had parts of their intestines removed because of obesity: They are cured of type 2 diabetes. In a recent NIH-led study of several thousand patients, roughly 85 percent achieved remissions. These patients, according to Castoe, apparently reactivate almost all of the same molecular signaling pathways that his group has been tracking in snakes when they break their fasts and regenerate their organs.

"If you want to summarize this stupid jargon I said in the last five minutes," Castoe said, after walking me through this molecular rigamarole of snake biochemistry, "pythons regenerate organs by avoiding getting type 2 diabetes. That's the point, literally. That's what we think is going on. It's nuts!"

In many scientific narratives, there's a point where exciting results can edge into "just so" stories. Findings are tantalizing; everything seems to fit; Valhalla is just over the next hill. The University of Texas researchers haven't conclusively proven anything yet. When Castoe first described the most recent research to me, it hadn't even passed peer review at a journal. But even if the findings require modifications, as they almost certainly will in a molecular odyssey whose surprises have unfolded like a Russian doll, the research has underlined the surprising, and promising, parallels between snakes and humans.

"What I think that means," Castoe said, "is that there is evidence for a pretty exciting, kind of backdoor, poorly known, but highly conserved regenerative capacity for vertebrates, that's in common between things like snakes and humans. We don't really know or understand it much, because it's very hard to observe. It's very rare. But it's there. The blueprints are there."

Castoe is an enthusiastic and bubbly scientist when it comes to talking about what might be considered the metabolic magic realism of snakes. Because of the complexity and invisibility of the dramatic molecular events he and his colleagues have uncovered, it's hard to imagine they would play well on a nature show. But the research has riveted experts in snake biology. "Todd Castoe is the most innovative person in our field right now, in my opinion," said herpetologist Joe Mendelson, who is completing a massive new textbook for the field. "What he's doing is just mind-blowing. And the work he did on the digestive system dynamics with Steve Secor—I remember when I first got tipped off on what they were doing, reading that and going, 'Seriously? This is what these animals are doing right in front of me, and I had no idea?'"

At this juncture in the metabolism story, however, Castoe becomes uncharacteristically circumspect—for very good reason. As he and his team gingerly noted in their 2022 *BMC Genomics* paper describing this implausible "switch," their findings about the interplay of growth and stress signals in snakes may have identified "potential molecules that hold promise for therapeutic control of regenerative responses." That's a cautious way of saying that python tricks might tell us how to regenerate human tissue.

Castoe elaborated a bit more, but not a lot, on what he called a "remarkably simple idea" when I spoke with him shortly after the 2022 paper came out. "Could it be as simple as manipulating one ion channel for another that changes fundamentally your ability, one, to not get diabetes, and two, [to] undergo regenerative growth? Don't get me wrong," he hastened to add. "The simplicity about talking immediately about drugs and therapies needs to be off the table. Because what we're talking about is playing with fire. You're talking about playing with cell proliferation. We want to be thoughtful and reasonable about saying we're not going to...there are reasons we need to be a little gun-shy." That said, he couldn't suppress his excitement about the idea that such modest molecular interventions could produce such dramatic, non-canonical, rule-breaking outcomes— at least in snakes. "That suggests that, yeah, [with] a couple of tiny tweaks, maybe we could get regeneration like this in a regular vertebrate like us."

No one is playing with fire yet, and there's a long, long way to go from Old World pythons, which evolved their metabolic tricks 70 million years ago, to something that might find a place in a medical clinic. But one final tidbit in this molecular odyssey, in an odd way, brings us all the way back to Asklepios. The metabolic signals in pythons that lead to stress tolerance and tissue regeneration involve molecules—mTOR and IGFs—that may sound vaguely familiar to regular readers of the Science section. That is because aging researchers (not to say scientists pursuing anti-aging and even "immortalizing" therapies) have been traveling the same biochemical roads for decades in their search for treatments that would slow down, arrest, or possibly even reverse the process of aging. Scientists have been exploring the role of these molecules in mice and nematodes and yeast, with frankly not a great deal to show in terms of medial applications for all that effort (and government coin). Maybe if we had granted snakes membership in the lab-animal club earlier, we might be further along in understanding the biomedical potential.

When I spoke to Bronwen Wickkiser, the expert on ancient Greece, she reminded me that Asklepios's life as a mortal ended when his healing arts became, in a sense, non-canonical. "Only the gods had the power to give and take life," she said, "and when Asklepios revived people from the dead, he got out of his lane, so to speak, and Zeus zapped him." Part of the enduring mythology about Asklepios, the god of healing, is that he may have acquired his prodigious healing powers when a snake whispered its secrets into his ear. Modern science is recapitulating this ancient transfer of biological wisdom, with snakes now whispering molecular secrets. But our contemporary demigods of molecular biology might revisit Pindar's account of Asklepios's ultimate fate as a healer: smote by that thunderbolt hurled by Zeus to erase the possibility of immortalization in the world of mortals. Even nature's stop signs are there for a reason.

Snake Road
Isola Tiberina, Rome, Italy

There is a stretch of pavement (not even a road) on an island in the middle of the Tiber River in the middle of Rome, which flanks a massive retaining wall of pitted, discolored marble, on which, about halfway up the wall, you can make out the faintest and sole remnant of pagan snake worship: the curving serpentine form of a snake in relief, crawling up the wall toward the heavens.

It is probably the least known, least celebrated, least visited scrap of sculpture in all of Rome—a curvy three-foot eversion of travertine, cracked and pitted and defaced by more than two millennia of weather and vandals and civic indifference. But it shows the unmistakable form of a snake encircling and ascending a staff. And if you step back, you can see that the thick slabs of marble on which the snake crawls were shaved and shaped, centuries ago, to suggest the hull of a small boat. This is the only fragment that remains of a story told by classic poets like Ovid, by Roman historians like Pliny the Elder and Livy, by travel writers like Augustus Hare and Georgina Masson, by art historians like Robert Hughes and countless others, all of whom have delighted in describing the arrival of a sacred snake in the third century BCE on this very spot.

At that time, the very air of Rome was "polluted by a fatal plague," according to Ovid in *The Metamorphoses*, book 15. Around 292 BCE, the desperate citizenry consulted local oracles, seeking relief from the raging pandemic. The oracles instructed the Romans to seek help from the son of Apollo, none other than Asklepios, and the Roman senators dispatched a delegation to sail by boat from Rome to Epidaurus, in Greek Peloponnese, where they would implore the god of healing to accompany them back to Rome. "The Greek response was mixed" (Ovid again), but as the Roman visitors slept in the sanctuary, Asklepios appeared to them in a dream and told them he would turn himself into a serpent and accompany them back to Rome. In the morning, the god—now in the form of a snake—announced his intention to accompany the visitors. He slithered down the steps of the sanctuary, paused to cast a final glance at the temple in Epidaurus, and made his way through a path strewn with flowers by his Greek admirers to the waiting Roman ship. Within a week, the ship had reached Italian waters and sailed up the western coast of the boot of present-day Italy to the mouth of the Tiber River.

In a wonderful scene lovingly depicted by Ovid, the snake-god rested his head on the stern of the boat as it passed well-wishers who rushed to line the banks of the river. The Vestal Virgins celebrated his arrival. The smell of incense from dozens of impromptu altars wafted over the boat. Countless Romans sacrificed animals in honor of the arriving deity. As the ship entered the part of the river that flows through the center of Rome, Ovid wrote, "The god / lifted himself on high and coiling round / the mast-head, moved his neck side to side / to find the place best suited for his shrine." As the boat neared the island, the serpent jumped off and swam ashore. When it climbed up the downstream bank of the only island within the city walls (the island now known as Isola Tiberina), it was an unmistakable signal to locate an Asklepion sanctuary at that spot. As he climbed ashore, Asklepios resumed the form of a god and "put an end to grief and pestilence." Ever since, the island has been a center of healing. To this day, two hospitals stand on tiny Isola Tiberina.

What makes this richly detailed myth even more momentous is what it portends. Rome's appropriation of the Greek god of healing symbolizes the eventual transfer of the center of power from the Hellenic world to the Roman world, with the snake as the essential messenger heralding this transfer. Which is why Rome, and Isola Tiberina in particular, is a fitting place to contemplate the fate of the snake. Upon that scrap of surviving Asklepian architecture, that faint residue of nautical imagery, with its weary marble and weather-worn serpent, was built—no, *built* is too neutral a term; was superimposed upon, or plunked—a Roman Catholic church, the church of San Bartolomeo, in the 12th century CE: an architectural and cultural metaphor for how organized religion stomped upon and crushed the Snake as a symbol of healing and helped shape cultural antipathies toward the demon serpent for centuries.

For proof of that, one need only walk along several more Roman roads and streets—the route covers a little more than two miles, and not quite two millennia after the arrival of the boat—to the Galleria Borghese museum. There, in a gallery on the second floor of the museum, one can view, take in, admire, wonder at, or recoil from the single most influential image in all of Western art in the vast oeuvre of snake hatred. It is one of Caravaggio's greatest masterpieces, *Madonna dei Palafrenieri*, a painting dating from 1606 that famously depicts a conspicuously buxom Virgin Mary lowering her bare and fearless foot to crush the head of a snake, while guiding the baby Jesus to do the same.

The painting, originally commissioned by the Vatican, lasted only a week on display there, perhaps (as some art historians have surmised) because the scarlet dress and bulging bosom of Mary were a bit too racy. But if you think of church-sponsored paintings as the Renaissance version of an instructional video, you begin to understand how morally necessary it was to learn to detest the snake in all its physical and symbolic manifestations. In teaching baby Jesus to stomp on the cunning serpent, the Virgin Mother is teaching all of Christendom how to think about snakes.

It is no accident that the sculpted travertine snake on Isola Tiberina has itself been partially decapitated, its wise and healing head chipped away to a flat empty space, a negation of figurative detail that recalls early depictions of the Buddha. To defy these teachings would entail transgression, subterfuge, rule breaking, a kind of social crypsis. Something that snakes, and snake-worshippers, in other parts of Italy and elsewhere, have proven to be very good at.

5

The Wasabi Connection

Sensation

T he first thing Elena Gracheva noticed about the building was
the absence of windows. It was still early on a South Texas
morning in February 2009, but the long, low-slung brick edifice
seemed almost prison-like, as if designed to keep whoever or whatever was inside from getting out into the world. The second thing
Gracheva noticed was the cheerful woman greeting her at the door.
Elda Sanchez, professor of chemistry at Texas A&M–Kingsville, was
also director of the facility, which happened to be one of the largest
serpentariums in the world. Known as the National Natural Toxins
Research Center, the building housed, at any given time, an average
of 450 highly venomous snakes. Several of the serpents were about
to donate their bodies to science—difficult, dangerous, and daring
science.

The third thing Gracheva noticed made the strongest impression
of all: the noise. "When you enter that building," she told me, "you
hear this…*tseeeeeeeee*." She made a dry, attenuated, bristly sound,

like wind rustling brittle leaves in a shingle oak. "We were working with rattlesnakes," Gracheva said, "and they were constantly rattling."

The "we" were three hard-core molecular biologists visiting from San Francisco—the kind of scientists who primarily traffic in genes, proteins, test tubes, centrifuges, and all the virtually invisible moving parts that power cells and physiological systems. The closest they usually got to wild animals were lab mice and fruit flies. Suddenly Gracheva and her colleagues, Gunther Hollopeter and Yvonne Kelly, were in a little room that housed, by Sanchez's estimate, about 65 western diamondback rattlesnakes (*Crotalus atrox*). As soon as anyone walked into the room, the agitated snakes started rattling. "Some people get really unnerved by the sound," Sanchez told me, "but then some people describe it as being a peaceful sound. Have you ever heard a rain stick? Kind of like that."

It probably didn't sound like a rain stick to the people whose weathered tombstones in Texas graveyards memorialize the fact that they died by fatal snakebite; Harry Greene mentioned a couple of those tombstones in a dispatch he once sent to the *Mason County News*. And it didn't sound like a rain stick to the molecular biologists, either, who were more accustomed to the click of their pipettes and the whir of blenders stirring up their lab happy-hour margaritas. Western diamondbacks had a reputation for being an especially aggressive species, Sanchez told the visiting scientists. The pit vipers lived up to their reputation, proving to be ornery, unpredictable, and dangerous—even *after* death.

Of all the satellite outposts of the National Institutes of Health in Bethesda, Maryland, the National Natural Toxins Research Center surely ranks among the most unusual. Located on a patch of Texas land that once formed part of the legendary King Ranch, the center harvests and distributes (for a fee) the venoms of some 600 deadly serpents to scientists conducting molecular research on toxins and their potential medical use. Qualified researchers can buy venom, purified and lyophilized (dried into powder), by the gram: $75 for a gram of western diamondback venom, $75 for the Trans-Pecos copperhead, $100 for the red spitting cobra, $1,000 for the massasauga, $2,000 for the bushmaster, and, over in the bargain bin, $56 for a gram of western cottonmouth venom. The center also carries venoms harvested from forest cobras, terciopelos, Brazilian lance-heads, Malayan pit vipers, coral snakes, cantils, Gaboon vipers, and roughly two dozen species of rattlesnakes.

The scaly residents of the serpentarium posed occupational hazards that were hard to ignore. Sanchez told me workers at the Kingsville facility had suffered six snakebites, gratefully none fatal, over the years, including one poor guy who got nailed three times in a single year. Gracheva and her two colleagues couldn't help but notice disquieting signs of the danger they were about to confront as they stole glances at members of the serpentarium staff. Some had disfigured fingers and hands, others had divot-like cavities in their limbs, yet another appeared partially paralyzed—all the result of accidental snakebites. They weren't in fruit fly country anymore.

Gracheva is a slender, energetic, gracious, yet ruthlessly efficient biologist who bristles with determination. Even back in her native Russia, she was known to go to unusual lengths to pursue her scientific projects. There was the time during the perestroika years in the 1990s—an era when Russia's scientific infrastructure had basically disintegrated—that she had to obtain lab animals on her own from a farm (eight chickens) and house them in her Moscow dorm room (along with four roommates) in preparation for an experiment. But in her science, she trafficked in more microscopic menageries, parsing out the molecules that allowed animals to make sense of their surrounding environment. She eventually obtained a PhD at the University of Illinois–Chicago, and, in the fall of 2008, landed a postdoctoral position at the University of California at San Francisco in the laboratory of David Julius, one of the leading researchers on sensory perception in animals. It was in that capacity that she had traveled to Kingsville, Texas, with Hollopeter and Kelly to conduct the first—and most dangerous—steps in a complicated experiment. They wanted to figure out how snakes with pit organs, the small little sensory divot on either side of the snout, employed those devices to create a thermal picture of the world. "Seeing" by temperature, as it were, as well as by light.

Snakes with pit organs have an ability to detect infrared radiation with a sensitivity that shames most human-made instrumentation. They can discern variations in temperature to a level of precision as small as 0.003 degree Celsius—three one-thousandths of a degree. The pit organ itself is a gossamer circlet of tissue, invisibly pimpled with thousands of exposed nerve endings that are hot-wired to the rattlesnake brain. Certain Old World pythons and boa constrictors possess pit organs, too, but the genus of animals with the most sensitive infrared detectors happened to be rattling around in that noisy room in Kingsville: pit vipers in general, and New World rattlesnakes in particular. The UCSF team was there to dissect the pit organs, tease out the delicate filigree of nerve tissue that connected the pit organ to the brain of the snakes, pack that fragile material in dry ice, and ship the precious cargo by overnight courier to San Francisco, where Gracheva would use the latest genetic tools to probe the molecular mysteries of infrared sensation.

In order to even begin their test-tube experiments, however, the UCSF molecular biologists had to venture far outside their comfort zone and deal with the western diamondbacks directly. They had to surgically remove the pit organ and the bundle of nerves that traveled from the pit to the brain, specifically a node of neural tissue known as the trigeminal ganglion; in addition, for scientific rigor and as a control, they needed to harvest a separate node of nerve tissue, buried in bone and running all along the spine of the reptiles, known as the dorsal root ganglion. Both nerve nodes were tiny and encased in very hard bone, so the dissections demanded both brute force and delicate microsurgery performed with the help of a microscope. Elda Sanchez and her staff at Kingsville prepared more than a dozen rattlesnakes for the California scientists. The snakes were milked of venom, anesthetized, euthanized, and then decapitated by a guillotine-like device. It fell to Gracheva to palpate, by hand, each rattlesnake to make sure the heart had stopped beating before the decapitation. She also had to carefully clip away the fangs, a task for which she fatalistically donned a double set of heavy gloves. "One or two," as she put it, "it really doesn't matter, the fangs are so sharp..."

You might think that dissecting the disembodied head of a dead pit viper would be relatively straightforward (albeit grueling) work. You would be wrong. As each of the California scientists learned in turn, it was like performing brain surgery on a grasshopper. Every time they touched the detached head of a euthanized snake, it jumped. Shards of fangs still poked out of the mouth. Yellow droplets of venom still glistened (at least a milliliter, by Gracheva's reckoning). If you see any venom, they were all told, whatever you do, don't touch it. Indeed, experts in venomous snakebites never tire of warning that many life-threatening envenomations occur *after* people have killed the snake.

As Hollopeter maneuvered the detached rattlesnake heads on

his impromptu surgical field, a diaper, he couldn't help but marvel at the huge muscles—"like a chicken leg"—surrounding the venom gland and powering the snake's bite, the better to inject poison. Then he began to probe the brain stem in search of the trigeminal ganglion, the tiny node of nerve cells that measured a mere six or seven millimeters in diameter—the size of a small pea. The diamondback's head—triangulated, like many vipers, into an expression of indifferent menace—rested on the diaper. Every time Hollopeter touched the bony skull to get at the neural tissue, the head started "going bananas," as he put it. "It's just a reflex, but you're digging around in the brain, so the head is basically jumping around on the table. Really strange!"

That was but one reason why, as the scientists would report later, rattlesnakes were "inconvenient subjects" to study. Why even bother? Because those arriviste molecular biologists were about to discover what herpetologists had known for centuries. Snakes, deadly pit vipers and harmless garter snakes alike, possess extraordinary sensory equipment.

~

"If you want new ideas," Gordon Burghardt is fond of saying, "read old books." One of the books that captivated Burghardt, a longtime herpetologist and professor at the University of Tennessee, was published in 1909, in German, by an Estonian biologist named Jakob von Uexkull. The "new" idea that Uexkull proposed, which has guided much of Burghardt's career research, was a concept called *umwelt*—the totality of sensory information that created an animal's conception of the external world and how to behave in it. "Our anthropocentric way of looking at things must retreat further and further," Uexkull argued (in a passage Burghardt translated himself), "and the standpoint of the animal must be the only decisive one."

Understanding the standpoint of the animal meant understanding sensory perception at the most basic level, which in snakes included the thermal detection of pit vipers, but even more their *chemical* sense of the world. Beginning in the 1960s, Burghardt began to investigate the *umwelt* of snakes by investigating the nature of their chemosensory awareness. He performed an experiment that still has scientists shaking their heads in admiration half a century later. Burghardt tested the feeding preferences of neonatal queen snakes (*Regina septemvittata*) moments after they emerged from the mother. He focused on queen snakes because they are known to be exceptionally picky eaters. Adults only eat freshwater crayfish, and only within an hour or two of the crayfish having shed their shells (or "molted"). Anything else, they send back to the kitchen. Burghardt waved cotton swabs drenched in extracts of various serpentine hors d'oeuvres—goldfish, crickets, earthworms, mice, even crayfish with hardened shells—in front of baby queen snakes. These snakes, which had never eaten a single meal and had never seen a Q-tip, passed on everything on the menu except one— the extract of freshly molted crayfish. Not only did the baby snakes attack the cotton swabs—they tried to swallow them! "Just still the most mind-blowing experiment," Harry Greene told me. Not all chemosensory sensation in snakes is innate, as subsequent research has shown, but they come to the table, as it were, with incredibly refined chemosensory tastes.

How animals sense the world around them has always fascinated biologists, to be sure, yet researchers have often succumbed to inadvertent tunnel vision in exploring sensory phenomena. For the most part, contemporary scientists stick to a small gang of "model organisms"—common laboratory animals—to probe the molecular secrets of sensory perception. The usual suspects include mice, rats, bacteria, yeast, fruit flies, and a lint-size nematode called *C. elegans*. (Gracheva's PhD research involved the nematode.) But many species other than the standard model organisms display unique and evolutionarily profound solutions to the problem of making sense of the world. That bats use sound waves to echolocate and map three-dimensional space is commonly known. Star-nosed moles use an unusually sensitive proboscis to discern food by touch. And snakes, despite their lack of eardrums and often poor vision, have evolved remarkable sensory properties to compensate.

Take that flickering tongue. Snakes possess the hardware for olfaction, the ability to detect airborne volatile odors, housed in their nostrils. But their snouts and tongues combine to form an even more sophisticated chemosensory detection system for larger, heavier molecules in the environment. When snakes flick their tongues, they're not just testing the air; the tongue is stirring up and fetching invisible, heavy, non-volatile molecules on or near the ground. Once the tongue retracts into the mouth, it delivers these molecules to two fancy side-by-side rooftop biochemical labs, one for each tine, lodged in the top of the palate that connects to the central nervous system. Known as the vomeronasal organ, this chamber of nerve tissue—common in reptiles, amphibians, and many mammals (perhaps including a version in humans, although that remains controversial)—assesses these non-volatile molecules, distinguishing the chemical signature of prey, predator, and potential mates, as well as transmitting pain signals.

In snakes, this chemosensory processing center is known as the Jacobson's organ, named after the Danish surgeon who reported its presence in humans in 1809 (although it might more fittingly be called "Ruysch's organ," for the Dutch anatomist who actually discovered it—including in snakes—nearly a century earlier). The entire vomeronasal system is so sensitive that some researchers have shown that the two separate tines of a snake's forked tongue gather molecular "data" with such precision that the difference between the left and right tine provides a chemosensory gradient that functions a bit like binocular vision—using molecules to create a three-dimensional perception of space. "Smelling in stereo" is the way evolutionary biologist Kurt Schwenk has described it. Ponder that for a moment: a 3-D sense of smell. That is what *umwelt* is all about.

"Snakes are the most chemically sensitive vertebrates we know," said Robert Mason, a biologist at Oregon State University who has dedicated his career to understanding the chemical language of snakes. "We now know," Mason has said, "that with a single tongue-flick, a male garter snake can determine not only whether another snake is a member of its own species, but also if it is a male or a female, a female from the male's own den versus another den, a large female versus a small female (larger females produce more young), and whether a female is likely to reproduce this year or store his sperm for a following year." That's a lot of information from one tongue-flick.

Neuroscientist Mimi Halpern did pioneering chemosensory research in the 1970s and 1980s at the State University of New York–Downstate demonstrating the primacy of the vomeronasal system in snakes. "I really wanted to understand how the brain used the senses to accomplish survival," she told me. She had no particular love of snakes but became enamored of garter snakes as a research organism when she discovered that fully one-quarter of their brains—a structure called the nucleus sphericus, analogous to the amygdala in mammals—was the sole destination of the vomeronasal system. In experiments with John Kubie, she showed that if snakes were allowed to tongue-flick earthworm extracts sprinkled with radioactive markers, traces of that radioactivity worked their way through the animal's neurological system. Halpern and Kubie even had garter snakes navigating simple mazes in the 1970s. "It wasn't high-level training," Kubie said, "but we laid down odor trails, and they do learn."

On the basis of decades of experiments, Halpern believes olfaction tees up the vomeronasal system to interrogate something important in the environment; volatile airborne molecules tease the snake with information, she told me, but the molecules detected by the vomeronasal system on or near the ground carry more critical information for survival. Despite these discoveries, when Halpern came up for tenure, her department chair questioned why a medical center (SUNY-Downstate includes Kings County Hospital, one of the largest in Brooklyn) should support research on snakes (she nevertheless received tenure).

As Halpern suspected, the vomeronasal organ is essential to snake survival. It allows the animals to find their way back to their winter dens, year after year, after roving for months and miles in the wild. It allows a rattlesnake to track down a rat or rabbit it has bitten; despite the jolt of venom, the fleeing prey can often attempt a futile effort at escape, traversing 100 yards or more, but with those flicks of the tongue and a special "tracker" protein included in the venom, the snake can methodically follow the breadcrumb molecules of its wounded prey and track down its meal. It allows males to find and assess females during mating season—despite all the human obsession with sexual pheromones, snakes don't detect them through their snouts but rather with their tongues.

And in snakes at least, the vomeronasal system is crucial for copulation and reproduction—that is, to life itself. In one clever experiment, Mason and his colleagues plugged up the olfactory hardware of garter snakes and found that the animals were able to mate just fine; when researchers blocked the vomeronasal system, however, the snakes couldn't detect pheromones, couldn't find mates, and couldn't reproduce anymore. Garter snakes can even differentiate the relative strength of sexual pheromones on opposite sides of a blade of grass. The ability to read these faint chemical signals is a matter of life and death for snakes, so they are incredibly astute at distinguishing friend from threat.

That idea received a harrowing test back in the 1930s. Grace Olive Wiley had a radical theory about the sensory perceptions of snakes. During her brief stint as curator of reptiles at the Brookfield Zoo outside Chicago, Wiley believed that snakes were so chemically astute that they could grow familiar with, even habituated to, their handlers if they were exposed to their specific chemical signature, as unique as a fingerprint. She later put her theory to the test in the most perilous way imaginable, according to a fascinating account in the *Bulletin of the Chicago Herpetological Society* by James Murphy and David Jacques. She would toss pieces of her own clothing, pre-worn and unwashed, into the cages of recently arrived venomous snakes so they could get used to her sensory signature and in a sense become tame in her chemical presence. While these "experiments" were not scientifically rigorous, the hypothesis-testing was dead serious, as the snakes she attempted to tame included cobras, rattlesnakes, puff adders, and Gaboon vipers. If Wiley's hypothesis was true, it would suggest that the snakes did not have an innate sense that she was a threat, but rather had *learned* that her scent was not threatening.

For a while, the theory appeared to hold true, as Wiley posed for a number of breathtaking, almost erotic photographs showing her embracing, nuzzling, practically necking with king cobras and rattlesnakes. Depending on your point of view, her hypothesis was tragically disproven or proven mortally true in the summer of 1948, when she was bitten by a newly arrived Indian cobra that had not yet been habituated to the scent on her clothing. Among her last words were, "He really didn't bite me, did he?" (He did, and she died hours later.)

It is not uncommon to hear modern herpetologists dismiss Wiley as at best eccentric or, as one snake expert told me, "nuts." But Mason, who has studied the sensory perceptions of snakes as deeply as anyone in the field for four decades, began to revise his opinion when I mentioned Wiley's theory. "I wasn't really aware of that," he confessed from his Corvallis laboratory, "but you know what? That really is a kind of brilliant idea. It's hard to say, especially in her case. But my prediction would be that that would work to some degree, that they would recognize her or recognize that smell as something that just is not a danger." Even Harry Greene surprised me by saying, "I think it's entirely plausible." The fact that such a wild idea might be plausible may be due in part to Grace Wiley's scientific intuition but also to an increased biochemical and molecular appreciation of just how perceptually discerning snakes can be.

Of all the interesting things that Gordon Burghardt has discovered about the vomeronasal system, perhaps his most startling observation is that we still don't even have the words to describe it. Olfaction—the sense of smell, the detection of volatile compounds in the air—has scent, odor, smell, aroma, stench, and every possible variation of stink. Whatever the vomeronasal system detects has no name, although 30 years ago, Burghardt proposed "vomodor," and you occasionally see the word in scientific papers. All we know is that it delivers large, heavy, probably more complex, and presumably more informative molecules to the animal brain. Without words, we can't even begin to imagine how snakes—or any other animals, for that matter—truly create a sensory picture of their world.

That is why research on the pit organ of rattlesnakes was so important and groundbreaking. It represented one of the first instances of a high-powered, molecular biology group stepping outside the narrow world of predictable lab-animal *umwelt* and trying to see how real-world vertebrates, not inbred clones, made sense of their environment. Herpetologists like Burghardt and Mason had shown for decades how exquisitely attuned snakes were to their world, both chemical and thermal. Here was an opportunity to see—or rather, begin to see—how those traits played out at the level of genes, molecules, and neural pathways.

But before the molecular biologists could get to the whir of their ultracentrifuges, they would have to deal with the "tseeeeee" of the snakes.

~

The molecular basis of smell, taste, touch, and pain sensation has fascinated biologists ever since they began to acquire the tools to rummage around in nerve cells and figure out how genes, molecules, and pathways convert a scent in the air, for example, to a sensation in the brain. Some of that research relied predictably on lab rats; Richard Axel of Columbia University and Linda Buck, now at the Fred Hutchinson Cancer Center in Seattle, shared a Nobel Prize in 2004 for research in the 1980s showing how rodents detected olfactory cues through specialized molecules on the surface of sensory cells known as receptors. One of the postdoctoral fellows in Axel's lab in the 1980s, David Julius, inherited precisely that approach when he set up his own independent lab among the University of California–San Francisco's high-powered molecular biology groups in the 1990s. But Julius had a hunch that different, weirder animals might tell different, weirder stories about how non-human organisms perceive the world.

If Julius's name is faintly familiar to foodies, it is because he made a big splash in the scientific world in 1997 when his lab group identified the molecule in mammals that conveyed the (delectable) pain of capsaicin, the fiery component in hot peppers that creates that burning sensation in the mouth from eating spicy food. The capsaicin receptor molecules were arrayed like sentries on the surface of nerve cells that specialize in the detection of heat; when those molecules encountered hot pepper, the sensory nerve cells fired off a message to the brain that essentially said, "That's spicy (and a little painful)!" The same molecule, however, also functioned as a thermal detector; it could register heat and thus pain such as when we touch a hot stove.

As it turns out, the capsaicin detector belonged to a larger family of proteins known as TRPs (for "transient receptor potential" channels, pronounced *trips*), which control the flow of ions into nerve cells. Invisible, silent, and ever so slightly freighted with an electrical charge, ions are the atomic protagonists of sensory perception because ions change the electrical charge of nerve cells, triggering them to fire and send a message to the brain. Each TRP protein seemed tuned to a particular sensory stimulus, whether heat, cold, menthol, pain. A few years later, Sven Jordt of the Julius lab identified the receptor that detected several other "household staples"— wasabi mustard and cannabinoids. By the early 2000s, armed with a powerful new set of technological tricks, biologists raced to figure out which receptors other animals used to sense their environment.

It was around that time that Julius began to consider non-traditional animals for experimentation, including snakes. As is often the case, inspiration came to him by chance.

"My kid was little at the time," he told me, referring to his young son, "and we were sitting down watching some television. One of these nature shows came on. It featured at some point—I can't remember exactly when—pit vipers! And I thought, *That's kind of interesting. I wonder if it's related to what we work on in some way.*" After a little laugh, he continued, "And you know, it was just one of those problems you can't turn away from once you start thinking about it."

In one sense, moving on from hot peppers to pit vipers was in keeping with Julius's long history of straying from the conventional. He used mice in his experiments, like many other neuroscience researchers, but he always wondered how other, "non-traditional" model organisms sensed their worlds: scorpions, sharks, spiders, vampire bats, skates, snakes, crocodilians. A former graduate student recalled him obtaining a couple of baby alligators for the lab because he wanted to know how the nervous systems of crocodilians employed a mechano-sensor in the jaw to detect faint ripples on the surface of water as they lay in wait for prey. "If someone comes into my lab and brings to the table an idea that 'This animal does x, y, or z,'" he told me, "then I'm willing to give it a shot."

Once he started thinking about the pit organ, Julius couldn't drop the idea. In 2005, he tracked down Elda Sanchez, forged a collaboration with the serpentarium in Kingsville, and tapped a graduate student, Gunther Hollopeter (affectionately known as "Fuzzball" because of his buzzcut), to go down to Texas to collect the nerve tissue of pit vipers. Hollopeter was the likeliest person in the lab to take a first crack at the rattlesnake project; his father, a small-town doctor in Idaho, had treated a number of snakebite victims in his time, and Gunther himself had performed minor surgeries on ranch animals. (He had also attained unwanted notoriety among colleagues for destroying a neighboring lab's margarita blender while trying to mince wasabi root for an experiment.) Hollopeter's first foray to Texas, in October 2005, hit a dry well and was dangerous to boot. He ultimately managed to gather enough material from the rattlesnake pit organs, but they had nothing to compare it with—neither other snakes nor other nerve tissue in the rattlesnakes. Hollopeter finished his PhD and moved on. Still, Julius wouldn't let go of the idea.

Nearly four years later, while still working in Chicago, Gracheva sent her CV to Julius in the hope of getting a spot in his lab. He never answered. She sent him her CV a second time. When he failed to respond to that, either, she decided to buy an airline ticket, fly to San Francisco, and deliver it in person. In 2008, while Julius was taking a break from writing a grant proposal, he found this young woman with a charming Russian accent hovering in the hallway outside his office. "I know who you are!" he cried. Gracheva pressed her CV into his hands again and begged to join his lab. The third time was the charm. "Finally I realized: This woman is very persistent!" he told me with a laugh. "And then I realized: She's also very smart." There was one catch. Julius made her an offer most people would be happy to refuse: working with venomous snakes instead of nematodes. "If you want to come to the lab," he told her, "why don't you try that project?"

"That sounds very interesting," Elena Gracheva lied. She was very fond of sensory experiments, not so fond of snakes. "She does not like snakes," Julius confirmed. "But she's pretty fearless otherwise."

Not long after, in February 2009, Gracheva found herself shopping for dry ice at the Walmart in Kingsville and heading over to the serpentarium to disentangle the facial nerves of dead—but still deadly—western diamondback rattlesnakes. Julius asked Fuzzball if he would be willing to go back to Kingsville and help Gracheva and Kelly with the dissections.

They again had to travel to Texas to do the preliminary work because, as Julius thunderously told Gracheva, "No rattlesnakes at UCSF!"

~

The pit organ in rattlesnakes—a tiny piece of membrane about a quarter inch in diameter—is one of the most exquisitely sensitive patches of sensory real estate anywhere in the animal kingdom. Similar to the way the human eardrum is a membrane buried deep within an anatomical cave (the ear), the pit membrane is a curtain of nerve tissue stretched like some cryptic solar antenna deep within the pit. Pythons and boa constrictors possess a dozen or more small, less sensitive, dimple-like "labial" pits around their mouths. Vipers possess only two pits, one on either side of the head between the nostril and the eye, but they are larger, deeper, and up to 10 times more sensitive than the pits in constrictors. Among pit vipers, the western diamondback rattlesnake's ability to detect infrared radiation exceeds that of all others. Inconvenient or not, it was the perfect place to look for infrared-detecting molecules.

But sensory detection always involves more than just the part of the body that bumps up against the external world. In the case of the pit organ, the membrane is infiltrated, indeed teems, with nerve cells that dispatch signals; these signals travel along filaments that extend from the cells like cables and gather like braids in a hypersensitive microscopic rope that threads back to the animal's brain. In animals up and down the phylogenetic ladder, from snakes to humans, the bundle of nerves that collects the messages from all over the face and head, poised to detect everything from a caress to a slap, is known as the trigeminal ganglion.

In mice, this little knot of tissue is tiny, about the size of a peppercorn. When the researchers from UCSF started probing for the trigeminal ganglion of their hip-hopping rattlesnake heads, they were in for a surprise. "It was *huge*," Hollopeter recalled. It was so big, in fact, that they had to double-check to make sure they had the right neural tissues. The herpeto-surgeons delicately traced nerve fibers all the way from the thick knot of tissue to the pit membrane to confirm it was indeed the trigeminal. In the other direction, the fibers traveled to a part of the snake brain known as the optic tectum. In neural tissue, size matters: Just as the huge nucleus sphericus in garter snake brains told Mimi Halpern the vomeronasal system was paramount, the fact that the trigeminal ganglion was so big in the rattlesnakes told the UCSF researchers that it must be linked to an especially important function.

As molecular biologists, Gracheva and her colleagues were ultimately on the prowl for molecules that somehow detected infrared radiation or heat and translated that into nerve impulses. The equivalent "translator molecules" in vision were proteins in the retina known as opsins, which respond to different wavelengths of light and spit out ions that send nerve impulses to the visual part of the brain, which then melds all the inputs into a Technicolor image of the world. To pinpoint a single molecule able to detect temperature, the experiment relied on new genetic technologies that were becoming available at precisely that moment, including a marvelous technique called RNA seq (pronounced *RNA seek*), which essentially allowed scientists to see every gene turned on in a single cell. Their plan was to gather as much nerve tissue from the rattlesnakes as possible, including both the pit membranes and the trigeminal ganglia, and send it all back to San Francisco. There they could apply this technology discovered in 2009 to understand a trick of biology evolved by pit vipers perhaps 35 million years earlier.

The new RNA seq method allowed Gracheva and her colleagues to perform an audacious form of microscopic eavesdropping. They could intercept, at the molecular level, every message being sent by the nerve cells embedded in the pit organ. In the metabolic hum of cells at work, these messages—biochemical work orders, if you will—are dispatched via a molecule known as messenger RNA. The RNA tells a cell what proteins to make, and the working hypothesis of the biologists was that the specialized nerve cells arrayed along the rattlesnake's pit membranes would be sending an overabundance of messages telling those reptilian cells to make a highly sensitive receptor molecule specifically "tuned" to detect infrared radiation. From humble bacteria (with their single cell) to us humans (with our 80 trillion cells), receptor proteins form the interface between the outer world and the inner world, between the environment "out there" and the nervous system of every sentient animal. This same general strategy had allowed scientists to identify the opsin molecules and capsaicin receptors. Rattlesnakes undoubtedly possessed some kind of special thermal receptor to detect infrared radiation. What was it?

Once back in San Francisco, Gracheva and Nicholas Ingolia unpacked the rattlesnake tissues and performed a laborious analysis of the RNA sequences isolated in the pit organ. They left no molecular stone unturned in their effort to identify the supersensitive rattlesnake molecule and prove beyond a reasonable doubt that it detected thermal radiation.

The analogy may seem a bit unusual, but the upshot of all this effort was redolent of the climactic scene in the Dr. Seuss story *Horton Hears a Who!* where every single Who on their tiny planet screams the same phrase at the same time to convince Horton the Elephant's doubters that the invisible creatures indeed exist. After the scientists collected all the RNA messages from the cells in the pit organ and sorted them out on a data plot, one message in particular resounded, as if all the cells embedded in the pit membrane were yelling the same genetic phrase. That genetic phrase (and its bespoke molecule) overwhelmingly dominated the cells lining the pit membrane; those same cells sent nerve fibers threading back to that gargantuan trigeminal ganglion. And the message, astonishingly, was a protein almost identical to one with which virtually every human being on the planet is familiar: the molecule that detects the distinct, painfully delicious piquancy of wasabi mustard. Known as the wasabi receptor, this molecule (the technical name is TRPA1, pronounced *Trip A1*) is deployed widely in the animal kingdom. It is especially good at instantaneously detecting noxious, and potentially dangerous, chemicals in the environment. Snakes deployed a slightly different version to detect temperature.

Why such exquisite sensitivity? Pit vipers are typically ambush hunters—they lie in wait, often at night, for prey to wander into striking range. Infrared detection tells the animal that its prey is warm-blooded and nearby, how big it is, and where it is going, even in the dark. If your diet is primarily rodents and other small furtive mammals, that's a pretty good device to have. After the experiment, Gracheva went to a pet shop in Berkeley with an infrared camera and took pictures of a live mouse standing next to a boa constrictor. It wasn't just that the mouse, in infrared, looked like the neon sign of a cheap roadside motel; it was that the darkest (and thus coolest) area in the entire image was the snout of the snake, precisely where the pit organs were located. "They have a system to cool down the pit," she explained, "so they can make the hunting strategy very effective." Other research has suggested that pit vipers may also employ the pits as a kind of ecological thermostat, using them to identify warmer or cooler spots in the surrounding environment to help them regulate body temperature.

When she first plotted out the data, working in the lab at midnight, Gracheva couldn't believe the results. There was a single dot in the data plot, a single overwhelmingly populous molecule. This doesn't usually happen in experiments. Data is typically messy. When she showed the plot to Julius the next morning, his reaction was similar: "It's unbelievable. It's too good to be true." But he added, "It was amazing to see this one dot right up there. Just beautiful." The fact that it was a close cousin—an orthologue—of the wasabi receptor was unexpected, so they repeated all their lab work to make sure of the result. What the UCSF scientists reported in 2010 is still true today: The molecule they plucked out of the pit organ of those jumpy Kingsville rattlesnakes is the most heat-sensitive receptor ever identified in any vertebrate species.

At the evolutionary level, the discovery was one more lesson that we are not so different from the seemingly different animals that we loathe; evolution sometimes functions like an auto parts store, where over the course of millions of years of evolutionary time, nature evolves a warehouse of molecules and different creatures pluck the same basic components off the shelf and adapt them to suit their specialized needs. Nature took this utilitarian part and repurposed it. In the case of pit vipers (and Old World boas and pythons, as Gracheva and her colleagues went on to show), the snakes essentially used an off-the-evolutionary-shelf molecule and adapted it for thermal detection. We humans use a very close cousin of that same molecule to savor sushi.

But there was one more surprise. In tracing the path of the nerve fibers leading out of the pit membrane, the UCSF researchers showed that the infrared signals threaded their way to the "optic tectum." As the word *optic* implies, this is the part of the snake brain that processes visual information, and what it suggested to researchers was that snakes with pit organs, rare among all known vertebrate animals (including us), create a single, integrated map of the external world that combines visual and thermal information. It's not two different parts of the brain talking to each other and overlaying their respective maps; it's a single, integrated image of the world designed to combine multiple sensory inputs into one somatosensory map.

David Julius likened this scenario to synesthesia in humans—cases where two different senses, such as sound and taste, are conflated, so that a person feels like they can "taste music" or "hear colors." Synesthesia, of course, is considered an exceptional neurological phenomenon in humans (the novelist Vladimir Nabokov and the physicist Richard Feynman, among many others, claimed to be synesthetes), and it is sometimes perceived as a dysfunction of the nervous system. In other organisms, however, it may be a product of physiological design, not an aberration but rather an exceptional sensory skill. "In terms of that direct integration of information," Julius said, "I think the snake is an unusual example of that."

Harvey B. Lillywhite, a herpetologist at the University of Florida, went even further. In his book *How Snakes Work*, he noted that it's not only infrared and visual information that converge in the pit viper's optic tectum, but also auditory and motor signals. "Thus," he wrote, "the infrared and visual information, along with other sensory inputs, merges to form imagery of the external world in a manner that we cannot well imagine."

Cannot well imagine. It seems we're still on the outside looking in when it comes to a serpent's *umwelt*. The mere suggestion that snakes can achieve a kind of synesthesia, not as a neurological oddity but as an enhanced way of experiencing the world, is indeed hard to wrap your head around. But it's something to keep in mind when people mention, almost always with ill-concealed disdain, the "primitive" reptilian brain.

〜

As major articles in *Nature* are wont to do, the pit viper research got a lot of attention, but it would also be fair to say that doing research on snakes was no longer a scientific backwater. Elda Sanchez went on to become president of the North American Society of Toxinology, which has highlighted recent research on new snakebite treatments. Gunther Hollopeter, who gingerly excavated nerve tissue from hopping diamondback heads on his dissection table, is now a professor at Cornell University, working with non-threatening nematodes. Yvonne Kelly, the technician who learned some of her earliest surgical chops dissecting rattlesnakes, recently finished a fellowship as a transplant surgeon at New York Presbyterian Columbia/Cornell hospital system. And David Julius invited the persistent and fearless Elena Gracheva to attend the ceremony in Stockholm in December 2022 when he accepted the 2021 Nobel Prize for Physiology or Medicine for his research on pain and heat sensation. Gracheva, continuing her lifelong curiosity about the unusual ways animals sense their worlds, now runs her own lab at Yale University. She studies how ground squirrels, another non-traditional model organism, manage to suppress their appetite and survive without eating for eight months during their winter hibernation. She currently is hunting for the mechanism by which squirrels, with the flip of an internal genetic switch, activate a form of self-induced reversible anorexia.

After she finished the rattlesnake experiments, Gracheva asked Julius if she could spend a day or two away from the lab—her first vacation since she had arrived in San Francisco. Julius ordered her to take two weeks off. Perhaps it was just coincidence, but she chose to go to Hawaii, the only state in the country with no native snake species (and one of only two states in America free of venomous species, Maine being the other).

The snakes managed to get in the final word, though. Shortly after the *Nature* paper came out, in April 2010, David Julius's son Philip wandered into their house in San Francisco's East Bay one day and said, "There's kind of a funny snake on the porch. It's hissing when I look at it."

Julius dashed out onto the deck and discovered that the intruder wasn't hissing. It was rattling. It was one of the many rattlesnakes that inhabit the dry hills flanking San Francisco Bay. Julius's first thought? "This is like the revenge of the snakes!" Or maybe it was just a different kind of ambassador snake. In many ancient cultures, snakes are considered messengers from nature, and maybe the message here was: "Despite all your sequencers and data plots and molecular magic, you're not even close to understanding our *umwelt*."

Snake Road
Via Orto Magliocco, Cocullo, Italy

There is a country road in the Abruzzo region of southern Italy, about two hours east of Isola Tiberina, that climbs the side of a picturesque valley in the Apennines, then curls to the right as it reaches a small plateau and arrives in the main piazza of the tiny village of Cocullo. If you try to drive up that road on May 1, the date of Cocullo's annual snake festival, you will have to park miles down the hill (many traffic police will be happy to direct you) and then walk through a gauntlet of roadside vendors selling everything from pastries and steaming chunks of porchetta to doormats, fava beans, rugs, amulets, household cleaning supplies, brooms, you name it. If you drive up any other day of the year—even, say, April 30—the road is wide open, sparsely traveled, and you might even find a parking spot right in front of the Church of San Domenico, which honors the itinerant 10th-century Benedictine monk who allegedly protected locals against snakebite, rabies, and toothaches.

On the eve of the festival, there's an elevated hubbub of the Sunday-afternoon crowd at the main café, but the real action is around the corner in a small modern building flanking the main square. This is the municipal building of Cocullo, population 211, which houses the mayor's office, village agencies, and, on the second floor, a little "Mostra Dei Serpenti," or Snake Exhibit, which features live specimens of the local fauna—a four-lined snake known as cervone (*Elaphe quatuorlineata*), a green whip snake (*Hierophis viridiflavus carbonarius*), and—wait for it—an Aesculapian rat snake (*Zamenis longissimus*). Not since Raymond Ditmars stashed his snakes in the city coroner's office have live serpents been so close to the seat of municipal power.

It didn't occur to me until much later that the juxtaposition of the church, the town hall, and Cocullo's lone café captured in microcosm a millennium-long tension among the church that has demonized snakes, the state that now protects them, and ordinary village residents who, with possible historical ties to earlier pagan cults in the vicinity, revere snakes in defiance of the other two institutions. This is especially true of the local breed of self-taught naturalists called *serpari* who prowl the steep rocky hills of the Abruzzo, collecting snakes. Not for food or sport or environmental control, but as the central players in a ritual that has local echoes in ancient pagan rites, was adopted by the Roman Catholic Church perhaps as early as the 11th century, and has persisted, according to some experts, for 500 years. The modern festival survives now as a tourist destination that swells the local population 100-fold to witness one of the strangest, most exuberant, and weird folk celebrations in Italy.

On the last day of April 2023, less than 24 hours before the festival, a steady stream of *serpari* climbed the steps of the town's municipal building to a second-floor conference room. Cloth bags with serpentine bulges dangled from their hands or were cinched to their belts. There were old-timers, to be sure, and young men, too, with hard chiseled faces and the assured, firm hands of experienced snake-handlers. There were also women, young children, entire families, all bearing snakes. They had captured the snakes in the surrounding countryside during early spring, a task particularly challenging that year because of a snowfall the previous week. Yet the scene that unfolded wasn't the guidebook version of the Festa dei Serpari, nor even the one I witnessed as a young, Rome-based journalist in 1977, when the event was less formal, less crowded, and perhaps more charming, given the absence of cell phone cinematographers, selfie sticks, and international media.

To begin with, as soon as a snake was taken out of its bag, a group of scientists converged on the animal. One scientist ran an electronic scanner along the length of the serpent, to see if it contained an identifying microchip (each had to be electronically registered for the following day's festival). Many did—they were, in their limbless way, veterans of the festival, even though they had been caught in the wild in the preceding few weeks; if not, a veterinarian would deftly insert one. Ernesto Filippi, a biologist with the Italian Ministry of Environment and one of country's leading reptile experts, measured the length of each snake, and then weighed them. Domenico Otranto, a veterinarian from Bari, took swabs from the mouth and vent of each snake to check for parasites—Otranto's team had recently identified a new pathogen in Moroccan cobras (*Naja haje*), and everyone is on the lookout for snake fungal disease these days. Two of Otranto's assistants took blood samples. Jairo Mendoza-Roldan, a native of Colombia, nimbly inserted a needle to draw blood (a splendid tattoo of a serpent ran the length of his forearm, which he joked could be made to look like a viper or a harmless species, depending on how he squinched his wrist). His colleague Nicole Szafranski, a visiting intern from the University of Tennessee, jotted down the data. "It's so cool," she said, "to be preserving an ancient folk ritual while still doing science."

"We just reached the 100th snake!" announced Gianpaolo Montinaro, who sat at a long table and entered all the data into a master registry. A murmur of approval rustled through the room.

Montinaro, an Italian-born ecologist based in Germany, played a major, although initially awkward, role in converting a century-old folk festival into a citizen-science event. Back in 2006, he had been hired by the World Wildlife Fund to conduct an amphibian-and-reptile survey in a nature preserve in Anversa degli Abruzzi, just down the valley road from Cocullo. During that project, Montinaro encountered some of the *serpari* from Cocullo catching snakes. What Montinaro knew— and the local *serpari* did not—was that collecting snakes in the nature preserve required not only a permit from the local regional government, which the locals had, but also a permit from Italy's Ministry of Environment to conform with European Union regulations, which they did not. "So there was a problem," Montinaro told me. This meant that, technically, the Cocullo festival had been using illicitly procured snakes for years. "From the middle of the 1990s, when the European Union laws were published, it was illegal," he said.

Montinaro met with the mayor, helped the local *serpari* obtain the necessary permits, and then realized that the festival offered unique possibilities for science and public education, both in the service of conservation. The event had been the academic fief of anthropologists dating back to at least the 1890s. Montinaro's thinking was, "Okay, we have snakes. Did no one think about the possibility of communicating something positive about snakes?" Meanwhile, the local snake hunters could provide useful ecological and conservation data: where they found their snakes, when, and in what condition.

New problem. The mayor was on board, but the *serpari* were not.

"No, I will not tell you where I collect my snakes!" he recalls being told. "Forget it! That's my secret!"

Fast-forward 17 years, and most of Cocullo's best-known *serpari* crowded into the snake registration room in the town hall. Montinaro photographed each serpent and logged all the data in a ledger spread out on the table in front of him. By the end of the day, Montinaro announced the final tally: 115 snakes (five different species in all) bestowed upon the festival by 30-odd snake-catchers. Once registered, the serpents were returned to their owners, who would bring them to the ceremony the following morning.

Italy has its share of idiosyncratic folk festivals, of course, but why snakes? And why here? Although the origins of Cocullo's festival are as shrouded in mist as the nearby mountains were on that chilly Sunday in April, there is good reason to believe that the local reverence for serpents arose in the pre-Christian era, largely due to a female goddess named Angizia, whose mythic story shares more than a few plot points with the legend of Asklepios. In the fourth century BCE, the Marsi, an ancient people, occupied territory in central Italy stretching from the mountains in the Abruzzo to modern-day Umbria. As described in the peerless 1911 version of the *Encyclopaedia Britannica*, "The Marsi were a hardy mountain people, famed for their simple habits and indomitable courage. It was said that the Romans had never triumphed over them or without them (Appian). They were renowned for their magicians, who had strange remedies for various diseases."

Chief among those magicians was Angizia (the name possibly derives from *angue*, Latin for "pertaining to a serpent"), and chief among the afflictions she cured was snakebite. She also, it was said, possessed the power to kill snakes by merely touching them. Whether goddess, magician, or witch, Angizia was revered by the Marsi. At roughly the same time that the cult of Asklepios exploded in Greece, the Italic embrace of Angizia became prominent in the early fourth century BCE. The most celebrated shrine to Angizia stood in a sacred grove near the present-day Italian town of Luco dei Marsi—a mere 25 miles west of Cocullo. It's not just that the fascination with snakes has a long local history; it's the independent temperament of the Marsi that seems to find a nearby latter-day echo in the *serpari* of Cocullo.

So there is a bracing mix of religion, residual paganism, citizen-science, and conservation in the current version of Cocullo's Festa dei Serpari, beginning with the special squad of *carabinieri* dedicated to wildlife protection who fan through the crowd on festival day to make sure all the participating serpents contain the required radio chip. At the 2023 festival, much anticipated after a three-year pandemic-occasioned hiatus, a steady rain fell throughout the morning—so steady that a rumor circulated among the cognoscenti that the event might be postponed by the bishop of Sulmona, who had the ultimate say on this ostensibly religious celebration. The rumor proved untrue, but the rain put a damper on both the size of the crowd (often up to 20,000 people) and the usual festive pre-parade musical warm-up. Still, the traditional Abruzzese bagpipe players gamely wheezed out songs in the rain, and the municipal band played marches, although the musicians mostly huddled under a portico of the village administration building.

When it appeared that the parade was about to commence, Gianpaolo Montinaro leaned over a railing and imparted a quick word of advice to me and my friend Riccardo Scalera. "I will give you one tip: Watch the *serpari*, and when you see them start to move toward the front of the church, follow them! Just follow them to the front."

Easier said than done, especially given the hundreds of umbrellas uncorked in the square.

Riccardo and I followed Marco Ognibene, who did not carry so much as wear a long *cervone* snugly wrapped around his neck like a beautiful four-lined scarf as he slithered through the crowd. Soon, as if on cue, a dozen *serpari*, all clutching handfuls of snakes, flanked the steps of the church, waiting like ambush predators. You could see a trickle of blood running down the hand of Maria Rita Zinatelli, a fearless young *serpara* who kept a firm grip on her feisty whip snakes. Suddenly, young girls in white dresses emerged from the church bearing special festival cakes, also known as cervone, over their heads (but under sheets of plastic). Finally the statue of San Domenico—wooden in both substance and expression, wearing a pained and bewildered look—wobbled into view, on the shoulders of several townspeople. Once down the steps, the men carrying the statue lowered it, and more than two dozen *serpari* surged toward the figure of the saint.

"*Indietro!*" shouted police officers. "Get back, get back!"

Still, the crowd pressed forward, and they let out a soggy cheer moments later when the statue was hoisted again. As the saint's head and body rose above the crowd, serpents covered, surrounded, encircled, slithered, and otherwise threaded their limbless forms around the pained face and halo of San Domenico—a veritable snake ball of *cervoni* hugging the haggard saint's head. The bitey coachwhips were not invited to participate in this phase of the procession.

As they have done for at least a century—the website of the "Istituto Luce," the archive of the Italian film studio Cinecittà has grainy black-and-white footage of the procession dating back to the 1920s—the townspeople of Cocullo paraded the statue around the main square, and up and down the streets of the village, threading their way among umbrellas, TV cameras, and selfie sticks. At the end of the parade, the *serpari* reclaimed their snakes and later released them, chips and all, back into the mountainous Abruzzese terrain at the very same sites where they had been captured. San Domenico retreated into the sanctuary, where he would magically dispense protection against snakebites, rabies, and toothaches for another year. Hundreds of umbrellas dispersed into the mist.

At a lunch hosted by the mayor after the sodden ceremony, an anthropologist named Claudio Corvino, who has witnessed the festival many times, made a point about the fragile reconciliation between pagan cultural practices and religious protocols. "San Domenico did not choose the *serpari*," he told me. "The *serpari* chose San Domenico." By which he meant that in order for these descendants of the fiercely independent Marsi to continue their vestigial pagan rite, they needed in a sense to use the church, and its local saint, as a cover. That was the theory of Alfonso di Nola, the Italian anthropologist considered the greatest expert on the ritual and Corvino's mentor. But it might also be that, in order to ensure popular support for the church among this historically transgressive population, the church opted to absorb, tolerate, and ultimately endorse a long-standing pagan ritual, dressed up though it may have been in liturgical livery.

Earlier in the day, as everyone stood in the rain wondering if the procession would take place or not, I chatted with one of the *serpari*, who was leaning against a wall to get away from the rain as tourists took turns holding one of his snakes. I posed a question that I had begun to ask at many stops along the snake roads I traveled: Are you finding fewer snakes in the nearby Abruzzese hills than you used to? Manuele Gentile, a handsome, rough-hewn young man who tended to avoid eye contact (but kept a very close eye on how people handled his snake), paused, glanced to his right, and finally gave a curt nod. True, it had snowed the previous week—less-than-optimal conditions for herping—but it wasn't just climate change that resulted in fewer snakes, Gentile said. "Also the *cinghiali*," he said. Wild boars. "They eat everything."

You could make the case that, in their literally headstrong rooting and rummaging, wild boars also destroy habitat, and they are, one way or another, yet another invasive species introduced by humans. But the point Gentile was making was that the animals simply devoured serpents, venomous and non-venomous alike. It was a reminder that in the stories of reptilian extinction that are writing themselves all over the globe, ecology is complex. Like politics, all habitat crises are local. Climate change is global. Both are ultimately human-made.

There were several somewhat surprising scientific postscripts to this pagan celebration, which add yet more complicating wrinkles to the ecology of snake–human interactions. Following the 2023 event, geophysicists from the Max Planck Institute of Animal Behavior in Germany reached an agreement with the Cocullo organizers to test the idea of using radio-implanted snakes as advance warning sentinels for earthquakes; based on earlier observations in China and elsewhere that snakes appear noticeably more active prior to major seismic events, scientists are now tracking the movements of five serpents as they slither through the Abruzzo mountainside. And in February 2024, the same scientists who crowded into that small conference room, measuring and swabbing the participating serpents, reported in the journal *PLoS Neglected Tropical Diseases* that 28.5 percent of the snakes collected for the 2023 festival bore signs of either mite-borne parasites or orally transmitted pathogens that could potentially spread diseases to humans. The scientific assessment of the health of the snakes, the authors wrote, may "generate conservation policies to benefit the human-snake interactions" and improve both serpent and public health in future iterations of the festival, which they deemed, in a respected scientific journal, a "sacro-profane ritual."

6

The Evolution of Pleasure

~~~~~

Reproduction

It was a typical spring morning in the Brennan lab. Shortly after 9:00 a.m. on a chilly but gorgeous April day in western Massachusetts, Patricia Brennan swiped her passkey into Clapp Hall, the venerable stone edifice housing the biology department at Mount Holyoke College, unlocked the heavy oaken door to her first-floor office, whisked by the large portrait of an emerald tree boa staring down from above her desk, and popped her head into the adjacent laboratory to greet Carmen San Diego, her pet corn snake (*Pantherophis guttatus*).

"Good morning, Carmen!" she cried.

Carmen earned her name, she explained, because she is shifty, squirms when handled, and "has escaped multiple times." Rohini, the ball python in an adjacent terrarium, was a bit less social this morning, not deigning to make an appearance. "These are all my ambassador squamates," she said—tame and user-friendly snakes that put both curious schoolchildren and trepidatious adults at ease.

A pale axolotl, the enigmatic salamander that can regenerate its limbs, floated in a nearby aquarium. Scribbled in black marker on a whiteboard on the far wall were the words SNAKE GENITALIA TASKS, with a catalog of species to be dissected, molded, and measured. Even by herpetological standards, it was an unusual to-do list.

If some lab members still had a little sleep in their eyes, it was because the previous evening, they had convened at Amethyst Farm, the Brennan family's sustainable hay and horse agricultural spread in nearby Amherst, to eavesdrop on the spring 2023 concert by the chortling male frogs known as spring peepers. These tree frogs inhabit wetlands on the property, and it had been a late night as the humans grooved to the sound of male frogs piercing the air with shrill mating calls.

A short while later, Rachel Keeffe, a postdoctoral fellow working in Brennan's lab, hovered over the lifeless, three-foot length of a recently thawed, beautifully banded Ethiopian mountain adder (*Bitis parviocula*). Keeffe delicately snipped through the belly scales and pried away filaments of tissue to expose the viscera of the venomous serpent (as a precaution, she had taped its head and mouth shut, even though the snake had been in a deep freeze for three years). In a nearby sink, a similarly defrosting Gaboon viper (*Bitis gabonica*), proud possessor of the serpent world's longest fangs and perhaps its boldest geometric patterning, awaited a similar postmortem investigation. Halfway up the midline of the adder's belly, Keeffe paused to look up at her visitor. "Oh, I should have asked—you're not squeamish, are you?" Let's just say I was grateful spring pollen had effectively plugged my nostrils.

As the snake lay open, Keeffe pointed out various organs: heart, lung, intestine, liver. When your external body plan is essentially a tube, your internal organs tend to be thin, elongated, and tightly packed inside a fuselage of 100 or more curvilinear ribs. "Hmm, a fang," Keeffe chuckled at one point, plucking an internalized tooth presumably swallowed by the snake. She eventually worked her way down to the female adder's reproductive organs, pointing out the vaginal pouch. She then deftly injected a vivid purple form of liquid silicone (the same kind used by dentists), waited about five minutes for the soft plastic to cure, and gently extracted a perfect, Barney-bright, three-dimensional mold of the female viper's sexual organs. The procedure resembled a kind of evolutionarily flavored variation on the Plastercasters, the 1960s-era groupies who, using a similar form of dental molding material, made casts of the genitalia of rock stars like Jimi Hendrix and the Beach Boys. In another corner of the room, an undergraduate named Grace aimed a fancy laser scanner to create a 3-D digital representation of the penis of a large Burmese python. In its specimen jar, it looked like an oversize, multi-lobed, mutant albino dill pickle.

When Brennan dropped by to see how the dissection was going, the two scientists marveled at the surprisingly small size of the male genitalia of the previous specimen Keeffe had molded that morning. "Oh, jeez," Brennan exclaimed. "I don't know if you can see these. Look how tiny they are!"

"That's gotta be a baby," Keeffe replied. Despite the smell of formaldehyde, the exchange had more the whiff of distaff locker room talk.

It probably exceeds the bounds of permissible irony, perhaps even polite conversation, to juxtapose this high-minded, quantitatively rigorous, and frankly graphic research on genitalia with the oaken, high-ceilinged rooms and soaring windows of the institution in which it took place, whose roots trace back to a 19th-century female-only seminary. But Room 120 is the epicenter of some of the most provocative reproductive biology research going on anywhere in the world, in any taxa of animal, with results that have landed Patty Brennan's research in the *New York Times* and *The New Yorker*, on cable television and NPR, and in hot water with conservative commentators who mocked her work, even as she headlines symposia at prestigious scientific meetings.

Apart from the public fascination, her research has surprising and profound implications for how the seemingly incompatible shapes of male and female sexual organs have evolved in vertebrates like snakes, birds, and mammals, with ramifications for the sexual health, habits, and antagonisms of humans. And this lab is also a cauldron of sociological upheaval in snake studies: Brennan was one of four scientists (all women) who authored a paper in December 2022 reporting that female snakes have two clitorises, known as hemiclitorises—a discovery that came more than a century after Victorian scientists (all men) began to report that male snakes each possess two penises, known as hemipenes, pronounced *hemi-PEENS*. As Brennan put it, "It's fine that it didn't occur to them, because not one single person can come up with all the questions and all the answers. What this really illustrates is why diversity in science is important. It becomes a problem if you exclude whole groups of people from being in science, right? That's when it's problematic."

Brennan is a petite and intrepid dynamo, dressed this morning in a bright-turquoise blouse, dark jeans, and stylish boots. Officially an evolutionary biologist, she unofficially comes across as equal parts flower child, evidence-based feminist, and old-school pass-the-scalpel anatomist who is, in the words of one prominent collaborator, "scientifically unstoppable." She speaks with the energy of an exuberant teenager, exhibits a brand of intellectual curiosity that is off the charts, and has been the protagonist of a scientific picaresque that has led, step by step, stop after stop, on what feels like an inexorable journey from an early love of whales to a sophisticated theory about genitals, snakes, and female sexual choice. It fits perfectly into this non-conventional scientific narrative that her husband quit his post as a postdoctoral researcher in behavioral ecology at Yale University to become a farmer, that the couple runs a CSA and Airbnb on the side, that she consults the price of hay on the commodity exchanges as diligently as the pages of *Science*. She is both celebrated and infamous for her area of interest, specifically: penises, vaginas, clitorises, copulation, rough snake sex.

At first blush (and there's very little blushing in these rooms), it might seem like an odd, almost fetishistic interest to catalog the precise, three-dimensional measure of the sex organs of serpents. But once you get past the blushing, it becomes apparent—as it did to Brennan two decades ago—that the shape or "morphology" of sexual organs varies dramatically between species, especially among snakes, where the genitalia display idiosyncratic bells, whistles, and gewgaws suggestive of sex-shop paraphernalia. And yet, as Brennan and her many distinguished collaborators have argued, these bizarre shapes also tell a story—an evolutionary story about the war between the sexes, about sexual antagonisms or compatibilities writ in the shape of male and female reproductive organs, and about some truly wild biological innovations by females to exert reproductive choice. Yes, even in snakes.

Given the Judeo-Christian tradition of demonizing serpents as symbols of temptation and sin, and the Freudian psychoanalytic tradition of viewing serpents as phallic symbols representing unconscious sexual desires, it is perhaps fitting that snake sex is a little bizarre. It's not just that males have two penises. They droop internally, like uninflated balloons, below the cloaca, the all-purpose vent that disperses urine, feces, and sperm; that's why male snakes tend to have longer tails than females—it's also a storage room for their sexual equipment. There is a confounding percentage of intersex individuals—snakes that possess anatomical features of both males and females. Some males, known as she-males, mimic females, manufacturing and emitting female sex pheromones to distract rivals. And some males are so good at this mimicry that they end up courting themselves, endlessly circling and tongue-flicking their own bodies.

And the sex can be rough. In certain species, the male hemipenes come outfitted with startling anatomical "ornamentation": hooks, barbs, calcified projections known as spines, ridges, "chalices," small flower-like petals as feathery as a foxglove, pits, protuberances, and bulbous rotund masses; sometimes the male organ is as forked as the snake's tongue, splitting into two distinct lobes. The females, for their part, can have multi-lobed vaginas. Or not. They often suffer injury and bleed from the rough appendages of the males. But as Brennan and colleagues showed about a decade ago, some females have evolved a few tricks of their own, including something that might be the envy of certain bipedal mammals, a trait called post-copulatory sexual selection, which is a fancy scientific way of saying, "Do I really want to keep this dude's sperm or not?"

"Snakes are weird," Brennan said, beaming. Things began to get really strange in the spring of 2021 when Brennan received an email—half scientific fan mail, half a plea for technical help—from a 23-year-old graduate student in Australia. Megan Folwell contacted Brennan because she wanted to do research on snake vaginas. "She was like, 'You know, I think what you're doing is super cool and so I would like to do some of that with snakes in Australia,'" Brennan recalled. "I was like, 'Great, let's collaborate!' Because of course there are a ton of species in Australia that we would not be able to get here."

Brennan and Folwell began a long-distance, pandemic-isolated collaboration, meeting through Zoom every month or so. One day early in the summer of 2021, Folwell asked Brennan whether she thought female snakes have the equivalent of a clitoris—a small nub of erectile tissue that becomes engorged during copulation.

Without knowing it at the time, by merely posing the question, Folwell had waded into decades of scientific controversy. Herpetologists had identified several lizard species sporting a pair of externally projected (or "eversible") hemiclitorises, so the structure had been reported in at least some reptiles. But a clitoris had never been convincingly reported in snakes. Folwell mentioned that she had started to comb through the scientific literature on female genitalia in snakes but confessed that many of the claims left her "confused." The question was especially touchy because the presence of a clitoris in snakes could suggest tactile sensitivity during copulation and perhaps even sensations of pleasure—things that no one had even dared suggest were a possible aspect of snake reproductive behavior. Brennan, who by then had dissected "a ton of snakes," pooh-poohed the idea.

"I've never seen them," she told Folwell on their Zoom call. "And from what I have read, they're not supposed to happen. Most people just think that they are not there."

"Okay," Folwell replied, and the issue seemed to be settled.

They scheduled a follow-up session a month or so later. It began with Folwell telling Brennan, "I want to show you some pictures..."

~~~

There is a widespread belief in the lay public, so strong that it barely rises to the level of afterthought, that snakes are unfeeling, asocial animals. How could they be social? They don't speak, bark, meow, howl, growl, snort, or otherwise communicate verbally. They don't do waggle dances like bees or create vast societal hierarchies like ants or other insects. They don't form monogamous couples like birds or travel in packs like wolves. The prevailing thought is that they are too simple, too biologically aloof, too reclusive to display anything resembling social behavior. This was even a prevalent sentiment among many herpetologists, according to Gordon Burghardt, who co-authored a 2021 academic book whose title alone refutes many of those biases: *The Secret Social Lives of Reptiles*. The volume is a compendium of reptilian behaviors suggesting that these animals are much more socially complex than historically believed.

Male and female cottonmouths reportedly form monogamous couples, at least on a small island in Florida. Harry Greene showed that female black-tailed rattlesnakes protect their young for at least a week after birth; Melissa Amarello and Jeff Smith more recently showed that rattlesnakes sometimes babysit for other females (but the moms are choosy, and won't let just any snake stay with the kids); and one of Emily Taylor's graduate students, Matt Holding, did a fascinating study in 2017 showing that when he physically moved male rattlesnakes 300 meters from their home territory during mating season, they rushed back faster to their home base than other dislocated snakes, in part to "protect" the females. These dislocated snakes even sprouted a bit of new neural tissue in their brains, which Holding believes reflected the importance of navigating quickly back to the females. Garter snakes can discern, through subtle chemical signals, whether a nearby snake shares the same hibernation den or not. Nowhere is sociality—in all its fraught manifestations—more apparent than in the realm of reproduction.

As anyone who's been lucky enough to see it in the wild, or

viewed the repository of YouTube snake porn online, the reproductive behavior of snakes is spellbindingly beautiful. They are like some preposterous M. C. Escher optical illusion come alive—cylinders entwining and braiding and disengaging, seemingly in defiance of physical and topological laws, and then reengaging again like dancing ropes (is it an accident that Escher's very last work of art before he died was called *Snakes*?). There are chin taps, feigned bites, "periscoping," undulating waves, kinetic braids. How fitting that their version of the act of transferring DNA from male to female itself resembles a double helix in motion. These acrobatic movements are all in the service of daunting physical mechanics accomplished without the aid of limbs. The act of mating can also make the animals vulnerable in the most heartrending ways: In one YouTube video I viewed (since removed), two copperheads are shown mating on what appears to be a suburban lawn in North Carolina, a mesmerizing three-minute pas de deux that is suddenly interrupted by a gruff off-camera voice saying, "They're copperheads, and they're having sex, and we gotta kill 'em," followed by the sound of seemingly endless blunt-trauma blows on a blank screen.

Unlike ducks, whose courtships and consummations last exactly 300 milliseconds as Brennan showed in a 2010 experiment on waterfowl, some snakes mate for hours; Greene wrote of a pair of western diamondbacks going at it for more than 25 hours. The basic mechanics involve the male everting one of his two hemipenes, inflating it with blood, and inserting it into the female cloaca. It sounds pretty conventional, as far as copulation goes, yet the diversity of snake genitalia and behaviors is what makes the lock-and-key version of intercourse a laughably inadequate metaphor. In a facetiously titled paper called "Are snakes right-handed?," Rick Shine and colleagues collected data showing that male garter snakes used the right-sided hemipenis more often than the left, with possible implications for reproductive success.

Snake sex can be downright orgiastic, like the famous spring-time snake balls in central Manitoba, where the mating behavior of tens of thousands of red-sided garter snakes (*Thamnophis sirtalis parietalis*) has become a tourist destination at the wildlife refuge in Narcisse, with a huge garter snake sculpture greeting visitors. It can be sexless in the case of some species, including pythons and cobras, where the females have reliably been shown to conceive on their own parthenogenetically, without insemination by males. It can be gender-defiant, in the form of the brahminy blind snake (*Rampho-typhlops braminus*), a species native to India and Southeast Asia in which all the individuals are female and all the offspring are clones of the mother. It can sometimes resemble aspects of a sperm bank, in the sense that females in certain species can internally store sperm from multiple males for a year or more. It can be reproductively (and physiologically) diabolical in some species where the males, after mating, create a physical plug in the female that stymies other males from depositing their sperm. It can sometimes seem to defy the laws of physics: University of Florida herpetologist Harvey Lil-lywhite once witnessed a male eastern yellow-bellied racer (*Coluber constrictor flaviventris*) dragging a female partially up a pine tree in mid-copulation by the strength of a single hook on one hemipenis.

It might even be—and this is the most astounding implica-tion of the recent results coming out of Brennan's Mount Holyoke laboratory—pleasurable for the females. Is it possible that snakes enjoy sex? And if so, why? Patty Brennan and her Australian collab-orators have been thinking about that a lot recently.

In many respects, Brennan was perfectly positioned, scientif-ically and temperamentally, to pursue these questions. Her career

path has the serendipity of a random walk yet follows an undeni-
ably organic logic. "Tortured" (her word) by three older sisters while
growing up in Bogotá, Colombia, Patricia Ligia Rodriguez discovered that she could hold her own against her siblings by fearlessly
embracing the creepy crawlers that her siblings shunned. "I was the
one who always liked to pick up all the creatures, all the frogs and all
the beetles, and throw them down my sisters' shirts," she said with a
mischievous grin.

Brennan's first scientific interest was marine mammals, and one
of her first scientific papers, in 1999, grew out of a one-year stint as
a member of the scientific staff on a 98-foot research sailboat ply-
ing the Galápagos Islands. That paper described a species of beaked
whale that had never previously been seen in the wild; she was
dumbstruck by the idea that an animal as enormous as a whale could
remain undiscovered for centuries, and it helped shape her research
philosophy ever since: looking for, and finding, things that others
had overlooked. "When you find that there are unknown whales in
the ocean," she told me, "it gives you this amazing window into all
of the things we don't know." That epiphany was part of "an absolutely magical year" in which she fell in love twice—with the idea of
doing scientific research as a career and with the computer geek on
the boat who maintained all their electronic gear, a guy named Bernard Brennan. He was interested in bioacoustics; she was interested
in dolphins and whales. They looked for a graduate school in the US
that would accept them both, which is how they ended up at Cornell
University in Ithaca, New York.

Brennan's mentors at Cornell steered her away from a PhD project on marine mammals (boats, gas, too expensive) and toward the
study of birds in Costa Rica. The office of Harry Greene, by then one
of the most famous herpetologists in the world, was right across the

hall in Ithaca, and the two of them often talked snakes. Snakes even barged into Brennan's initial avian research project, which focused on an ancient species of neotropical bird—a plump, shy, peculiar, hen-like relative of ostriches and emus called the great tinamou. "What drew me to them is that the males do all the incubation and all the parental care," Brennan said. "The females are just completely emancipated. And I was like, 'That's so cool!'" During the long reproductive season in the tropics, the females would constantly churn out big blue eggs—in part because one of their main predators, the tropical chicken snake (*Spilotes pullatus*), would burrow up from below their nests and eat the eggs *while the male was incubating them.*

The understandably skittish behavior of these birds led to the next epiphany, in the early 2000s, when Brennan was observing two tinamous copulating through her binoculars and saw this enormous white...*thing* sticking out from the male. "I started looking at it," she recalled, "and he started pulling it in, and I realized it was a penis! I was *so* surprised. I had been at Cornell for, I don't know, four years, and I had never heard that birds had external genitalia."

It was not a total surprise that she had never seen one before. Roughly 97 percent of bird species don't have penises or, more specifically, lost them at some point during evolution. And that struck her as a worthwhile thing to investigate. "That's how I started looking at genitalia," she told me. "That was it. It was like, looking at these tinamous having sex, and wondering why I'm not hearing about genitalia in birds." Since the road less traveled also runs through scientific disciplines, she returned to Ithaca, wrote up a grant proposal to study bird genitalia, and received funding from the National Science Foundation. "The reason that nobody was talking about bird genitalia is because nobody was *doing* anything with bird genitalia," she said, "which was really shocking to me."

· Brennan quickly forged alliances with two scientific power-houses in the bird world (as one scientist told me, "*Everyone* wants to collaborate with Patty"). She first established a research partnership at the University of Sheffield in England with evolutionary biologist Tim Birkhead, one of the world's experts on avian sperm. Then Yale University accepted her husband as a postdoctoral fellow, and since she was splitting her time between England and New Haven, Brennan set up a collaboration with Richard Prum, a world-famous ornithologist at Yale known for, among many things, his book *The Evolution of Beauty*. While working in England, she made the discovery that won her invitations to talk shows, coverage in newspaper articles, invective in the conservative press, and eventually the eternal admiration of her scientific peers.

She went to a farm near Sheffield, bought two male ducks, brought them back to the lab to dissect, and discovered that the males possessed a penis shaped like a corkscrew, turning in a counterclockwise direction. Like everyone else in ornithology, she knew that there was a lot of forced copulation in ducks; unlike everyone else in ornithology, she immediately wondered what the female genitalia looked like. The next day, she went back to the farm, bought two female ducks, and brought them back to the lab to dissect. "I was sure there had to be something going on with the females," she recalled.

There sure was. Brennan patiently and delicately peeled away multiple folds of female duck anatomy to uncover—"all wrapped up, like a little present"—the bird's vagina. It, too, was shaped like a corkscrew—but it turned in the opposite, clockwise direction. Rarely had evolution made sexual antagonism so mechanically explicit.

How did male ducks evolutionarily respond to this feminine twist in the story? Using a high-speed video camera at a California breeding operation, Brennan and Prum went on to show that male ducks externalize their 20-centimeter penis (just shy of eight inches) and ejaculate explosively in one-third of a second. "Basically, like a ballistic missile," Brennan likes to say. Their video of the duck ejaculation has garnered nearly 3 million views on YouTube. That finding became part of a larger body of work with a clear message. "In species where the males force-copulate on females," she said, "the males have very long penises, and the females have very convoluted vaginas. But in species where there is no forced copulation, the males have a short penis, and the females have a very simple vagina. So it's showing this really very nice co-evolution between the male and the female that's sexually antagonistic."

That 2010 paper by Brennan and Prum on ballistic duck ejaculation landed with quite a splash in the media pond: massive coverage in the popular press, an appearance on the John Oliver show, even a mention on *Saturday Night Live*'s "Weekend Update." But not all the attention was positive. Fox News conducted an online poll of its viewers about Brennan's research; 88.7 percent agreed that studying duck genitalia was a waste of taxpayer money. Tom ("Snowball") Coburn, a Republican from Oklahoma, derided Brennan's duck study on the floor of the US Senate. When Sean Hannity wondered mockingly on Fox News, "Don't we really need to know about duck genitalia, Tucker Carlson?" Carlson mockingly replied, "I know more than I want to know already!"

One of the reasons Brennan's bird genitalia research is so beloved, even by herpetologists, is that she refused to back down in the face of this political bullying. She and colleagues fired off op-ed pieces to *Slate*, *Current Biology*, and other outlets defending the value of basic research. Prum defended her in the pages of *The New Yorker*.

While the conservative media continued to snigger, Brennan

toyed with a much more provocative scientific idea: Maybe the shape of genitalia might reveal a back-and-forth evolutionary argument between males and females not just in birds, but in other vertebrates like mammals and reptiles. It was around that time that Greg Watkins-Colwell, senior manager of herpetological collections at Yale's Peabody Museum of Natural History, mentioned to Brennan in passing, "You know, if you think duck genitalia are cool, you really should look at snake genitalia."

Not long after, Brennan was doing just that, heading to one of the most famous red-light districts in all of snakedom: rural Manitoba.

～～

The moment Brennan learned that male garter snakes deposited a so-called chastity belt plug in females to prevent them from mating with other snakes, she immediately began to think up experiments. She first heard about the plugs at a talk by Christopher Friesen in January 2011 at a meeting in Salt Lake City. Every spring tens of thousands of garter snakes emerge all at once from underground limestone dens in Manitoba, about 55 miles north of Winnipeg, and almost immediately begin to congregate in massive copulatory orgies known as mating balls—not so much a spherical ball, however, as a kinetic warp and woof of sexual frenzy, a living undulating textile in brown and yellow and red stripes jiggling in a weave of reproduction. At the time, Friesen was a graduate student at Oregon State University, working in the lab of the snake chemosensory expert Bob Mason, and at the Utah meeting, he described the war of the sexes, garter-snake-style, with hooks, plugs, and mesmerizing undulations. Even before Friesen finished speaking, Brennan made moves to launch a collaboration.

"She was in my ear from the row behind me immediately when

I sat down after my talk," Friesen recalled. That exchange was followed by a boisterous lunch at a bistro where their enthusiastic talk about snakes and genitalia became so loud that other diners stopped eating and stared at them. Brennan had just received a grant to study male genitalia in snakes, and garter snake orgies offered a unique outdoor laboratory. Three months later, she was up in Manitoba.

The frigid prairie of central Manitoba, where the daily low temperatures average -20 degrees Fahrenheit in midwinter, might at first seem like an odd place to document a hot war between the sexes, serpent edition, and it certainly did not appear to be a situation likely to shed much light on something as grandiose as the timeless evolutionary issue of female choice in vertebrates. But it was one of the most famous sites in snake science.

"The reason the project in Manitoba has been so successful," Brennan explained, "is because you have these giant mating balls, right? So you have a *huge* sample size right there. Most people who study snakes—you have to go out at night to find them. And how the hell do you find them? They don't make sounds. They're cryptically colored. They don't move. It's really one of the biggest challenges of working with snakes, is really doing fieldwork in species where you can't find them in significant numbers." Numbers were not a problem in Manitoba. There were so many snakes that the researchers could barely move without stepping on them. Rick Shine, the Australian snake expert who has also collaborated with Mason's group, called the experience out of this world. "You're in an area the size of your living room with 10,000 sex-crazed male snakes crawling around and completely ignoring you. Part of it is, it's not just the numbers. It's the fact that the snakes don't care that you're there."

For a genus that has never enjoyed the racy celebrity of venomous serpents, the humble garter snake has been a fount of fascinating knowledge, reproductive and otherwise. Bob Mason has visited the Manitoba dens every spring since 1983 (with the exception of two recent pandemic years) and, along with a stellar troupe of graduate students, has asked shrewd scientific questions and gotten surprising answers.

It begins with the otherworldliness of the site. "I think it's one of the wonders of the natural world," Mason told me. The unique limestone geology has created a series of underground dens where the snakes hibernate for eight months during the frigid Canadian winters; those freezing temperatures, however, have created freeze–thaw cycles that have cracked the limestone so that the ceilings of some of the dens have collapsed, forming a kind of sinkhole where the snakes gather in the spring. You walk through a grove of aspens, Mason said, and along a path carved in the prairie, and suddenly you hear a strange sound. "You can actually hear them before you see them," he continued. "Of course, snakes don't make any sounds, right? The most a snake can do is hiss, and most snakes don't even do that. But what you hear is their scales rubbing against the other guys. And it's a very—I don't want to be too woo-wooey—but it's a very ethereal, kind of rustling sound. It's not off-putting. It's like, '*That's* a sound I've never heard before.'" Then you look down into a squarish depression, about the size of someone's living room, and, as Mason likes to put it, "It's just a sea of living spaghetti."

Male garter snakes emerge from their dens before females, so by the time a female emerges, she barely has time to rub the metaphorical sleep out of her eyes before she is approached, surrounded, courted, and essentially assaulted by up to three dozen would-be suitors. Some research has suggested that the males nearly suffocate their mating partners into submission using "caudo-cephalic waves" (essentially muscular contractions from tail to head), which induce oxygen deprivation in the female, causing stress and inducing her to open her cloaca. Size apparently doesn't matter: "small males with small hemipenes are just as successful," Friesen and colleagues have reported. But duration of copulation *does* matter.

Since most female snakes experience, even seek, multiple paternity, the male garter snakes try to trademark their semen by depositing a gelatinous plug in the female's vent after mating, which presumably prevents other males in the Manitoba conga line from depositing their sperm. But continuing research by Friesen and Mason has shown that the plug has more cunning male-centric purposes: As the plug dissolves over the course of several days, it releases even more sperm, and the scent of the sperm acts as an anti-aphrodisiac to suppress the urge of other males to mate. The plug also likely releases a chemical that dilates the female's oviducts, improving the odds of fertilization. The longer the copulation, the larger the plug left by the male and the higher the odds of insemination. The only way females can exercise choice, it seems, is through some form of physical rejection: either rotating her body to rebuff the male or literally squeezing out the male genital with the powerful muscles surrounding her vaginal pouch. To counter female rejection, male garter snakes evolved hooks to hold on to the female and prolong the time of copulation.

Under the watchful eyes of the emerald tree boa photograph up on the wall of her office, Brennan called up a striking photo on her computer from her initial trip to Manitoba. The image, dated May 2011, had been taken at a den near Inwood, and it showed a male red-sided garter snake copulating with a female. The hard, bony, curved appendage at the base of each hemipenis has been likened, even in the scientific literature, to a grappling hook. At the point where this barbed organ entered the female, there was a lot of red, and it wasn't part of the snakes' coloration or camouflage. It was blood. It was a vivid, non-microscopic, non-molecular, non-genetic marker of physical damage. And that of course piqued Brennan's interest. "If you're going in there and you're harming the female," she said, "that's usually a signature of sexual conflict."

As Brennan scrolled through more images of female snakes, she said, "The females are bleeding everywhere, right? That's his hemipenis going right in. And you see all the blood. Obviously, blood during sex is not good for females because that means injury." So Brennan, Mason, and Friesen cooked up an experiment in the field that would disrupt the duration of garter snake sex. It involved anesthesia, scissors, stopwatches, and what the researchers euphemistically called "genital manipulation." It also involved some improvisations in their field experiments when their initial research plan didn't quite pan out. "But that's okay because we would just shift on the fly and come up with even better experiments," Mason recalled. "And Patty is just a *whiz* at that."

First, they clipped off the "grappling hook" from the hemipenes of half the males—these snakes, they reassuringly reported, displayed "no deficit in courtship ability" after surgery—while leaving the hooks on the rest. Then they timed each copulation to plus-or-minus 10 seconds. They measured and weighed each copulatory plug. And because they did it in Manitoba, with dozens of crazed snakes as research subjects, they had the kind of numbers that lent statistical power to whatever they found. When all was said and done, the length of copulation was much shorter in males who'd lost their grappling hook, and the plug was smaller. Clearly, the hook conferred a reproductive advantage to the males. Conversely, when the researchers anesthetized the reproductive opening of the females, with the intent of dulling potential pain, but also diminishing the female's ability to squeeze out the hemipenis, the duration of copulation was longer. Crudely put, sex was shorter when the males lost their penile grappling hook, and sex was longer when the females experienced less pain during the interaction. This provided classic biological evidence of sexual conflict between males and females. "So the females basically actively are trying to terminate copulation to get the optimum that's good for them, and the males are using the spines to counter that," Brennan said.

Snakes offer a particularly vivid window on sexual conflict and evolution because of a word that has a lot of biological as well as sociological resonance these days: diversity. "There is a lot of variation in snakes," Brennan said, specifically referring to their reproductive anatomy. Intersex snakes, for example, possess male gonads and a female reproductive tract; others seem to have fusions of male and female organs. Even a cursory review of the literature suggested that snakes had a dizzyingly broad array of genital shapes and sizes. The male hemipenes could be cylindrical, forked, or bi-lobed, with all those hooks and spines. The female vaginas could either match the males (roughly cylindrical or bi-lobed), or, like the duck vagina, assume a shape that made it extremely hard to penetrate and inseminate. And in response to forced copulations, females in some species had clearly evolved reactive behaviors and morphological defenses to thwart those assaults. For studying evolution and sexual conflict in snakes, Brennan lacked only one final piece: a technique that would allow scientists to measure, with quantitative three-dimensional rigor, the precise shape of their warring genitalia.

It was around this time, as Brennan was contemplating next moves, that a colleague mentioned an obscure medical paper by Paula Pendergrass, an obstetrician-gynecologist at the Women's Health Research Group in Durham, North Carolina. Hypothesizing that the shape of a woman's vagina could influence everything from childbearing success to feminine hygiene product design, Pendergrass discovered she could make molds of the vaginas of patient volunteers using dental-grade silicone. The project was short-lived. As science writer Rose Eveleth discovered two decades later, people found Pendergrass's research off-putting and embarrassing; some even called her a "dirty old woman," and researchers considered the work scientifically uninformative. Brennan, however, leapt at the chance to try it in animals. Brennan dug up an early Pendergrass paper, adapted the technique, and soon added CT scanning and laser-directed automatic landmark measurements to create "cloud shapes"—three-dimensional, digitized representations of snake genitalia that could be mapped, categorized, and compared. This "3-D morphometric" technique suddenly allowed incredibly precise comparisons. It was also perhaps one of the first times that humans served as a model organism for reptile health.

In an early test of the technique, Brennan and colleague Dara Orbach dissected the reproductive tract of a dolphin and created a 3-D mold of a dolphin vagina. By this time, Brennan had become a virtuoso dissector, which was important because the vagina of this marine mammal was hidden, almost guarded, in a complex matrix of folds—more evidence, in yet another species with forced copulation, of females evolving genital structures to stymie unwanted male behavior. For decades, scientists believed the elaborate folds surrounding the dolphin vagina were there to keep salt water out of the reproductive tract. What they were really doing, Brennan and Orbach discovered, "is that they stop the penis from going into the vagina." That discovery, published in 2022, made waves not just because it created a 3-D model of dolphin genitalia; it was the first indication that female dolphins had a clitoris. And once they showed that the 3-D method worked in dolphins, Brennan and her team became the Plastercasters of the animal world. They did ducks, bats, chickens, alpacas, sharks, seals, skates. And snakes. "We're doing these dolphins and figuring out these folds," she recalled. "But all throughout this, I had snakes in the back of my mind. Snakes really required that 3-D component because their genitals are super complex, with all these spines and things," she said. "I was like, 'Okay, now we can do snakes!'"

She had dozens of snake specimens in three basement freezers in Clapp Hall, and started to dissect as many different species of snakes as possible, making casts of their genitalia and creating a growing library of three-dimensional molds—computerized, landmarked, and prepped for statistical analysis. The 3-D shapes began to tell intriguing evolutionary stories. During a deep dive into a genus of water snake called *Nerodia*, she discovered that the species with a more ancient lineage (*Nerodia rhombifer*) had huge penile spines, while two more recently evolved species (*N. fasciata* and *N. sipedon*) had lost those prickly appendages. Why? Brennan hypothesized that in this genus of snakes, the male genitalia have evolved to be less harmful to females—a process that, at least in snakes, but perhaps in other vertebrates, may happen much more quickly than scientists traditionally thought possible. "Genitalia are actually quite labile," Brennan said. "They can evolve different features fairly quickly. It doesn't take as long as you think it might take. And I think that partly is because they are directly involved in this fundamental process of copulation, which is an integral part of reproduction."

But if reproduction is essential to fitness of the individual, wouldn't female genitalia that discourage insemination be an evolutionary non sequitur? And if copulation is so harmful and unpleasant to females, why wouldn't they develop an aversion to sex? That was the bigger, much more puzzling question implied by all the hooks and folds of genital morphology. Like all living creatures, male and female snakes obviously agreed that reproduction was necessary. But, as Brennan told one interviewer, they sometimes "disagree on the details."

"Imagine you're a female snake," Brennan told me, "and you're mating with a male and this male is *digging* into your vaginal mucosa and causing bleeding and injury, and presumably even pain, even though we actually don't know; that's something we're looking into. Is this painful for females or is it that they don't have sensory receptors that would detect pain in the vagina? We don't know. But if this is somehow harmful, noxious, painful, why doesn't this result in the evolution of aversion, right? Females should avoid sex, but clearly, they're stuck. They can't really avoid sex because they need sperm to reproduce. So it's kind of this weird pickle that we find ourselves in. It's like sometimes sex can be harmful and painful, but you still need to have it. So then how do you bypass that evolution of aversion?

"Of course," Brennan continued, not missing a beat, "one possibility is that actually there might be another way that sex can be stimulating and pleasurable, and that could be through the presence of a clitoris, a functional clitoris."

Right about the time that Brennan was wrestling with this paradox, that young graduate student in Australia got in touch. Like Brennan, Megan Folwell had independently pursued a similar idea about creating 3-D molds of snake genitalia but had run into a roadblock—she couldn't get her molds to set. That's when she happened to notice the dolphin clitoris paper. Before long, Folwell and Brennan were dissecting the genitalia of as many snakes as they could.

It is not exactly clear when science officially discovered the reproductive organ of male snakes. In 1886, Hans Gadow, a Prussian zoologist, reported in the *Proceedings of the Royal Society* that lizards and snakes had a double penis (he did not use the term *hemipenes*); there was even a citation in an 1833 German-language text mentioning "Morphologie Der Hemipenes." Unofficially, Bob Mason surmised that hemipenes were probably discovered by whoever first cut open the tail of a male snake; they were that obvious. But by the 1890s, research on the male organ had become mainstream, even cutting-edge. Our old friend Edward Drinker Cope had not only described 235 different snake hemipenes in an 1896 paper but led a "brilliant attempt" (in the words of latter-day herpetologists) to use hemipenes as *the* anatomical feature on which to base the classification of every species of snake. For the better part of a century, no one seems to have wondered if there was an equivalent type of erectile tissue in female snakes.

Megan Folwell began to wonder about that in 2020, when a project she planned to pursue on elephants fell through and she switched at the last minute to snakes. She ended up working with Kate Sanders, a well-respected reproductive biologist at the University of Adelaide, investigating parthenogenesis and multiple paternity in sea snakes. The research convinced the 23-year-old Folwell that "snakes are *not* straightforward. They have a variation in every single aspect of their mating and reproduction." When Folwell embarked on her PhD project in the spring of 2021, her focus was relatively narrow; similar to Brennan's work, she performed snake dissections with the aim of creating three-dimensional molds of snake vaginas. She couldn't get the molds to set, but she stumbled onto something altogether different in the midst of that frustration.

In April 2021, while dissecting a female taipan (the extraordinarily venomous snake common in Australasia), Folwell uncovered a triangle of tissue near the snake's cloaca suggestive of a clitoris. That didn't rise to the level of a discovery, but it was an intriguing hint. "It was while researching that that I started thinking about the other genitals of snakes," she said, "and why I heard nothing about the clitoris. So I went looking into the literature." That preliminary dip into the scientific literature showed that descriptions of snake genitalia had been historically muddied by the presence of intersex snakes, with their confusing mix of both male and female anatomical features. Some of the research struck her as flat-out wrong. As she looked further back in time, she found Gadow's century-old descriptions of two male snake penises. Reports on female snake genitalia were, as Folwell and her colleagues would later note, "conspicuously overlooked."

Those twin frustrations, historical and technical, prompted Folwell to reach out to Brennan, but a likely third unspoken frustration came when Brennan initially scoffed at the idea of a clitoris in female snakes. Brennan was nonetheless impressed with Folwell's scholarship and asked her to give a presentation on everything she had learned in her literature review at the annual meeting of the Society of Integrative and Comparative Biology. By then, Folwell had documented no less than 10 mistaken or improperly cited articles on the presence, absence, or nature of the snake clitoris.

As Folwell told the scientists at the meeting, it was widely believed that female snakes lacked hemiclitores, which was puzzling because with the exception of birds, every other animal with a distinctive embryonic sac (or amnion), including reptiles and mammals, had one. Moreover, in reptiles like lizards, this tissue typically developed in females from the same embryonic starting point as the hemipenes in males, suggesting a common developmental origin. As Folwell and her colleagues later pointed out, one 2017 paper mistakenly identified scent glands as hemiclitores; another research group speculated that clitorises, although present, were non-functional vestigial organs; and yet others reported their presence in individuals that displayed both male and female characteristics.

"Everyone was like, 'Well, we can't find it. People have gone looking,'" Folwell told me. "That's fair enough, because depending on what species you're looking at, and especially if you're dissecting them the way that we dissect the tail for a male to find hemipenes, you would completely miss the structure. And because it's so variable, and sometimes really tiny and quite discrete, it would sometimes be luck of the draw. If you don't know what you're looking for, you would definitely miss it."

This knack for finding something that others hadn't even bothered to look for, a central thread running through Patty Brennan's scientific career, was now picked up by Folwell as she continued to dissect Australian snakes. But luck, as she noted, played a role, too. The variety of snake species native to Australia was extensive, and Folwell had 73 snakes in a freezer at the University of Adelaide. But she very luckily started with the common death adder (*Acanthophis antarcticus*).

It is an uncommonly beautiful snake, with alternating bands of ginger and vanilla running down the back; it has the look of a viper, with a flat triangular head, but is actually a relative of cobras and mambas. It is also extremely deadly. "Honestly, I think they're gorgeous," Folwell said. "They're such cool creatures that I think that's also what drew me to them." It turned out to be a fortunate (and non-fatal) attraction, because adult death adders are large, even chunky snakes—big enough to raise the odds of finding small structures like hemiclitorises if they existed. Like Brennan, Folwell came at the problem from a different angle.

For decades, vertebrate biologists had typically dissected snakes with a vertical line down the middle of the ventral (or belly) surface of the tail. That was efficient for finding male genitalia, not so much for finding the analogous female tissue. "I knew that if [there] were to be a hemiclitoris, that's internal," she said. "It would likely be a lot smaller, more fragile. So instead of dissection through the middle, like most people do, I dissected down the side." After gently separating skin from the cloaca, she carefully cut through some muscle. "And then saw the hemiclitoris." It was a small, distinct, triangular wedge of tissue located just below the cloacal vent of the female, between the two scent glands located in the tail.

But how on earth could you prove that this negligible nub, this *hint* of reptilian tissue, was in fact the snake version of a clitoris? Old-school histology, for a start. Folwell took microscopically thin slices of the tissue and stained them with a dye that reveals the presence of blood vessels; the clitoral structure lit up with a thatch of vasculature. But it was when she stained the tissue with a silver chemical that she got really excited; silver staining reveals nerve fibers, and nerves threaded through the tissue like worms in a compost pile. That suggested function (the tissue *did* something) and, more important, sensation. To enrich the microscopic panorama even more, she used a high-powered form of CT scanning to achieve high resolution. "Every step of this, as soon as I found out a new little piece," Folwell said, "it was like a little dance, a little celebration."

By August 2021, when she had one of her regular Zoom calls with Brennan, Folwell had assembled enough images with CT scans and histology stains from the death adder dissection to present her case. "I want to show you some pictures," she began.

Brennan was blown away. "She was like, 'What do you think? Do you think that...?' And I was like, 'Those are *definitely* clitorises.' It was crazy."

Encouraged by these preliminary findings, they went on to perform dissections in females from nine different snake species. The "focus species" was the death adder, because Folwell had two females, one male, and a juvenile down in the freezer. But they looked at eight other species in all: a Mexican ground viper (also known as a cantil; *Agkistrodon bilineatus*), puff adder (*Bitis arietans*), Norman's keelback (*Helicops polylepis*), Honduran milk snake (*Lampropeltis abnorma*), carpet python (*Morelia spilota*), Collett's black cobra (*Pseudechis colletti*), pygmy mulga snake (*Pseudechis weigeli*), and Ingram's brown snake (*Pseudonaja ingrami*). In every species, they found the serpentine version of the clitoris. Not that it was easy. They were often minuscule structures—ranging in size from one millimeter (about the size of the tip of a pen, even in fully mature adults) to more than seven millimeters (roughly the size of a thumbnail). "Again, it's the kind of thing where if you're not looking for them, you don't find them," Brennan marveled. "In some cases, they are so tiny, I don't even know how she found them."

Brennan suggested trying one additional stain. This one would reveal the presence of collagen—the kind of elastic padding they would expect to see in an erectile tissue like a clitoris. Collagen, too, was there. The composite picture meant the structure Folwell initially identified in the female death adder was an erectile tissue fed by blood vessels and rich with nerve fibers. "That's pretty suggestive, right?" Brennan said brightly. "That this fills up with blood, and they've got all these nerve endings, ergo, it's functioning for something. And because it's a clitoris, the 'something' is quite likely to be pleasure."

Put another way, after a century-plus of obsession with male genitalia by German anatomists like Gadow, "titans" of American science like E. D. Cope, and celebrated reptile curators like Raymond Ditmars and all the other "snake guys," we only learned in the last couple years from the "snake gals" that female snakes might actually have evolved a genital structure that allows them to enjoy copulation—despite, or perhaps because of, forced copulation with males that have as many penile doohickeys as a Swiss Army knife.

"I was super excited about it for many, many reasons," Brennan told me. "Because it's again one of those examples of something that hadn't been described because the people asking the questions were not very diverse. So as soon as you widen the diversity of who's asking the questions, you start getting more interesting questions and more interesting answers! But also, because it solved that problem in my mind of females evolving an aversion to sex if it is painful or harmful, because they are probably being stimulated even as they are being harmed, and so that can probably bypass the possibility that females might not want to have sex—which wouldn't make any evolutionary sense. So this is cool for that particular reason."

Not that the scientific world swooned. Two journals turned down the paper; one reviewer flatly dismissed the possibility that a female snake could have a clitoris. Brennan, still miffed a year later, fumed: "But look at the pictures, look at the data, dude!" The report ultimately came out in December 2022 in the *Proceedings of the Royal Society Biology*—completing an ironic and incredibly slow-to-close circle, as it was the same journal in which Hans Gadow's original paper hinting at hemipenes appeared in 1886. More than a few herpetologists have suggested to me that the novelty of the 2022 paper may have been overstated, overlooking earlier reports suggestive of hemiclitorises that appeared in the scientific literature 15 or 20 years ago. But as Harry Greene told me, "A lot of guy herpetologists never paid attention to it." And Brennan was having none of it. "People were basically saying, 'Oh, yeah, lizards have very well-described hemiclitorises.' But citations of snake hemiclitorises? Nothing. If it's not written down and published in the literature, it doesn't exist!"

If you go back to the earlier German reference to hemipenes, it took almost 200 years to close the circle, but roughly a nanosecond for the word to spread. The internet blew up as outlets from *The Atlantic Monthly* to the BBC carried accounts of the clitoris discovery. Once again, Brennan's work received more press coverage in a week than most scientists experience over an entire career.

This may seem like an extraordinarily long account of an ordinary scientific observation that just happens to have happened in snakes. But the sum of a decade of work on snake genitalia has now demonstrated that female serpents have evolved incredibly cunning mechanisms to exercise reproductive choice—even *after* copulation. They can constrict vaginal muscles to squeeze out hemipenes. They can close the gate to their oviducts, preventing insemination of their eggs. They can create a more acidic genital microenvironment that effectively kills unwanted sperm. And Patty Brennan would like you to consider the ramifications—*all* the ramifications—of this remarkable, and politically vilified, scientific saga.

"What can we learn from this?" she said on the morning I visited her lab. "At a very basic level, it's how females can co-evolve with males to reassert their reproductive autonomy. Females are often, and have often been portrayed as, passive, as somehow victims; they're waiting for the males to fight it out so that they can then mate with the male. That is an inaccurate portrayal of females, and complexity in females.

"So on one hand, a big part of it, is to say, 'Hey, look, yeah, males evolve traits to control females, to manipulate females.' But then females co-evolve traits that *prevent* males from succeeding in those strategies. Because they're not passive vessels. They are active evolutionary agents. And I think that we need to hear that message more because people commit the naturalistic fallacy all the time, whether we like it or not, and justify what's right or wrong based on what we see in nature. So, if we say, 'No, actually in nature, females are quite active. They evolve these cool responses. They protect themselves. They get away from things they don't like.' That's an important message, I think.

"But with snakes in particular," she continued, "I think there is another opportunity that's kind of cool that has to do with the evolution of spines in hemipenes. It turns out that spines are super common in most vertebrates. Some sharks have spines, and some mammals have spines. Some snakes have spines. And those spines are going to generate a mechanical response in the female tissue. And I think that understanding that, the sort of material properties of female vaginas, how vaginas may protect themselves against this mechanical harm, can actually potentially be important medically. We don't have very good models to study those questions in humans. The human vagina is subject to a lot of mechanical damage, during parturition especially, but also sometimes during copulation. Right now, we use rats as a model for humans. I think rats are definitely not quite as good as other systems might be. I'm curious about what we can learn from snakes how females are managing that potential damage. And I think that's kind of exciting."

Snake Road
Fifth Avenue, New York City

There is a road in New York City, a famous one-way street, that flanks Central Park, where the teenage Raymond Ditmars caught the first snakes in what would ultimately grow into the founding collection of reptiles at the Bronx Zoo. You won't find many snakes (if any) in Central Park today, but Fifth Avenue also passes the entrance to the Metropolitan Museum of Art, an institution whose holdings cut across hemispheres, belief systems, animal spirits, time itself. Anyone who has spent time perusing the jewelry collection in the Egyptian art wing, with its glistening gold serpentine pins and amulets, might conclude that there are more snakes in the Met than at the Bronx Zoo.

Some of the objects on display hint at the project envisioned by snake-lovers who want to rewrite the Garden of Eden narrative and reduce the ophidiophobia it has engendered. The art of a recent exhibition and the story attached to it goes back some 2,500 years, to a different kind of fig tree—the Bodhi tree (*Ficus religiosa*)—that rose in a forest in what is present-day Nepal. There was a famous snake there, too.

When a dissolute young prince named Siddhartha Gautama decided to renounce all worldly desires and seek enlightenment in the fifth century BCE, he faced many challenges. He had to escape from the 60,000 wives in his harem, according to ancient texts, traverse seven walls and seven moats, and elude the 500 guards ordered by his royal father, leader of a tribal clan, to prevent his escape. He had to slip through the locked gates, threaded with ringing bells, surrounding the royal kingdom in what is now western Nepal. And once outside, he had to avoid detection, like a cryptic serpent, by the legions of countrymen seeking to find him and bring him back. Gautama wandered as a mendicant and seeker until he found an isolated patch of grass under a Bodhi tree in Uruvela, a small village in northern India, where he commenced 47 days of non-stop meditation in his effort to achieve enlightenment.

At one point during this extended meditation, torrential rains battered the region, causing floodwaters to rise. The water inched up closer and closer to the fig tree, endangering the recently enlightened Siddhartha Gautama (no longer a prince, now the Buddha). At this moment of peril, according to an ancient sutta, when the newly enlightened Buddha was so deep in thought that he remained oblivious to the worldly danger threatening to engulf him, a snake spirit known as a *naga* materialized in order to protect him. This particular *naga*, known as Mucalinda, the king of serpents, assumed the shape of a cobra, encircled the meditating Buddha in seven protective coils and expanded its hood over the head of the Enlightened One to shelter him until the storm abated. Snake deities have long figured prominently in Asian cultures as symbols of the power of nature. But the flood around the Bodhi tree, and Mucalinda's role in ensuring the safety of Gautama, is one of the most famous episodes in the story of the Buddha—memorialized in the celebrated Mucalinda Sutta.

Mucalinda came to New York in the summer of 2023. In a second-floor gallery of the Met, at Fifth Avenue and 82nd Street, an exhibit called *Tree & Serpent: Early Buddhist Art in India, 200 BCE–400 CE* honored this confluence of nature, myth, human enlightenment, and serpent beneficence in a series of quietly dynamic sculptures mark-ing the origins of Buddhist art. Entering the first room of the exhibit, visitors encountered a beautifully ornate column of carved sandstone that depicted—no, *venerated*—Mucalinda, rendered as five cobra-like snake spirits arching up and linked in an animate umbrella of protection sheltering nothing more than an empty rectangle of space. That is because in early Buddhist art, as curators explained, the emp-ty space or vacant throne signifies the Buddha, who could be depict-ed only as a blank rectangle.

As Holland Cotter, the estimable art critic of the *New York Times*, noted, in words as pertinent to snake conservation as spiritual cele-bration, Buddhism is "a permission-giving faith, offering us, as it does, myriad ways to save our souls, including through practices of gener-osity." But it is also, he continued, "a faith of ethical absolutes, a major one being: stop killing—your fellow beings, meaning all living things, and the earth, which has a consciousness of its own."

That sandstone sculpture of Mucalinda, dating from the second century BCE, was discovered only in the 1960s in a village in eastern Maharashtra state, outside Mumbai, and it held pride of place in the first room of *Tree & Serpent*. When I visited, I stumbled unplanned onto a tour of the exhibit led by the art historian who organized the exhibit, John Guy, the Met's curator of South and Southeast Asian art. Standing in front of the Mucalinda sculpture, Guy pointed out that snakes, then as now, were often feared on the Indian subcontinent because of the human toll of snakebite, adding that early Buddhist art sought to convert that fearful perception into something different, replacing predator with protector, and trepidation with veneration.

Following the death of the Buddha around 400 BCE, preexisting shrines dedicated to nature spirits and demigods in India ("cult deities") were, according to the show, "commandeered and repurposed" as sacred Buddhist monasteries throughout South Asia and marked the emergence of the earliest Buddhist art. "Snake (*naga*) shrines were often chosen as sites for new monasteries," one exhibit text reads. These sacred sites, known as stupas, housed artifacts such as gems and jewelry touched by the Buddha. Surviving fragments of these holy sites show the snake deities as spiritual sentries, "the supreme protector of the relics."

In a piece of limestone sculpture known as a drum panel from the Amaravati Great Stupa in Guntar, dating to the second half of the first century CE (roughly 500 years after the death of the Buddha), a massive five-headed cobra, fanning out like an animate version of barbed wire, rises to guard the entrance to the shrine. "Two iconographic devices," Guy wrote in the exhibit catalog, "were privileged above all others in the adornment of the southern *stupa*: the branching *bodhi* tree and the snake (*naga*)." Another fig tree, another snake, a dramatically different storyline. Vedic texts dating back perhaps to 500 BCE mention the presence of cults "devoted to the veneration of deified snakes…"

To the tens of millions of people who fear snakes, including millions in present-day India, where annual snakebite fatalities are greater than in any other country, it might seem folkloristic to venerate cobras and vipers and other creatures of similar venomous disposition. But part of the wisdom of Buddhism, an essential part of its generosity, is the obligation to show compassion to the disenfranchised, embrace the loathsome, venerate the detestable. As one label in *Tree & Serpent* put it, "The teachings of the Buddha radiate from this art:

his simple message of compassion for all living beings and respect for the habitat that we share."

Six months before the Met exhibition had even opened, I had a conversation with longtime herpetologist Gordon Burghardt on almost the same exact theme as the show: snakes, trees, and reverence. He was making the point that all religions up to the early 1600s insisted on the veneration of nature, noting that the 19th-century writer J.G.R. Forlong, author of a massive history of religion, argued that "the three origins of all religions were trees, serpents, and fertility." Plucking another new idea from an old book, Burghardt continued, "Snakes have this awe and wonder that are associated with them, and trees, we're finding out, are super important to our survival and ecology. The nature connection disappeared by the early 1600s. But in all the South American, African, American Indian religions, snakes are really important." Because the two most important organisms in religions are trees and snakes, he said, we need a new religion that integrates and celebrates both (he proposed calling it Treepents, but we can probably do better than that).

Many people believe that reverence and science stubbornly stick to different sides of the street, but a recent BBC *Planet Earth* video suggests that the spirituality Burghardt envisioned is already alive and well in parts of contemporary India. The three-minute clip shows women and children in a small village in West Bengal walking barefoot—and, more to the point, without apparent concern—around a six-foot cobra as the snake slithers along a village street, chases down prey in their midst, and even pokes around in their dwellings. The villagers, according to the narrator, venerate the local snakes and allow them to wander freely underfoot. "Scientists believe that this tolerance had led to a change in the snakes' behavior—they move more slowly and are less likely to strike when disturbed," according to the BBC's Instagram account.

Two other objects in *Tree & Serpent*, seemingly minor, stuck with me. One was a six-inch-tall copper alloy figurine of the Roman god Poseidon, dating from the first century CE. The other was a small ivory figurine of an Indian courtesan—"bejeweled and garlanded but otherwise naked"—known as a yaksi dating from the same time period. Neither figure includes anything suggestive of a snake, but the Roman artifact was excavated in Maharashtra, India, and the Buddhist artifact was discovered in Pompeii, Italy. Which suggests not only a vigorous commerce of artistic objects and goods between the pre-Christian era of Rome and the Buddhist era in Southeast Asia, but perhaps also a vigorous affirmation of the idea, shared by both pagan and Indic cultures, that snakes deserve not just tolerance, but veneration.

7

No Legs? No Problem

Locomotion

T he problem with most papers in the scientific literature is the absence of drama. There's plenty of drama in research, of course, but it rarely finds its way into the official publication. This is true of scientific research in general, but it is especially true of a high-profile experiment about a decade ago that analyzed the motion of a very unusual snake—dramatic, it must be added, not least because the researchers aspired to measure the tiniest of physical forces in a species of snake that happened to be venomous.

In the spring of 2014, a small army of researchers from the Georgia Institute of Technology, Carnegie Mellon University, and Zoo Atlanta published a landmark paper in the journal *Science* deconstructing the unusual movement of a rattlesnake known as a sidewinder (*Crotalus cerastes*). Sidewinders, nocturnal denizens of the desert Southwest, have a unique form of locomotion that defies logic and compass: While the head points in one direction, the body moves sideways, at a 32-degree angle to the direction the head is facing. This form of locomotion allows the snakes to travel lightly over sloping, sandy, unstable terrain, but it is a mesmerizing, confounding riff on slithering; as one member of the research team put it, "if you look too long at a sidewinder, you'll go mad."

In a cleverly designed series of experiments, the researchers discovered some surprising new aspects of the mechanics and physics of this distinct form of movement. The paper got a fair amount of attention, with ample coverage in both the scientific and lay press, including a nice video on the *New York Times* website. And then, like news of all sorts, the findings subsided under the mass of newer news. Still, that initial paper has spawned more than half a dozen follow-up studies over the last decade and has even led, in an indirect way, to a redefinition of what it actually means to "slither."

But the initial drama lies in what didn't make it into the paper.

To start, there was the loud, rancorous cell phone argument that broke out as Hamidreza Marvi, the lead author on the eventual study, stood in line for final boarding of a plane at Hartsfield-Jackson Atlanta International Airport for a trip to Arizona to catch the rattlesnakes to be used in the experiment. "Whoever's on the other end is just *screaming* at him, in Farsi, so I have no idea what anyone's talking about," recalled Joseph Mendelson, one of Marvi's traveling companions. Mendelson could tell Marvi was agitated, but insisted he hang up because the flight was going to leave. It was then that Marvi, born and educated in Iran, sheepishly admitted, "That was my mother. She just figured out the real reason we're going to Arizona." Mendelson burst out laughing. "I was like, Okay, Hamid didn't tell anyone in his family that he was going to catch venomous snakes in the desert. And his mom just connected the dots and figured it out and went ballistic on him." Despite the fervent parental pleadings, Marvi boarded the plane, with no regrets. "I *loved* that trip," said Marvi, now a professor at Arizona State University.

There was the moment, later that evening, when the excited scientists swept their flashlights over the desert terrain outside Yuma, searching for their research subjects in a particular GPS location where local snake hunters had spotted sidewinders the previous night. Unbeknownst to the out-of-state scientists, the GPS coordinates were about equidistant between the US border with Mexico and a high-security Arizona state correctional facility. "And guess what?" Marvi recalled with a sheepish grin. The scientists suddenly found themselves temporarily surrounded by border patrol officers and local law enforcement; they'd neglected to alert authorities about their nocturnal collection expedition.

Then there was the awkward conversation back at Georgia Tech, in which Daniel I. Goldman, a professor of physics and the senior scientist on the sidewinder project, had to explain to university administrators that if the school insisted on its policy that no students could in any way be in the vicinity of venomous animals, on or off campus, the researchers might have to return $1 million in grant money to the US government. A hasty compromise allowed the scientists to throw together an ad hoc experimental lab on the grounds of the Atlanta zoo.

And finally there was the instigation for the entire enterprise in the first place: the epic failure several years earlier of a state-of-the-art, snake-inspired robot named Elizabeth that had been carted from Pittsburgh to a seaside archaeological site in Egypt. Roughly 4,000 years earlier, ancient Egyptians had created the "harbor of the pharaohs" on the Red Sea at Mersa Gawasis, where boats of Lebanese cedar plied Middle Eastern trade routes and returned with cargoes of gold, ivory, incense, and obsidian. Egyptian traders had carved eight storage caves out of fossilized coral lining the harbor at Mersa, and a team of Boston University archaeologists had explored the site. But several caves were deemed too unstable and dangerous to risk human exploration. Who knew what treasures lay in the unexplored caves?

Enter Elizabeth. A team of robotics experts at Carnegie Mellon, led by Howie Choset, had designed and programmed Elizabeth to move like a sidewinding rattlesnake, which effortlessly glides up sandy slopes in its native environment. But during a field test in Egypt in July 2011, they learned a brutal lesson about robots in the real world. Inside the cave, with its gently sloping sand, the robot flopped, literally and figuratively. She rolled over, she slid, she floundered, she sank in the sand—and the reviews were withering. "Failing miserably," in the words of biomechanics expert Jake Socha, seemingly writing Elizabeth's obituary. "Their limbless robots were *terrible* in terms of locomotion," said one of Choset's collaborators. Real sidewinders clearly knew something the roboticists did not.

In a field like herpetology, where for decades published papers usually had only one or two authors, generally field biologists, the ensuing collaboration typified a sea change in snake science— big groups, multidisciplinary, international. Choset reached out to Goldman, a physicist with an expertise in sand (technically, "granular media") at Georgia Tech. One of Goldman's colleagues at Georgia Tech was Mendelson, a herpetologist who, in addition to being head of research at Zoo Atlanta, had hunted sidewinders when he was a teenager growing up in San Diego County, California. By the time they added graduate students like Marvi, postdoctoral fellows like Henry Astley, physicists like Goldman, herpetologists, snake-handlers, roboticists, and mechanical engineers to the international research team, they had more collaborators than Medusa had snakes in her hair.

If you happen to see an ordinary snake gracefully slithering through the grass, you might pause to admire the elegant motion without giving much thought to how it happens. In fact, trying to explain serpent motion might seem like a sacrilege, unweaving a scaly rainbow. But it is a testament to the complexity, sophistication, and versatility of snake locomotion that it took mechanical engineers, mathematicians, physicists, evolutionary biologists, biomimicry experts, roboticists, and electrical engineers to *begin* to understand sidewinding and, in a larger sense, shed new light on snake locomotion. This sprawling research team wanted to know how sidewinders knew something that a bunch of PhDs struggled to figure out: how to negotiate a sandy, sloping terrain. If not exactly another example of breaking the rules, it surely defied logic. As Socha, a biologist in the mechanical engineering department at Virginia Tech, crisply framed the challenge: "Having no limbs to push off should make the matter worse, yet the snakes make it look simple. How do they do it?"

After their brief skirmish with law enforcement in Arizona, the Georgia-based scientists returned to Atlanta with six sidewinder rattlesnakes, roughly 400 pounds of genuine Arizona desert sand, and hopes not only of answering Socha's question but also of creating a new, improved, and much more sand-savvy version of Elizabeth. And Elizabeth, sand-challenged though she was, would help them learn a very important lesson about snake evolution.

~

Snakes don't do math or physics, of course, but they *know* math and physics—arguably better than our finest basic and applied scientists. They may even, according to a recent study that grew out of the original sidewinder research, have an intuitive understanding of quantum mechanics. Who cares? In addition to all the scientists, there's the National Science Foundation, the US Army Research Office, the Defense Advanced Research Projects Agency (DARPA), NASA, and more than a few biomimetic start-up companies throughout the world, all of whom have invested millions of dollars, and countless investigator-years, trying to replicate snake locomotion, only to discover a humbling truth. At least to date, snakes do it better. Much better.

Snakes are masters of a multitude of gaits. We usually associate the word *gait* with walking, but it scientifically refers to any mode of locomotion that navigates a terrestrial environment. Bruce Jayne, the elder statesman of snake locomotion, having researched the trait in his University of Cincinnati lab for nearly four decades, recently argued that snakes have 11 distinct and unique ways of propelling their bodies through the world, although they more or less fall into four general gaits. One of those modes, known as lateral undulation, is what we casually call slithering—the sinuous, curvy flow of motion that uses lateral body waves, a complex lattice of muscle and bone, and tiny morsels of Newtonian friction to propel forward. It is a seemingly simple three-part concerto of nerve impulses, muscle contractions, and forceful pushing to the side in a body-length wave that is endlessly repeated, propelling the snake's body forward.

In 2009, a group of scientists led by David Hu, then at New York University's Courant Institute of Mathematical Sciences, published what they called a "theoretical study," based on experiments using live Pueblan milk snakes (*Lampropeltis triangulum campbelli*) on a smooth fiberboard surface, that purported to explain how snakes slither. They even proposed a mathematical formula to capture the act of slithering:

$$Fr\ddot{X} = \bar{f}_{fric} + \bar{f}_{int},$$

It almost seems like aesthetic malpractice to reduce such graceful and elegant movement into a frieze of mathematical symbols representing acceleration, friction, and something called the Froude number. The snakes don't know what all those terms mean, but they do know, in a miracle of sensory awareness, neurological computation, and practical physics, how to navigate an uneven, constantly changing terrain without losing their purchase on the ground or the ability to move forward—except, it turns out, on a perfectly smooth surface like fiberboard.

It should come as no surprise that scientists have spent years trying to understand how snakes accomplish their rich repertoire of movement without limbs. The literature is thick with papers proposing formulas that explain the four main modes of snake locomotion: lateral undulation (slithering); rectilinear (straightforward crawling, or "belly rippling," as heavy-bodied species do); concertina (where the snake drags itself forward in serial contractions that resemble the expansion and contraction of an accordion, typically seen in species that navigate narrow spaces like tree limbs or tunnels); and sidewinding (where the snake's forward movement is actually at a sharp angle to the axis of the body).

What makes this agility especially impressive is that snakes started out with legs in the deep evolutionary past and decided, roughly 85 million years ago, to turn them back in. You can still see vestigial pelvic bones in CT scans of some snakes today, and recent research has shown that, almost like ghost limbs, vertebral nerves still fire at the site of these missing legs, even though there are no legs to move. This renunciation of legs is commonly characterized as snakes "losing" their limbs—a loss that religious scholars attribute to a famous passage in the Hebrew Bible that recounts how God condemned the snake in the Garden of Eden to crawl on its belly in eternal limblessness after it tempted Eve with the apple of knowledge.

On the religious side of the ledger, the loss of limbs was a punishment. If snakes could talk, however, they might describe it as a blessing. Bruce Jayne certainly sees it that way. "I think one way of looking at things is that there's simply many environments for which legs get in the way," he said. "They stick out from the other parts of the body. In many cases, even if a vertebrate folds its limbs back, they still stick out from the body wall, even if the limbs are held parallel to the long axis of the body. If you're trying to squeeze through tiny spaces, if you're trying to burrow through things, the legs can actually be a hindrance."

By losing their legs, snakes gained something enormously important: the ability to go almost anywhere. Temperature, not terrain, is their limiting factor, and because of that they have evolved to become exquisitely attuned to whatever geophysical challenges they find themselves in. They are the ultimate all-terrain vehicles. Unlike the world traversed by quadrupeds (most mammals) and bipeds (us), snakes negotiate a dizzying array of surfaces and obstacles—dirt, sand, water, mud, muck, rock ledges, poles, crevices, burrows, trees, forests, deserts, oceans, swamps. They crawl. They swim. They climb. They fly (or, more precisely, glide). And they do all those things *without limbs*. How do they do it? This quest to understand snake locomotion has produced an unstated, long-running, and amusing sideshow pitting real snakes against robots—a kind of science-based reality-show competition on who can do it better.

As Elizabeth attempted to show on her ill-fated trip to Egypt, snake-inspired robots could theoretically enter dangerous or inaccessible archaeological sites. They could be used in search-and-rescue operations after a natural disaster. They could be used to inspect difficult-to-access or potentially dangerous structures, from the hulls of naval vessels to nuclear power plants. They could be used medically, to reach small or inaccessible parts of the body during surgery. The military, according to Choset, was even interested in using snake robots for espionage or in hostage situations. "Staring at a bunch of snakes, people think you're crazy," Goldman said. "But they're beautiful science, and in fact potentially humanity-changing engineering."

Researchers usually love the animals they study—but not always. Choset loved robots, but snakes? Not so much. "I can't stand them," he told me. But he was willing to learn from them to build a better robot. He wasn't the first to try.

~

Snake robots, like snakes themselves, have lineages, a kind of genealogical family tree. Elizabeth's godfather was a Japanese engineer named Shigeo Hirose, and snake robotics got its start, in a manner of speaking, in a Tokyo restaurant in the 1970s. Hirose doesn't make snake robots anymore, but he is still considered one of the pioneers in the field of biomimicry—the idea of adapting traits observed in living organisms to design mechanical devices.

The idea of creating a robot that would imitate a snake came to Hirose when he was a graduate student at the Tokyo Institute of Technology. Like the legions that have followed him, Hirose was smitten by the possibilities of creating what he called a "soft machine" by mimicking a snake. "It's a very slender body, just a string," he told oral historians from the Institute of Electrical and Electronics Engineers in 2011, "but it can move or it can coil around an object or it can move from branch to branch. So it's a very versatile machine if we see it from the engineering standpoint." When Hirose learned in the early 1970s that a restaurant in the Shibuya neighborhood of Tokyo had "snake dishes" on its menu, he bought half a dozen live Japanese four-lined rat snakes from the proprietors, took them back to his lab, and began to study their movement from an engineering point of view.

Snakes may be very versatile, but that versatility makes them very hard to imitate mechanically. They have close to 20 degrees of freedom—20 different articulations, or directions they can move; the human arm, by contrast, has seven degrees of freedom (seven independent variables of movement) from the shoulder to the wrist. And what often gets overlooked, whether it's 7 or 20 degrees of freedom, is that it takes a sophisticated neural apparatus to coordinate all that movement. Beginning in 1972, Hirose spent four years trying to perfect a snake robot, creating five prototypes. Capturing the motion of snakes was so difficult that he essentially gave up and switched to spiders in 1976. It might have been viewed as an omen. It was not. (Serpents have lasted longer on the Tokyo restaurant scene than in Hirose's original experiments; patrons of the Snake Center Cafe now pay about 1,650 yen to have servers bring a live tame snake to their table to handle while they sip their coffee or tea.)

Nobody else picked up on Hirose's research until the 1990s, when Joel Burdick, a professor at California Institute of Technology, encouraged several of his graduate students to see if they could design a device with moving parts that could navigate over changing terrain like a snake. They didn't make a huge amount of progress, but one of Burdick's graduate students, Howie Choset, was, like Hirose, smitten by the snake-robot idea. "What makes them interesting is that they have many, many joints," Choset said. "A snake has 200 vertebrae in its back, on average, so you have lots of joints, lots of degrees of freedom that you have to coordinate." With so many degrees of freedom, according to Choset, there's no way that a human could neurally or intuitively control that many variables. "You need some sort of underlying theory or math," he said, "to really understand what it means to coordinate those joints, so the right motion happens." With that challenge foremost in mind, Choset joined the faculty at Carnegie Mellon University in 1996. He now co-directs its huge, world-famous, 1,000-member Robotics Institute. His journey all started with snake-inspired robots.

There was one nagging problem. Snakes did all those snake things better. The mechanical engineers struggled to understand exactly how snakes integrated an incredibly complex suite of physiological activities—ground and obstacle sensing, neuromuscular control, mastery of physical propulsive forces, huge muscular exertion in slender, nimble packages—to maneuver through environments that never resembled smooth fiberboard. It required more than new math. Biomimicry required intense observational study of real animals in real environments to make any progress in designing human-made devices.

The sidewinder project grew out of this biomechanical cul-de-sac. Goldman's team at Georgia Tech strove to understand and then incorporate unique animal traits in the design of their mechanical devices. When David Hu, the mathematician who was lead author on the NYU "slithering" study, joined the faculty of Georgia Tech in 2008, he and Goldman teamed up to search for inspiration in a menagerie of living, skittering creatures: sea turtles, sandfish lizards, cockroaches, even fire ants. "I was into things in the desert, David was into snakes," Goldman recalled. "We said, 'What combines those interests?'" And then, just around that time, Joe Mendelson invited the two researchers to take a tour of the snake collection at Zoo Atlanta for ideas. "We saw the sidewinders," Goldman recalled, "and David and I started talking and saying, 'That would be pretty cool to study.'" At about the same time, Choset reached out to Goldman for help in solving Elizabeth's "sand problem." An experiment began to take shape.

By then, Hamid Marvi had become a convert to the field of biomimicry. "I felt like there's *so much* you could do with that approach," Marvi told me. "We have so many ingenious designs in nature, and on the other hand we have so many problems in engineering that we have been struggling to find solutions to. And if you look closely, you can find solutions to most of those engineering problems in nature. They have evolved over millions of years, so you have a really good starting point. You don't have to start from scratch." Hu, Marvi's graduate adviser, gave him a kind of biomimetic ultimatum: ants or snakes. Marvi chose snakes. Soon after, he was standing in the Arizona desert at night—at a university-mandated distance of six feet away from any venomous serpents.

The "venomous" part of the experiment was "a logistical nightmare," in the words of Mendelson, the senior herpetologist of the group. "I wish garter snakes could sidewind," he lamented. "This would have been so much easier!" Since they couldn't do the experiments at Georgia Tech, he and his zoo colleagues built a three-by-six-foot custom-designed sandbox with high plastic walls in a shed at the zoo and added high-tech features for the experiment; they could adjust the surface of their imported Arizona sand, from perfectly flat to slopes of 10 and 20 degrees. After each trial, they could restore the sand to an absolutely pristine, uniform, flat surface through a sophisticated airflow system—what one scientist likened to "an air-hockey table on steroids."

The initial phase was observational but delicate. The scientists had to mark little dots, very carefully, along the bodies of the sidewinders, so they could train three high-speed cameras on the snakes as they moved through the sand; then, using the motion-capture data from the snakes, they could, as Choset put it, "reverse-engineer what we think the core math is." There's no poetry in mechanical studies; locomotion, as engineers define it, is a "properly coordinated sequence of 'self-deformations' (internal shape changes) that generate thrust to overcome drag forces (self-propulsion) via interactions with substrates." Translated into serpent-speak, a snake contorts its body in a coordinated series of motions and pushes against the terrain it's navigating through to propel itself forward.

Despite all the hardware and jargon, the scientists were still left speechless by the ineffable magic of the sidewinder's movement. "I've spent ridiculous amounts of time watching these things with Hamid, with Henry Astley, all of us in our little roasting sidewinder shed," Mendelson told me. "We can deconstruct it and describe it analytically, but I have never found a way to describe it verbally that does it justice. So I simply put on the footage and let the tape roll and go, 'Watch this.' This is as mesmerizing as Michael Jackson's moonwalk. It's like, wow, he's clearly moving to the left, but I'll be damned, his feet look like they're moving to the right. How is he *doing* that? I can't describe that any better than I can actually describe sidewinders, even though apparently, I'm one of the world's experts in sidewinding. I'll be damned if I can put words to it."

As anyone who has trudged up a sand dune knows, sand is a tricky surface to traverse. As long as there isn't too much weight focused on one spot, it acts like a solid surface, but when weight reaches a certain threshold, the sand gives way, behaving more like a fluid than a solid, and the weight sinks. That's why physicists prefer to refer to it as a granular medium, and why they spend so much time studying it. It gets even more complicated on a slope, where sand turns "liquid" more easily.

For comparison purposes, the researchers gave Elizabeth a turn in the sandbox and tested the ability of another dozen or so snakes in the Zoo Atlanta collection to negotiate sand, most of them venomous—copperheads, cottonmouths, and Mexican vipers, among others. "The reactions of the snakes varied," said Astley, a cheerful engineer with two branches of a long auburn beard spilling from his face and a competing tassel of ponytail in back. "Some of the species just kept trying until they got tired, others would give up and just sit there, and a few of the more aggressive or high-strung species would get quite worked up and thrash around quite a bit. Rattlesnakes are great because they wear their hearts on their sleeve, or more specifically their tail, so you can easily tell when they're irritated."

Although it was organized as a rigorous experiment, you could also think of it as an informal sandbox Olympics, and you didn't need to be a mechanical engineer to predict the results. Elizabeth reprised her debacle in the desert. She did okay on a flat surface, but with the sidewinding skills programmed into her by the engineers, she flailed on the 10-degree slope, either slipping or rolling downhill. The sidewinder rattlesnakes handled any slope up to 20 degrees with aplomb—in fact, the vipers were able to climb a slope just shy of 27 degrees, which is around the point where sand essentially becomes a fluid. "If you breathe on a 30-degree slope of sand, it can avalanche," Goldman noted. And because the researchers had filmed the snakes in action, they were able to translate those images into an explanation.

Viewing the film in slow motion, the mechanical engineers realized that the snakes send two waves of motion through their bodies. The horizontal, side-to-side wave was not a surprise; a snake loops its body almost like a lasso and throws it uphill to grab onto the sand. The other wave, up and down, was more of a surprise. At the beginning of each "step," the snake plunges its head into the sand, which seems to act like an anchor, and then sends the two waves through its body. The horizontal wave allows the snake to arch portions of its body forward—"forward" in this case meaning sideways, or orthogonal, to the head. The vertical wave, which is not simultaneous but rather in a delayed phase, ripples through the body up and down like a slow-motion strictly vertical jump rope, allowing the snake to control the amount of its body surface in contact with sand at any one time, which is crucial to success. "Only if you put high-speed markers on them and if you peer under them do you quantify their lifting," Goldman explained. "Too little, and you don't keep the sand solid enough and you potentially pitch over; too much, and you can't pick your body up over the next bit of sand." Perhaps most remarkable, after all these contortions and gyrations, the sidewinders leave a misleading autograph behind them in the sand: a series of straight lines.

The key to the movement, it turned out, was not the *side*-winding but the up-and-down wave. As the researchers analyzed the film, they learned that the snakes were masters at fractionally adjusting the amount of their bodies that touched the surface of the sand depending on the slope. More of a slope, more body contact with the sand; less slope, less body contacting the surface. The sidewinders made these minuscule, neuromuscular calculations on the fly, adjusting the amount of their body's surface area that comes in contact with the sand in milliseconds. Besides being instantaneous, these tiny adjustments were remarkably precise, shifting for example from 16.2 percent of body length to 16.4 percent as the slope changed. In a meter-long snake, a 0.2 percent adjustment translates into changing contact with the sand, on the fly, by two millimeters.

Hamid Marvi and his colleagues deduced lots of formulas and figures to describe this form of locomotion, but the tape doesn't lie: The motion is a living, breathing optical illusion. Translating all the variables into equations is nice for robotics, but it doesn't begin to capture the miracle of this form of motion in the wild, with varying surfaces and shifting slopes and random obstacles, all of which the snakes negotiate in real time using a serpentine computer light-years beyond what any robot programmer can currently achieve.

In fact, that's where this large, multidisciplinary team probably could have used one more member: a neuroscientist. A lot of the academic discussion about snake locomotion focuses on muscle, bone, tendon, skin, and the interface between a snake's scales and the environment, be it land, sea, or air. What often gets left out—and what makes snake movement so complex and sophisticated—is that there must also be some exquisitely sensitive neurosensory system that tells the animal that the terrain has changed, and that the animal must therefore adjust its gait—a gait, it bears remembering, that sometimes extends for 6, 8, 15, or 20 feet behind the head. This ability to change gait, in space and at speed, in an irregular environment typically strewn with obstacles, is virtually without parallel in the animal world; it would be like a car whose transmission not only changed gears from automatic to all-wheel-drive in a microsecond but also sensed a change in road surface and instantaneously switched from tires to tank treads or hydrofoil dynamics.

This is another instance where the reluctance to study non-traditional animals deprives us of novel, possibly revolutionary biological insights. No one has undertaken the admittedly challenging, but potentially revelatory, step of studying how the brains of snakes orchestrate this remarkable ability. Astley, a former graduate student of Jayne who now runs his own laboratory at the University of Akron, frequently gives lectures on snake locomotion, addressing this shortcoming head-on: "I used this all the time when I was on the market to become a professor—'I've prepared a talk on all the work that has ever been published on the higher-level neural control of snake locomotion.' Short pause. 'And that's everything we know about the higher-level neural control of snake locomotion.' Literally, there's nothing."

All in all, that's a lot of explanation for a form of animal locomotion that most people will never see. But the physical findings in that little shed in Georgia didn't stop with the 2014 *Science* paper. In a novel application of robotics, scientists used Elizabeth the robot as an instrument to test hypotheses about what the snakes were doing. Incapable of speech though they both were, the sidewinders and the robot commenced a dialogue in the sandbox that has continued for a decade; they've turned out to be very chatty, and the conversation has generated another half a dozen papers. The robot engineers used what they discovered from the sidewinders to "retune" Elizabeth, and by the time they finished all their tweaks, Elizabeth was capable of climbing sand on a slope. All she needed was an opportunity to show what she could do in the real world. An opportunity came a couple of years later when tragedy struck Mexico City.

⌒

On September 19, 2017, a 7.1 magnitude earthquake destroyed large portions of Mexico City. In the aftermath, the Mexican Red Cross reached out to the robotics team at Carnegie Mellon for help in search-and-rescue operations. Three CMU roboticists packed up

Elizabeth and a second snake-bot, hopped on a plane, and arrived in Mexico two days later, on September 21, with the best of intentions and, as they say in robotics, less-than-optimal outcomes.

During their time in Mexico, the robot team spent most of its time at Red Cross headquarters, according to several subsequent accounts by the Pittsburgh-based engineers, waiting to be pressed into service. Over their three-day stay, they deployed the snake-bots at exactly one site. They sent one of the snake-bots on two sorties (for a total of 10 minutes) into a small, pancaked apartment building where there were fears that three people might have been trapped. On one of those brief forays, Elizabeth managed to reach an other-wise inaccessible area, but it turned out to be an empty water cistern "which could not contain survivors." In terms of search and rescue, that was it. The Carnegie Mellon team showed up at several other sites, but the snake-bot delegation was not exactly rushed into ser-vice. As the robot team later characterized their on-site reception, "We waited for between a few minutes and a few hours before either deploying or being dismissed by the site coordinator."

As in all field tests, the engineers learned quite a few things to tweak once they got back to the lab. They learned that it was hard to see their laptop screens in bright sunlight (an aid worker had to hold a plastic bucket over the computer). The robot could not easily be operated by rescue workers, so the biorobotics researchers had to run things themselves. The robot operator, who maneuvered Elizabeth with a joystick, had trouble communicating with the "tether man-ager," who was feeding a long cable into the small, snake-size gap in the rubble, due to the noise of the surrounding excavation and dem-olition operations. And they learned that adding thermal imaging, chemical sensors (for gas), range sensors, microphones, and speakers would be helpful. Except for the audio, all those things come more or less as standard equipment on most serpent models. It was one more humbling experience for the snake-bots. In a sense, Choset conceded defeat when he told *Science* magazine, "We need to do a lot more, so we're that much more respectful of what a real snake can do."

It hasn't been all bad news. Roughly 1,500 heart disease and throat cancer patients at the University of Pittsburgh Medical Center and hospitals abroad benefited from a tiny, snake-inspired robot that allowed delicate surgical excisions of tumors, after the Food and Drug Administration approved the device for experimental use in 2015. Choset and surgeon Marco Zenati developed the robot and formed a company, Medrobotics, to market it. But the company lost a patent infringement lawsuit in 2020, ultimately forcing it to go into bankruptcy. "Is it a success that we offered it to so many people," Choset said, "or is it a failure because the company just didn't survive?" It was a reminder that the biomimicry business, like nature itself, can also be red in tooth and claw.

In a perverse way, the mission of the snake-bot engineers resembles a Borges story. As they try to incorporate more and more biological features from snakes into robots—more joints, more "actuators" (motors impersonating muscles), more feedback loops—the robots drift further and further away from the graceful creatures they seek to emulate, becoming so bulky, so clunky, and so sensately crude that they no longer resemble snakes at all but rather mechanical grotesques, caricatures of a beautiful living animal. Nobody would care about aesthetics if snake-bots worked, but the progress has been slow. Goldman doesn't mince words when talking about the capabilities of snake robots these days. "They're just terrible," he has said. He has moved on from snakes to robots based on centipedes—snake-like in their longitudinal form, but with legs.

Perhaps the final indignity for Elizabeth in this long saga, from the sands of Egypt to the rubble of Mexico City, is that her handlers "denominated" her—in a publication that emerged after the earthquake adventure, she was no longer Elizabeth but simply the Unified Snake Robot or U-snake. Despite being stripped of her name, it was clearly Elizabeth. Howie Choset admitted as much, and indeed delivered a kind of eulogy for the second generation of snake-bots when he was quoted in a 2017 CMU press release noting that the robot used in Mexico City was more than 10 years old and "on its last legs, even though it doesn't have any."

Despite all her struggles in the field, Elizabeth nonetheless bestowed an enormous gift on science. She has shone brilliant evolutionary light, in a series of follow-up experiments, on the deeper biological significance of sidewinding in particular, and of snake locomotion in general.

～～

The Georgia-based researchers continued to churn out, to their surprise, many more findings on sidewinder locomotion after 2014. They showed how the scales added a microscopically small but crucial force to sidewinding locomotion. How the snakes simplified, and in a sense automated, their sidewinding by use of repetitive neuromuscular templates, as if they had developed their own onboard algorithms trained by real-world experience. "None of that at all was in the plan when this project started," Mendelson told me. "This project started very much as a biomimicry—let's build a better robot! And then we stumbled across this idea that 'Wait a minute, the robot is actually helping us learn more about the evolution of animals.'" And as they learned more about the evolution of sidewinding, the robots helped them address what might be called—at least in Mendelson's view—the Attenborough Problem.

Science writers, science popularizers, and even scientists themself sometimes commit the Darwinian sin of saying an animal is "perfectly adapted" to the task at hand, be it locomotion, thermal sensing, or some other distinctive trait. The corollary term for mechanical engineers is *perfectly optimized*, meaning they've tweaked their device every which way to obtain the absolute best possible performance. No one complains about claims that robots are "perfectly optimized" (although snake-bots clearly have a long way to go). Evolutionary biologists, on the other hand, curdle at the notion that any animal has achieved a final state of perfection in the performance of a given trait.

"We realized that the corollary for *perfectly optimized* in evolutionary biology is this really dangerous phrase of *perfectly adapted*," Mendelson told me. "People—even, omigod, David Attenborough— use this phrase in documentaries all the time. But evolutionary biologists run for the hills. They avoid that phrase like the plague, because it's untestable. Evolution doesn't perfectly adapt anything. Evolution just keeps things that function in the gene pool. So we realized that if we get this robot to the point that it's perfectly optimized, and it's doing exactly what the snake does, and every parameter that we can possibly tweak reduces performance…Wait a minute, are these sidewinders perfectly adapted? Maybe there is not a better way to move on sand. Maybe there is not a better way to sidewind."

Mendelson paused as he was saying this, surprised to hear himself uttering scientific blasphemy. "I still can't bring myself to actually use the phrase *perfectly adapted*," he said. "I think that got beaten out of me in grad school. But using these robots, the physicists and biologists moved forward with this as a study system and realized that the robots are hypothesis-testing tools on animal functionality."

Elizabeth may have floundered in the field, but the snake-bot has

performed nobly in illuminating—indeed, honoring—snake locomotion. She established that sidewinders have achieved an evolutionary sweet spot when it comes to limbless locomotion. Perfect?
No animal is a perfect evolutionary machine in environments that
are constantly changing. But perfectly adapted? In terms of their
current habitat, Elizabeth would say yes. And that's exactly what the
scientists said, too, at a scientific conference in 2021.

And one additional surprise came out of Goldman's lab. Studying the motion of another desert-dwelling reptile, the shovel-nosed
snake (*Chionactis occipitalis*), graduate student Perrin Schiebel
reached an astonishing conclusion about how the snake uses its
spade-like skull to essentially burrow into and swim through sand.
"She discovered an absolutely gorgeous phenomenon," Goldman
said. "Those snakes basically mimic quantum mechanical parti-
cles. It turns out the snake is both a wave and a particle at the same
time."

There is an academic coda to the sidewinder research that indirectly curls all the way back to the informal term of locomotion
known as slithering. In the spring of 2015, Henry Astley, the postdoc
who participated in the sidewinder experiments, took a walk with his
wife and dogs in Little Mulberry Park in Atlanta when he happened
to spot a large black rat snake (*Pantherophis alleghaniensis*) along
their path. Astley took a video of the snake as it shot away from the
humans and noticed that the animal's motion superficially seemed
like lateral undulation (slithering), but that it incorporated a vertical,
up-and-down undulation as well. As he analyzed the impromptu
video later on, "I noticed that it was often moving straight or nearly
so, with far too few shallow lateral bends to be effective lateral undulation, but far too fast to be using rectilinear. It was at that point I
noticed it using vertical irregularities in the ground, most notably a

tree root." That got him thinking that perhaps the very definition of slithering was in need of a rethink.

Fast-forward eight years to a basement laboratory at the University of Akron, where Astley is now a professor in the biology department. With Penny, his seven-foot Taiwanese beauty snake (*Elaphe taeniura*), draped like a shawl on his shoulders, Astley showed me what has been the conventional rig for studying snake locomotion: an eight-by-eight-foot square box with several hundred upright wooden pegs and 12 motion-capture cameras overhead. For decades, researchers have deconstructed lateral undulation by watching snakes make their way through these human-made obstacle courses. But Astley's setup was in the midst of some high-tech remodeling. In place of the wooden pegs, he and his students are installing sensitive electronic detectors that can measure both lateral forces that push sideways against the detectors and also vertical forces pushing up against the ground as their test animals, usually corn snakes, slither through the thicket of wired pegs. Why?

Because, he believes, slithering is no longer just a matter of side-to-side undulations. "It really is a fully three-dimensional behavior," he said. "So I've been slowly, quietly pushing *slithering* to replace *lateral undulation*. Because it's not just lateral. It's also up and down." Astley insists that his insight is not a direct outgrowth of the sidewinder study. But you might say that both the rattlesnakes and a robot named Elizabeth set the stage for a new understanding of the mechanics—and meaning—of slithering itself.

Snake Road

"B Road," near Aiken, South Carolina

There is a two-lane asphalt road in South Carolina, but most people can't visit it, much less drive on it, unless they have national security clearance. "B Road" was the site of an unusual experiment that showed why crossing a road poses an existential danger for snakes.

The experiment, conceived by an ecological biologist named Kimberly Andrews, took place on the Savannah River Site, a 290-square-mile patch of wetlands, forest, and sandhills near the town of Aiken. The land, which is controlled by the Department of Energy, happens to be the place where the US government processes tritium and plutonium for use in nuclear weapons, and also disposes of nuclear waste, which is why it's off limits to casual human visitors. But it's pretty hospitable to animals (including 35 species of snakes), and in 2003 it was the setting of an experiment that answered a question few people probably thought needed asking: Why does a snake cross the road?

To answer that seemingly narrow research question, the scientists recruited nine local species of snakes; assembled an elaborate field setup with buckets, poles, and an observational blind; and threw in a 2002 Chevy Silverado 1500 pickup truck. In a larger sense, however, it was one of the earlier studies in a long-neglected area of environmental science now known as road ecology, which seeks to quantify the harm a road inflicts on animals that live in the surrounding habitat—harm that obviously begins with getting run over, but less obviously includes fragmentation of the environment, disruption of the natural behavior of the animals that occupy that habitat (be it foraging, migration, or reproduction), and, in a big-picture sense, all the future habitat destruction that roads inevitably inaugurate and accelerate.

The South Carolina experiment grew out of a master's thesis conducted by Andrews, an Australian biologist whose interest in road ecology dated back to the 1980s, when it didn't even rise to the level of a niche academic discipline. As part of her PhD work, she held a research position at the Savannah River Ecology Laboratory (SREL), which among other things monitors the health of plants and animals on the Savannah River Site. She worked with Whit Gibbons, the South Carolina herpetologist and nature writer who conducted research at the SREL for many years; the two of them collected nine local species of snakes—everything from a diminutive ring-necked snake (*Diadophis punctatus*) to Ben Franklin's favorite serpent, a timber rattlesnake (*Crotalus horridus*)—and watched what happened when the animals encountered a road. They wanted to know two things: how snakes behave as they approach a road, and how snakes behave near a road when a two-ton pickup truck blows by at 35 miles per hour.

The answer to the first question, not surprisingly, was: It varies. Smaller species such as the ring-necked snake and the southeastern crown snake (*Tantilla coronata*) seemed intimidated when they encountered pavement, almost never crossing. They avoided the roads, Andrews and Gibbons surmised, not because it exposed them to motor vehicles, but because it made them especially vulnerable to predators, mainly avian. Cottonmouths (*Agkistrodon piscivorus*) and southern water snakes (*Nerodia fasciata*) would often start to cross, have second thoughts, and then retreat to the side of the road where they started. Faster snakes, like the eastern racer (*Coluber constrictor*) and corn snake (*Pantherophis guttatus*), seemed less intimidated by the road. Most of the time, bigger and slow-moving snakes, like the timber rattler and the hognose snake (*Heterodon platirhinos*), didn't even try to cross. But in a subtly surprising observation, *every* snake that crossed the road did so at a 90-degree angle—in other words, even though their bellies hugged the ground, they somehow perceived (or intuited) the shortest distance to the other side and made a perpendicular beeline to safety. Was it vision or chemosensory clues? "When they cross a road, you *seldom* find a snake going longitudinally or even diagonally," Gibbons said. "They go straight. How do they do that? They're not supposed to be able to see that well. How do they know where the other side of the road is?"

Add a fast-moving vehicle, and the reactions varied again. Andrews and Gibbons tested just three species in this scenario, and only the timber rattlesnakes appeared to sense the oncoming vehicle ahead of time, exhibiting an "immobilization response." Two other species, the racer and the rat snake, froze, too, but only as the truck roared by. Oddly, the fastest snake in the group, the eastern racer, was the most reluctant to resume movement after the vehicle passed.

You can almost hear the ghost of Senator William Proxmire (D-Wisconsin) clearing his throat before bestowing one of his 1970s "Golden Fleece" awards for frivolous use of government funding. But snakes (and their fellow herps, especially turtles and amphibians) have an essential—and complicated—ecological relationship with roads. For nearly a century, almost as long as there has been a mass market for automobiles and a network of highways, scientists have observed that snakes often used the retained heat of paved roads to thermoregulate, especially at night. This aided the metabolism of the animals, of course, but it also exposed them to vehicular mortality and predation—including human predators who went out driving after dark to collect specimens for hobby and sport.

Roads even altered the trajectory of snake science. Laurence Klauber, CEO and chairman of the San Diego Gas and Electric Company and the elder (albeit amateur) statesman of rattlesnake studies in the US, practically invented the technique of "road cruising" to collect snake specimens in the 1920s. But as his early nighttime cruises into eastern San Diego County revealed, snakes believed to be incredibly rare, like the leaf-nosed snake (*Phyllorhynchus* spp.), began to show up regularly as roadkill; the "new technique," as Klauber described it, revealed leaf-nosed snakes to be one of the most plentiful species in the desert Southwest. Klauber's 1939 article "Night collecting on the desert with ecological statistics" in the *Bulletin of the Zoological Society of San Diego* grew out of nighttime expeditions near Scissors Crossing, in Anza Borrego Desert State Park, especially along a 20-mile stretch of California State Highway 78, which in the world of herpetology might as well be US 1, because it inaugurated the road cruising habit.

Roadkill might seem like a minor factor in overall snake conservation, but it poses a greater threat than many people realize. A study that came out at the end of 2023 concluded that vehicle collisions caused the majority of injuries to animals brought to wildlife rehabilitation centers in the US, based on data collected between 1975 and 2019. "Reptiles suffered the highest proportion of human-caused rehab admissions," according to an account of the research in the

Washington Post. In Australia, Rick Shine uncovered a complicated ecological story behind decreasing populations of common death adders (*Acanthophis antarcticus*). The bodies of many of the snakes littered the highway, and the hypothesis going into the study was that the snake populations had declined along a road in the Northern Territory because the adders had eaten poisonous cane toads, an invasive species in Australia. But when the researchers, as a control, compared the death rate of the snakes along nearby highways, they were surprised to find no dip in population. "The decline on the main highway was almost certainly due to the trucks that were roaring along and squashing snakes," Shine told me. "So the roadkill is actually a big deal. If you get a road with heavy traffic, it's going to be very difficult for a snake to survive long-term."

In Australia, roads also play a critical role in the ecological disruption caused by invasive species. "The invasive cane toads love to run down roads," Shine said. "The toads in the invasion front have evolved to be incredibly fast movers, and you can move a helluva lot quicker along a road than you can through the bush. You also bring in all sorts of invasive weeds and all kinds of other problems. Many of the terrible creatures that are causing such problems in Australia, like cats and foxes and so forth, just love roads, and they love the disturbed ecosystems on the sides of roads."

Highways ultimately pose a kind of holistic existential ecological threat. Roads mark the first incision of the human scalpel on any environment. Roads intersect routes that snakes use to find mates and prey. Roads force adaptations of behavior. And roads seem to incite a peculiar, species-specific form of road rage. Given the importance of the ecological dynamic between snakes and roads, you might think there would have been more systematic study, but besides the Andrews and Gibbons experiment, there has been only a handful of studies that I could find over the last two decades, notably including the collaboration between Bob Mason, lead researcher of the Manitoba garter snake dens, and Shine. But road ecology in its most expansive meaning is essential to the survival of species in general, reptiles in particular. Which brings us back, in an unexpected way, to that tract of government land in South Carolina.

It turns out that some of the most hospitable and protected habitat for snakes in all of America, and probably in the world, is associated with the military—army bases, air bases, training camps, weapons testing facilities, and similar off-limits government installations. Department of Defense land holdings amount to 26 million acres; that represents only 3 percent of US federal landholdings, but includes "more rare, threatened, and endangered species per acre than any other federal land managing agency," including national parks and Bureau of Land Management holdings, according to a 2015 DOD strategic plan for reptile and amphibian conservation.

What this means, in terms of coast-to-coast habitat hospitality, is that from Vandenburg Air Base on the central coast of California to the marine training base at Parris Island in South Carolina, there are millions of acres of federal military land subject to special environmental protection. In large part, that is due to a congressional bill most people have never heard of, called the Sikes Act, which requires every military facility in the country to have a wildlife management plan subject to the approval of local state fish and wildlife departments as well as the federal Environmental Protection Agency. All that real estate is rigorously monitored and protected, according to Robert Lovich, who co-directs herpetological management for the military, because viola-tions of federal environmental laws could compromise military train-ing and weapons testing. Since 2009, Lovich and colleague Chris Petersen have directed the Defense Department's Partners in Am-phibian and Reptile Conservation. As Lovich put it, "We unwittingly became the stewards of America's greatest biodiversity."

The military's environmental initiative has added some unexpect-ed wrinkles to snake biology. While monitoring a population of east-ern massasauga rattlesnakes (*Sistrurus catenatus*) near the National Guard's largest training area, Camp Grayling in northern Michigan, biologists discovered the first cases of snake fungal disease in that endangered species. Similar to the fungus that has decimated am-phibian populations, the pathogen infecting snakes, *Ophidiomyces ophiodiicola*, produces severe wasting and desiccation, according to Matthew Allender, the University of Illinois biologist who was part of the team that first reported the Michigan snake fungal disease out-break in 2015. Scientists have also been monitoring a SFD outbreak among eastern diamondback rattlesnakes (*Crotalus adamanteus*) at the US Marine Corps base on Parris Island.

The fungus causing the disease is not new. Allender has found evidence of the parasite in preserved museum specimens dating back to the early 1900s, and he thinks its pathology in snakes may loosely mirror the "cytokine storm"—an overreaction of the immune response to infection—that characterized many recent human cases of Covid. Whatever the exact cause of the pathology, even fervent snake-haters would have a hard time looking at images of the strick-en serpents—swollen faces, scabby skin, dislodged jaws, dissolved bones—without feeling empathy.

Stewardship of endangered species cuts both ways for the military. The spread of invasive Burmese pythons in Florida, for example, threatens to bring environmental chaos to the doorstep of military facilities in the southern part of the state thanks to not just the snakes themselves but also the way these snakes threaten other endangered or at-risk species, which the military has an obligation to protect under the Sikes Act. The pythons, for example, have already spread south to Key Largo, threatening the endangered Key Largo woodrat and the Key Largo cotton mouse. The responsibility for protecting those non-reptile species, Lovich said, could potentially compromise the training and testing mission of the navy, which is the largest landowner in the Florida Keys.

But one of the best historical examples of environmental chaos growing out of road construction happened a little farther north in Florida. In 1923, the industrial barons Henry Ford and Thomas Edison saluted a group of trail-blazing motorists as they set off from Fort Myers to carve out the initial route of the Tamiami Trail, an east–west highway bisecting South Florida. At the time, there were roughly a million Ford Model T's in circulation and approximately 2 million miles of roads in the United States, only 10 percent of them paved. A century later, there are roughly 283 million private and commercial vehicles sharing roughly 4.1 million miles of roadways, 90 percent of them paved. Those are the ingredients for a lot of potential "animal-vehicle conflicts," as road ecologists call them, but it's not just a matter of how a snake crosses a road or how many never make it across. It is that each road leads to something that will degrade the habitat even more: a mining camp, agricultural fields, a lumber operation in the forest, side roads that lead to subdivisions, industrial parks, shopping centers, parking lots. If each new road cuts into the landscape like a scalpel, civilization pries that incision open, just like a surgeon, and enlarges the wound, to the point where landscape and habitat will never fully recover.

The Tamiami Trail might be Exhibit A in the repercussions of a process often characterized as progress. Long before Burmese pythons got to Florida, construction of the Tamiami Trail devastated the ecology of the Everglades, according to both local indigenous people and conservationists. In connecting Tampa and Miami, the road blocked the natural north-to-south flow of water in the Everglades, effectively damming the wetlands and dramatically altering the habitat. The initial 1924 construction made an incision that engineers only now, a century later, are attempting to repair. Multiply this by 1 million, or 4 million, miles of highway in the continental US alone, and you begin to see the scope of habitat destruction and ecological disruption.

Although it is only anecdotal, I was surprised by how many prominent herpetologists (professional and amateur alike) mentioned how much more difficult it has become to find snakes over the course of just a generation. Bruce Jayne, the snake locomotion expert at the University of Cincinnati, mentioned it. Joe Mendelson, head of research at Zoo Atlanta, cited eastern indigo snakes (*Drymarchon couperi*) as a prime example. "They are just absolutely gone in most of their distribution," he told me. "That's in Florida, Georgia, and Alabama, Louisiana, everywhere that they occurred. And it's clearly a case of habitat disruption." Rick Shine noted the same thing in Australia—although he insisted that we need a standardized scientific survey to establish the exact state of snake populations in the world before jumping to any conclusions. The International Union for

Conservation of Nature (IUCN) plans to update its current "red list" of endangered snake species by 2025.

In the meantime, Shine, who is keenly aware of the way snakes defy convenient narratives, points out that not every habitat disruption story has an unhappy ending. In Australia, carpet pythons (*Morelia spilota*) have responded to the suburban sprawl outside some of the main cities by taking up residence in a new ecological niche: the attics and ceiling spaces in the roofs of new homes, which also attract opossums, one of their preferred food items. "It's also clearly true that there are a few snakes that can exploit disturbed habitats that we create and that are doing very well," he said. "Surprisingly well in many cases. So it's not all doom and gloom. There are winners as well as losers."

8

The Python Queen
of South Florida

~~~~~

Adaptation

W hen it comes to Burmese pythons, the monstrously large
snakes that have infested the Florida Everglades and are
invisibly slithering their way ever farther north, the word *cryptic* is
a dual-purpose adjective. In the language of herpetology, it refers
to the way an animal's coloration and behavior make it hard to see.
In the everyday language the rest of us use, *cryptic* most often suggests something puzzling, elusive, hard to understand, and ultimately unpredictable. You'll find the word in many scientific articles
about snakes, but its sheer power to suggest existential uncertainty is
nowhere more humbling than when you're standing somewhere in
the vast expanse of South Florida's "sea of grass," where non-native
Burmese pythons have made themselves very comfortably at home
since at least the 1980s.

Like their trademark camouflage coloration (irregular chocolate-brown splotches on a tan background), their very presence is cryptic—how many there are, where they are, how far north they have already spread, what their future range might ultimately be, and what that further spread might mean for implausibly distant ecosystems in (to hear some scientists tell it) the Carolinas to the north and parts of the Pacific Northwest. The technical meaning of *cryptic*—hard to find—renders the future of the pythons, and our future with them, cryptic in the larger sense: unpredictable and almost impossible to understand.

The irony of searching in the enormous Greater Everglades Ecosystem for a cryptic, "invasive" species occurred to me as I was standing on a raised plank of wood in the rear of a Nissan Frontier pickup truck as it slowly rumbled down a ghostly levee road just north of Everglades National Park one January evening. Standing next to me, sweeping a flashlight beam into a thicket of darkness, was Donna Kalil, a legendary snake whisperer and contract python hunter in South Florida.

I had met Kalil and fellow hunter Marcos (Marc) Rodriguez at sunset in the parking lot of the Miccosukee Gaming Center, a Native-run casino on the western outskirts of Miami. We left the neon behind and drove about 20 miles farther west on US 41, deep into the heart of the Everglades. US 41 is none other than the Tamiami Trail, that venerable two-lane highway completed in 1928 to link Tampa with Miami, and the road has a bit part, too, in Florida's python story. After crossing a little bridge just past the Miccosukee reservation, we turned right onto a narrow dirt road that brought us to a locked white gate, blocking access to a levee road known as L-28. Kalil, like approximately 100 python hunters in South Florida, possessed "the key"—the key that unlocks this and dozens of other gates, allowing access to the best hunting grounds. It did not occur to me until later that the gamblers at the casino probably had better odds of winning than we had of finding a python.

At the time we went out, on a Thursday night in January 2023, Kalil, along with her team of hunters, the "Everglades Avengers," had already yanked 735 pythons out of the South Florida environment and finished first in one of the state's annual "Python Challenge" events. As we rattled along a narrow levee road whose limestone soil cast an eerie glow in the headlights, Kalil confided that she wasn't about to quit until she had dragged a thousand invasive snakes out of the Everglades ecosystem. The remark attested to the ambition and skill of this 60-year-old former PTA president, former real estate agent, and full-time python obsessive who has been called the Python-Hunting Queen of South Florida.

There have been countless articles, videos, documentaries, TV shows, and even streaming sitcoms (*Killing It*) documenting the spread of these unusually large snakes in the Everglades. "There is no such thing as a bad python story," an editor at the *Miami Herald* once told me. They have decimated native mammal populations in the Everglades; rabbits, raccoons, opossums, bobcats, foxes, muskrats, squirrels, and rats have essentially disappeared, and the snakes are now working their way through wading birds. They can ingest deer, pigs, humans (at least in their South Asian habitat; no human prey in the wild so far in Florida), even alligators. A famous 2005 photograph of a dead alligator that partially ate its way out of the stomach of a large and equally dead python became an instant, iconic image of an ecosystem gone mad. That photograph inspired Kalil, who was a real estate agent at the time, to try her hand at snake hunting. "With these pythons," she said, "when I found out there was a problem out here, it's like, 'Wow, I've caught snakes all my life. I have a skill set that can help with the situation.' Every single one we remove is a plus for native animals that they would have eaten otherwise had we left them in the environment."

Before setting out that evening, I asked Kalil's supervisor, Michael Kirkland, if he had any tips for a first-time python hunter. Kirkland, who heads invasive species management for the South Florida Water Management District, urged me to train my eye by studying pictures. "To spot a python," he said, "you have to get the search image in your head first. Most people would just walk right past one of these things in the field because they blend in so well, are so well camouflaged." In other words, cryptic. Kalil's advice was a little more blunt: "This is one of the skills that I've mastered growing up—you're either predator or prey out here in the wild. And you don't ever want to act like prey," she added with a little laugh. "So when you're out here, we are the predators. We are hunting the hunters."

Even though January is not considered prime hunting season, we had reason to be cautiously optimistic. The week before I joined her in the field, Kalil had a banner day, snagging four big pythons, including a 13-footer; three of the captures were females, and among them she counted 100 eggs (she has been known to use python eggs to bake Christmas cookies). "This woman is incredible, she gets the snakes to come to her," said Rodriguez, who met Kalil when she was PTA president and he was PTA treasurer at the middle school their kids attended in Miami. He drove the Nissan about five miles an hour—"road cruising" in a way Californian Laurence Klauber could never have imagined—while we searched for anything pythonic in the darkness.

The drill was simple. I searched out to the left, Kalil searched out to the right. If we thought we saw something, we would yell "Python! Stop! Go back!" Rodriguez would brake, fling the truck in reverse, and we would inch back to take a closer look at whatever caught our eye. As we set off down the levee road, Kalil spotted a banded water snake in the canal nearby. Good omen? Perhaps, but she also paused to tick off a list of snakes we were better off not seeing: eastern diamondback rattlesnakes, cottonmouths, pygmy rattlesnakes, and coral snakes. "Those are the four we could run across out here," she warned.

As Rodriguez's truck rattled down the levee road, the word *cryptic* kept popping into my head. In a 2011 book, biologists Michael E. Dorcas and John D. Willson conceded that there was no accurate figure for how many invasive pythons were in Florida, although they speculated that the population "may very well number in the hundreds of thousands." A decade or so later, Kirkland was a little more cryptic. "We just don't know," he said, adding, "We're hoping we'll have a better answer for you in five years or so." Part of the reason for this protracted and maddening uncertainty is that Florida's Burmese pythons are incredibly hard to find and live in an extremely inaccessible landscape.

At one point, Kalil shouted "Stop!" and we backed up. Something shiny had caught her eye along the edge of the canal. We all directed light on the spot. It turned out to be the light reflecting off an empty bottle on the edge of the canal.

As we rumbled along the dirt levee road in Rodriguez's pickup (Kalil's pioneering 1998 Ford Expedition Eddie Bauer Edition, with its customized python-leather upholstery, was in the shop for some well-earned rehab), I couldn't help thinking about a little-known experiment—more anecdotal than scientific—that Dorcas and Willson described in their book. It put the challenge of this entire enterprise in perspective. In the summer of 2009, researchers placed 10 large male Burmese pythons in a fenced-in, 25-by-31-meter (roughly

81-by-101-foot) "semi-natural" enclosure in South Carolina; the open-air, walled-in pen had a small pond, a few trees, some bushes, and a sprinkling of dry underbrush and debris, but it was smaller than a lot of suburban backyards. The snakes, by contrast, were not small; they measured between 6½ and 11 feet.

The formal purpose of the experiment was to see how well the pythons might adapt to a slightly colder climate than in Florida. But as a kind of impromptu contest, researchers added what they thought would be a fun exercise. They recruited 19 observers—seven of them considered expert or experienced python hunters—and allowed them one at a time into the enclosure. Each snake hunter had 30 minutes to locate as many pythons as they could. Only 2 of 19 people managed to find a single snake, and five so-called experts failed to detect a single python. "Many areas of South Florida are largely inaccessible and contain heavily vegetated or aquatic habitats that afford poor visibility," the researchers later noted. "Thus, it is reasonable to assume that detection [probability] of wild pythons in Florida is well below 1% much of the time."

Day or night, python hunting is like fishing: You can spend hours and hours in the field only to return home empty-handed. But the hours are never idle. Lulls in the action are an invitation to yack about personal history, technique, family, the past, the ones that got away. Kalil told me she was a military brat; her father was in the air force and the family moved around every couple years, so she grew up in Venezuela, Oklahoma, and South Florida.

She described herself as a "*definitely* weird" kid; snakes not only fascinated her but seemed to offer company, social cachet. There was the pretty little snake six-year-old Kalil was playing with on the floor of her family's home in Venezuela when her nanny entered the room, screamed, and beat what she now believes to have been a venomous red-tailed coral snake (*Micrurus mipartitus*) to death. There was the corn snake that got loose in their North Miami home when she was nine. "My mom had birds at the time, and it ended up eating one of her birds," Kalil told me. "She was not happy. She said, 'Donna, you can't have snakes anymore.'" There was the six-foot yellow rat snake that she hid in her bedroom shortly after receiving the no-more-snakes edict. And there was the four-foot Florida king snake (*Lampropeltis floridana*) curled on top of her head when she removed her cowboy hat in high school to introduce herself to the boy who would become her husband. As she likes to put it, "He knew what he was getting into."

The one that got away? That would be the night in 2018 when Kalil got her hands around the neck of an extremely large python in the water just off a levee near where we were driving. "I circled around behind her in the water and grabbed her behind the neck," Kalil said. "I was straddling her, and then she just sort of bucked and threw me up and aside. Tried to follow her in the water, but she was gone." Using a video of the encounter recorded by a colleague to mark distances, they later estimated that the snake was about 17 feet long.

My big moment came when I spotted a suspicious form at the bottom of the levee, in a swampy thicket of vegetation. The dark form was motionless, yet almost perfectly sinuous; in fact, the curves were the only thing that distinguished it from the surrounding underbrush. Rodriguez dutifully inched the truck backward and directed his high-powered flashlight in the direction I had pointed, until the beam of light settled on…a duplicitously serpentine tree branch. "Better to check than miss one," Rodriguez said, humoring me. "Coming from the outside, I'm sure you've seen about a thousand stick-snakes, log-snakes, branch-snakes. Everything starts looking like snakes and pythons."

About an hour into our cruise, we encountered another group of contract hunters heading back out. "Is that Donna?" a woman's voice floated out of the darkness. "It *is* Donna!"

Everyone knows Kalil, and there was a mini reunion of the key-holders, with hugs, on the levee road. The main theme of conversation was meteorological: A brief cold front was moving through, with overcast skies and temperatures already dropping into the mid-60s. "This is the worst two months to get a snake, with this weather," said Dan, another contract hunter. "It's dropping down. You can feel it."

We continued up the chalky white levee road for another couple of miles, found a turnabout, and headed back to the Tamiami Trail. We did not see, much less catch, a single python. True, the temperatures were in the mid-60s. But I couldn't help thinking again about the long-ago South Carolina python-spotting contest.

The South Carolina enclosure in which the exercise took place measured exactly 8,181 square feet, or 0.000293453 square mile. Everglades National Park alone is 2,410 square miles; the South Florida Water Management District, larger still, comprises 18,000 square miles—roughly 61 million times larger. If a trained pair of expert human eyes couldn't find one python in an 81-by-101-foot enclosure, you can imagine how absurdly difficult, how utterly impossible it is to find pythons in the wild, much less conduct a reliable census.

That is what *cryptic* means in the context of ecological disaster.

~

This is not a widely shared opinion, I suspect, but the word *invasive* often strikes me as a brilliant marketing term, coined by humans to shift attention away from their own stupidity and instead redirect it toward species that had no vote on the matter. Invasion implies decision and cognition, a conscious and premeditated aggression intended to suborn an adjoining territory, as in Russia invading Ukraine. "Invasive species," by contrast, make all their decisions on a smaller, narrower, more reactive scale. The first priority? How to survive in the foreign ecosystem into which they've been unceremoniously dumped, whether by commerce (exotic wildlife), illicit trade (animal smuggling), pirates and whalers (the black rats and goats, for example, that overran the Galápagos Islands since the 16th century), dubious modern global trade in things like used tires (Asian tiger mosquito), delusional agricultural pest management strategies (cane toads in Australia), bilge water from ships (zebra mussels in the Great Lakes), leisure-class importations for sport hunting (wild rabbits in Australia), horticultural laissez-faire (Japanese honeysuckle everywhere), negligence (allowing non-native pets to escape), or craven irresponsibility (deliberately releasing Burmese pythons and at least six other non-native reptile species into the Florida wilds). These animals and plants did not choose to invade, and their order of Darwinian business is basic in the extreme—what to eat, how to survive environmental hardships, how to reproduce and sustain a population. When people sanctimoniously decry the spread of invasive creatures whose invasions were facilitated by human ignorance, greed, laziness, or negligence, it makes me wonder: What were they supposed to do? Go on a hunger strike?

If the "invading" animals are resilient and lucky and genetically nimble and, yes, cryptic to avenging eyes, they have a chance to survive and proliferate. The Burmese pythons have all those things going for them. And, as even the contract hunters will tell you, they are magnificent creatures. Creatures that are, as Kirkland put it, "just excellent at what they do."

The full origin story of the Burmese python (*Python molurus bivittatus*) invasion of South Florida will probably always remain as cryptic and mysterious as its subject, but the "most plausible" scenario—laid out in a comprehensive overview published in 2023 by the US Geological Survey, produced by first-author Jacquelyn Guzy and her 36 collaborators—involves a combination of a runaway pet trade, abandoned pets, roadkill, good genes, eye-popping genetics, and tenacious biological resilience, with perhaps a bit of folklore and conspiracy theory sprinkled in.

The story begins in 1979, when an unsuspecting motorist traveling along US Highway 41 just north of Everglades National Park ran over a 3.58-meter (almost 12-foot) adult Burmese python. That snake, recovered from the same road I took with Donna and Marc on the way to our nighttime hunt, was the first more or less contemporary sign of pythons in the wilds of South Florida (although the *Tampa Daily Times* reported a python sighting in the Everglades way back in July 1912). What happened after the 1979 roadkill incident is a little murkier.

Everglades National Park (ENP) personnel apparently spotted Burmese pythons on multiple occasions in the 1980s, specifically in the mangrove swamps and saline glades in the southwesternmost corner of the park, nearly 50 miles south of the 1979 roadkill on the Tamiami Trail. But because no records of those observations can be found, the reports are considered unconfirmed. Nonetheless, when Burmese pythons were observed more frequently in the national park between 1995 and 2000, most were spotted in the same general area, in the southwest corner of the park near the towns of West Lake and Flamingo, while others were identified roughly 20 to 40 miles farther north.

As Guzy and colleagues wrote, "These observations suggest that multiple generations of Burmese pythons were present in ENP by

2000 or earlier and that the population occupied a large geographic area." Probably earlier. "You don't start actually seeing these animals until they're well established," said Kirkland. "We can't say what was going on previously, but just to see the population explosion in the mid- and later 1990s, there must have been a reproducing population well before that. I don't know when that started, but I think we can say with confidence that it was well before that."

How did they get there? Everyone from university scientists to hunters like Donna Kalil is pretty much agreed on this point: The "founders" of this ecological disaster were fugitive serpents from the pet trade, either intentionally or unintentionally dumped in the Everglades. Between the late 1970s and 2011, according to stunning (and probably conservative) government statistics, at least 294,000 Burmese pythons and Indian pythons were imported into the United States from Thailand, Myanmar, and Vietnam—and that doesn't include countless more pythons bred in captivity by US hobbyists.

And it might not have taken a lot of snakes to get the invasion going. "It wasn't an act of God," said University of Florida biologist Frank Mazzotti, who has been tracking invasive species in Florida for at least a quarter century. "It was an act of people. But not a lot of people." He believes that "this *all* could have started from one female who was let go and was storing sperm and was able to lay eggs over several years." Female snakes can store active sperm for years—up to eight years in one documented case—and essentially self-fertilize at a time of their choosing.

There is a persistent suggestion—bordering on a conspiracy theory—that the Florida population exploded, if not started, when a large reptile breeding facility south of Miami was destroyed during Hurricane Andrew in January 1992. "I get a wee bit tired—and by 'wee bit,' I mean the exact opposite—of the [story about] Hurricane

Andrew spewing petri dishes of baby pythons into the Everglades, and if not for the hurricane, we would never have gotten the python problem," Mazzotti told me. "That's simply not true. It's been well documented that there were a very small number at first, and we know that because the original genetic diversity was very low. The other thing is, it's not likely that hordes of pet owners drove into the Everglades and let their snakes loose. More like one or two." Of the Hurricane Andrew hypothesis, Kirkland said, "We have no evidence of breeding facilities being destroyed. It's all just hearsay and rumors."

Then there's the outright conspiracy theories. One is that the government deliberately released snakes into the wild to create public alarm that would in turn spur legislators to pass regulations banning private citizens from keeping snakes as pets. Conspiracy theories being very much in season, this belief mirrors a rampant myth that has circulated in Europe at least since the 1970s, according to Italian invasive species expert Riccardo Scalera, that vipers and other venomous snakes have been deliberately spread by government agents in helicopters in order to encourage stricter environmental laws.

For the sake of argument, let's say it started with one released snake. Imagine for a moment that you are Python Zero, dumped by some distraught pet owner who bought a cute and beautiful exotic snake, kept it well fed, and then discovered how unwieldy (and expensive) it is to maintain a reptile that can add eight inches of length *per month* during its main growth spurt. What did abandonment feel like from the snake's point of view? I posed that question to Harry Greene, who famously enjoys reptilian thought experiments. He pointed out that the abandoned snake was no naif. Its phenomenal chemosensory apparatus would quickly begin to make sense of even a strange environment. "So it's coming to that [the new environment] with some innate knowledge," he said. "You know, it's coming to that with a somewhat innately constrained range of odors it will follow when it's hungry. It's going to be constrained on odors that it will follow when it matures sexually, and it's either a male looking for females or [the opposite]. And then it's also going to come with some capacity and propensity to modify its behavior in response to experience." That last observation is a sly scientific way of saying: They can *learn*. "And this last part?" Greene added. "I think all of us didn't expect that 20, 30, 40 years ago."

It was bad enough that they were intelligent, but there was an additional genetic complication. At some point, either in the hands of captive breeders in Florida or, more likely, back in their native Asia, the Burmese pythons bred (or "hybridized," as geneticists put it) with the related Indian python (*Python molurus molurus*). Scientists in Florida had already detected molecular evidence of this hybridization, which led to the grim speculation that genetic mixing of these two lineages "may provide the Florida population with novel genomes to facilitate adaptation to a wider range of conditions."

Once they were established, there was no going back, and this is the other, often overlooked human-made part of the story. Beginning in the late 1880s, agricultural entrepreneurs built hundreds of miles of drainage canals in the Everglades, converting an estimated 50 percent of the wetlands into farmland. Decades later, the state of Florida enlarged this vast network of canals and levees in and around the Everglades, for both agricultural and water management purposes; they control the flow of water out of Lake Okeechobee in central Florida as it slowly moves south through the Everglades and also cleanses the water before it reaches the protected areas. To visitors and tourists, those canals and swamps may look like placid out-of-the-way waterways, hospitable venues for alligators and Florida's dazzling array of avian wildlife. To the pythons, they looked like interstate highways. The snakes need high ground to mate and build nests, which is why the contract hunters mostly stick to the levee roads, but it's a bit like the drunk looking for his missing car keys under the streetlamp; for most of the year, most of the snakes are far from the roads—covert, hidden, and on the move while traveling through the water.

The animal has proven to be incredibly proficient at adapting to its new environment while remaining hideously elusive—not easy for a snake that can weigh upward of 200 pounds. They can go months, even more than a year, without a meal. Juveniles grow explosively, adding four feet of length in a six-month growth spurt. They can survive extended exposure to salt water—pythons have been spotted in Biscayne Bay up to 15 miles offshore. They can regulate their body temperature remarkably well, despite being "cold-blooded"; studies have shown they maintain internal temperatures around 85 degrees Fahrenheit. They can survive cold snaps behaviorally, by taking shelter in gopher tortoise or armadillo burrows, and brooding females, through a process known as shivering thermogenesis, can generate heat to keep both developing embryos and young hatchlings warm during cold weather. And they apparently see much better than scientists originally thought. Rocky Parker, one of the co-authors of the USGS study, recalled how a python detected him visually from a distance of 15 feet and quickly disappeared: "That's another reason that they're so cryptic—they are able to see threats at a great distance and change their behavior accordingly," he said. They have astonishing homing abilities; scientists have documented cases where radio-tagged pythons were captured, relocated, and made their way nearly 50 miles back to the exact location where they were originally caught, and Kirkland told me a few relocated pythons found their way back after traveling a staggering 200 miles. ("Burmese pythons are capable of homing at a scale previously undocumented for any species of snake," according to the 2023 report, and "presumably" use polarized light, celestial or magnetic cues to orient and navigate.) And, most important in their war with hunters, they are virtually impossible to see in the wild. On multiple occasions, biologists tracking snakes implanted with radio transmitters have discovered they're literally standing on top of the animal without realizing it.

By the time wildlife experts comprehended the scope of the problem, the problem had already spread like a slow-moving and irreversible watermark bleeding through a map of the southern third of the state. By 2003, pythons were found throughout Everglades National Park; within three years, they had moved north into Big Cypress National Preserve and east to Interstate 75, on the northwestern edge of Miami-Dade County; by 2010, they had spread north through much of Broward County and on into Palm Beach County; a year after that, hatchlings were spotted outside Naples, on Florida's Gulf Coast (genetics later suggested that a different founding population of pythons had spread on the western side of the state); by 2014, DNA evidence suggested they were present in Loxahatchie National Wildlife Refuge, on the doorstep of West Palm Beach, although actual snakes wouldn't be spotted there for another two years; by 2018, they had also spread south, to Key Largo and the upper Florida Keys. Back on Highway 41, where it all began, a roadkill hatchling in 2009 symbolized how entrenched the population had become. "There are probably more pythons in Florida now than there ever have been, both in terms of numbers and locations," Mazzotti told me in 2022. "Eradication is no longer an objective."

And the two big questions that dominated scientific and

public conversations a decade ago remain unanswered: How many are there? And how far might they spread?

The likelihood of a python being spotted by a human observer, according to government scientists, is technically less than 5 percent—but practically speaking, the "detection probability" is much worse than that. As the USGS report put it, "Pythons have essentially zero detectability unless they are in limited areas that can be searched." So even though, beginning in 2017, the South Florida Water Management District and the Florida Fish and Wildlife Conservation Commission each deputized 25 citizen–python hunters (background checks, real-time 24-hour satellite tracking of each hunter in the field, permits to euthanize the captured animals), and officials claim that more than 20,000 animals have been removed from the Everglades ecosystem since 2000, no one is willing to venture an on-the-record guess as to the overall population.

"You hear all kinds of numbers thrown out there," said Kirkland, "but the real answer is: We just don't know." The USGS report concurred: "There are no reliable estimates of python abundance or density," adding, "although there have been multiple reports in the media, an exact number is impossible to know." Margaret (Maggie) E. Hunter, who has been using cutting-edge genetic tools to track the spread of the snakes, told me, "So, the Number—it's kind of this Golden Chalice that everybody wants. But because we can't detect them, it's very, very hard to get it." The Number—critical to policy, management, and control—cannot be determined, much less uttered.

Meanwhile the governor of Florida, through press releases, touts the state's annual "Python Challenge," with prize money of $30,000 in 2023 for hunters who catch the most pythons and the biggest python, as making a dent in the population. Experts like Mazzotti are a bit blunter, noting that the hunting contests are good for public education, not so good for controlling the python population. Others, off the record, dismiss the event as a public relations stunt and "a circus."

If we're unlikely to get an official number (even a bad one) for another five years, according to Kirkland, what about the question of range? Can Burmese pythons spread beyond South Florida, and if so, how far? There was, mostly out of public view, an academic battle royale over this very question about 15 years ago. Two US Geological Survey scientists, Robert Reed and Gordon Rodda, studied the range of Burmese pythons in their native Southeast Asia, where the snakes inhabit ecological niches ranging from tropical jungle to the foothills of the Himalayas. Using that information, and pairing it with a climate-based assessment of the United States, they created a "niche model" suggesting that the Florida python population could theoretically spread into a much broader swath of the US—as far north as Virginia, south along the Gulf states and Texas, and as far west as California.

Press coverage was instantaneous and sensational. Scientific blowback was equally intense. A separate group, headed by Alexander Pyron at George Washington University and Frank Burbrink of the American Museum of Natural History, attacked the Reed and Rodda model as flawed. "That was one of the most famous brawls in herpetology," one scientist told me. "It was just rough. They got lambasted and just dragged through the mud for years." The models from both groups, oddly enough, agreed on an improbably distant destination for invasive pythons: the Pacific Northwest.

And then, right in the middle of this fierce debate, an unexpected meteorological event may have provided an ice-cold crystal ball revealing what the future may hold. The picture was not what many experts thought at the time.

⌇

In early January 2010, a historic and prolonged deep freeze swept across the southeast United States, reaching all the way into the subtropical Everglades. Temperatures hovered around 10 degrees Celsius (50 degrees Fahrenheit) for 48 hours; on January 11, thermometers in South Florida dipped as low as -4 Celsius (24.8 degrees Fahrenheit). Most people remember it, if at all, for the frozen iguanas that dropped out of trees and photos of citrus trees encased in icicles, like some fugitive Minnesota winter carnival smuggled into the Deep South. But to wildlife and invasive species experts, the Big Freeze marked the start of the Big Unplanned Experiment.

The immediate impact on the Burmese python population was clear. Carcasses of dead snakes littered roads; frozen specimens turned up in underground burrows; farther north in South Carolina, in the infamous "Where's Waldo" python enclosure, all 10 snakes perished during the regional cold snap. Researchers attributed the mass die-off to "maladaptive behavior," meaning many snakes tried to bask aboveground in the sun despite the frigid temperatures rather than seeking thermal shelter in underground or aquatic burrows. Python "removals"—captures by hunters, which served as a rough indication of the general population—had peaked in 2009 in the national park but plummeted almost five-fold in the following two or three years. It all looked like good news, at first.

But population numbers were still lacking, and "models" are still just models. Genes are where the rubber of biology meets the road of environmental challenge, and this is when the geneticists entered the story. They were less interested in the many snakes that had died, and more interested in the few that had survived. Like all snakes, Burmese pythons are ectotherms—they rely on warmth from the environment because they do not generate their own metabolic heat—so they have to develop biological tricks in their behavior or in their physiological resilience in the face of life-threatening cold to survive freeze events that do not occur in their native range. As the USGS overview put it, "Some portion of the southern Florida population survived" the 2010 event, "and these snakes and their offspring make up the current population." By 2014, the number of python removals in Everglades National Park had returned to pre-freeze levels. In genetic parlance, the 2010 freeze was a "bottleneck event"— only a few squeezed through and survived. But the ones that did, in the biblical sense, went forth and multiplied.

Beginning around the year 2015, several researchers studying python genomes to understand their metabolism became curious about the aftereffects of the big freeze. Daren Card and Todd Castoe of the University of Texas–Arlington teamed up with Maggie Hunter of the USGS office in Fort Lauderdale to look for evidence of what is called rapid adaptation, which might be thought of as an up-tempo version of Darwinian evolution. They examined the DNA of these survivor snakes to see if there were any genetic clues as to why certain pythons were able to withstand an extended freeze; more precisely, they compared the DNA of pythons that lived before the freeze event with the DNA of pythons that had survived to see if they could identify any differences that might explain the resilience of the survivors at the molecular level. The short and disturbing (although not definitive) answer was: yes.

It turns out that the survivors seemed to share genetic changes in areas of their genomes known to control thermoregulatory behavior and metabolism. "We saw a lot of things that just fortuitously overlapped with a lot of the same sorts of pathways and genes that we were studying in parallel, in a much more controlled fashion, using more lab experiments in Todd's work on Burmese pythons," Card told me. "As we delved into them, we started to see a lot of genes that are involved in things like thermal tolerance." The findings, published in 2018, suggested that the survivors shared variations in their genetic makeup that seemed to confer greater cold toler-ance and greater metabolic flexibility—two traits that Castoe's lab had been investigating since 2011. These snakes were more inclined behaviorally to seek shelter in underground refugia to outlast the cold—a good adaptation to the environmental reality of occasional freezes. And the metabolic changes seemed to favor a behavior that encouraged smaller and more frequent meals—a good adaptation to the ecological reality that, having already decimated populations of large mammals in the Everglades, the pythons might need to modify their diet.

This is not definitive news—the study was small, no one (surpris-ingly) has followed up on it, and, as Card put it, researchers still only have a view from 30,000 feet of what happened on the genetic level in the pythons that survived the Big Freeze. Whatever it is, the genetic changes—or, more accurately, the *selection* for genes that enhanced survival—appears to have happened very quickly. And that is, pos-sibly, very bad news. It suggests that the pythons are on the move, genetically as well as geographically.

The implication is that the 2010 freeze acted as a huge selection event, as evolutionary biologists put it—an environmental stress so dire, so extreme that it has the effect of rapidly winnowing out indi-viduals holding a bad genetic hand and "selecting" the lucky ones that hold winning genetic hands. The freeze culled out individuals that were susceptible to cold temperatures and selected individuals that possessed, at the genetic level, some form of cold-hardiness. Those genes were presumably passed down—immediately, profligately, and of course cryptically—to their hundreds of offspring.

The suggestion of rapid adaptation in the Burmese python population in Florida again contradicts our traditional notions of evolution as a glacial process of genetic selection and refinement that requires millennia, eons, and geological epochs. "We typically think of evolution as occurring over generally quite long time scales, on the order of several generations at the low end up to potentially thousands to millions of years," Card said. "I think with a lot of the tools that we've developed more recently, especially in things like genomics, people have taken a harder look at how quickly evolution can occur...And generally, when you see such an extreme thing occurring, it really suggests that there's very strong selection. Something's happening." In fact, Castoe believes the importation of so many pythons from so many different regions of Asia—all packing in their biological luggage a wide variety of genetic variants, known as alleles, for their trip to North America—set the table for a rapid genetic adaptation. As Castoe put it: "If you've got a good amount of genetic variation, given strong enough selection, this can happen in a heartbeat. 'I'm not waiting for anything! If I've got the allele, what the hell am I waiting for? I ain't waiting for nothing!'"

At a time when roughly 40 percent of Americans do not accept the notion of evolution, the pythons that survived the Big Freeze in Florida appear to believe in it 100 percent. The take-home genomics message from the snakes is that evolution is real, it's apparently happening at blindingly fast speed, and it argues that the 2010 cold snap may have created a subset of pythons better able to survive cold temperatures—and thus better adapted to spread beyond the northern boundaries of its current range. Python captures may have plummeted in the several years after the freeze, suggesting the overall population had taken a big hit. But captures quickly rebounded, to the point that by 2023, the USGS review suggested that the long-term effect of the 2010 freeze may well have been exactly the opposite of what appeared to be the case a decade earlier.

The 2018 freeze genomics paper did not lay to rest the arguments over the potential expanded range of the Burmese pythons, but another new genetic technology might. Over the last decade, Maggie Hunter at USGS has led efforts to trace the spread of the pythons by looking for traces of their DNA left in the environment—in shed skins, mucus, carcasses, even in their poop. Like genetic fingerprints, these scraps of DNA find their way into water, which can be easily collected and tested. The technique, known as environmental DNA or eDNA, lets molecular ecologists like Hunter detect python DNA in areas not easily accessible to the road-hugging hunters, allowing biologists to infer their spread without actually having to see a snake in the wild. If they've been there, they tend to leave a genetic calling card.

"So that 1 percent detection with visual sighting?" Hunter told me. "We have raised it—depending on the habitat—up to 90 to 100 percent detection with environmental DNA." Using this method, scientists in Florida have been able to track the northward march of the pythons. And they have been marching. "We have surveyed the Everglades and all across South Florida, and we're working our way up now—Lake Okeechobee, Kissimmee Slough," she said. "And we are getting detections higher and higher north."

There is nothing to celebrate about an introduced, non-native species that has decimated the native fauna of the Everglades. But there might be a little corner in a naturalist's heart that admires the adaptability of a creature that arrived in Florida as a commodity, was dumped as heartlessly as a puppy abandoned in the middle of nowhere, slithered off into a hostile and foreign environment, and nonetheless adjusted to this rude and cruel dislocation by surviving (the first Darwinian rule of law), reproducing, adapting, expanding, adapting some more, all the while remaining virtually impossible to see.

Gordon Burghardt, one of the elder statesmen of American herpetology, couldn't resist expressing a little serpent schadenfreude when I asked him about how things have played out in Florida. "What I find interesting with some of these invasive animals like the pythons," he said, "is it belies the idea that these animals are dumb and stupid and can't adapt their behavior and so on. Here they're wiping out *mammals*! Supposedly," he added with a little laugh, "those are the smart animals."

⁓

"Oh, boy...Yikes!"

It was sometime around noon during the following day's python hunting in the Everglades that Donna Kalil, behind the wheel of Marc Rodriguez's truck, brought the vehicle to a forced halt. We were ostensibly on a levee road, but there was no discernible path in front of us, not even tire tracks. Just dense underbrush and surprisingly thick saplings growing straight up in the middle of the road, with a steep drop-off to a canal on the left, a steep drop-off to a swamp on the right.

"Let's see if I can get around this stuff," Kalil said, easing forward.

Following our nighttime expedition on the levee road near the Tamiami Trail, I met Kalil, Rodriguez, and a volunteer hunter named Tanya Toutant for a daytime round of python hunting. This time we were much farther north—about 70 miles northwest of downtown Miami—in a place called the Holey Land Wildlife Management Area and the adjoining Rotenberger Wildlife Management Area. The two areas comprised more than 60,000 flat, desolate acres of saw grass marsh. Donna consulted an interactive map app on her phone that showed white dots marking the location of every capture in the Everglades. There was reason to be optimistic; Donna had caught her four big pythons in Holey Land the previous week, and several large pythons had been taken in this area around the same time in 2022. In another sense, it was forbidding territory, for python hunting or anything else. An online guide for bird-watchers in Holey Land warned that this remote network of neglected levee roads was "confusing," advised a full tank of gas and lots of water before driving into it, and pointed out that the best access was not by car but via airboat. Visitors were also warned to be vigilant about seasonal hunters.

The weather that Friday wasn't much fit for man nor beast, scaly or otherwise. Low clouds made the flat Florida landscape feel even flatter. There was no hint of sun, and the wind blew brisk and cold. Kalil took one look around and pronounced the snake-hunting conditions "horrible."

She typically consults a special radar app when she gets up at five o'clock each morning before deciding where to head. "You really just have to be a meteorologist yourself," she grumbled, dismissing mainstream media weather reports. The horrible weather did not dampen her determination, however. "One of the things I love saying is, 'I've never caught in conditions like this,'" she confided as we set out along another levee road, L-5. "I'd love to catch in conditions like this. So I'm up for the challenge to find a python today…I love proving myself wrong."

The words had barely left her lips when a cold, wind-whipped rain began to spritz the windshield. It was still raining when Kalil pointed out the spot where she had bagged four pythons the previous week. "This is where we caught 250 pounds of snakes in one day," she crowed. "I was very proud of catching right on this road, because five years ago, there weren't any pythons here. So they're moving north. We stopped a hundred pythons from moving north by removing those three females," she said, including their eggs as part of the haul.

But the going was tough. We were about 30 miles south of Lake Okeechobee, in "STA 3"—a stormwater treatment area, which was a vast sink of tainted water that had to be naturally filtered before it could flow (or, more accurately, ooze horizontally) into the Everglades farther south. It was a forlorn landscape, especially when we reached the overgrown part of the road. Kalil and Rodriguez disappeared on foot into the thicket, which looked impassable. One of the problems with levee roads is that there are precious few turnouts. If you reach an impasse, you may have to drive in reverse for miles on a 10-foot-wide dirt road, with a steep tumble into water on either side. The philosophy of sticking to roads to track down pythons is a little abiological; as the recent government report pointed out, about 97 percent of python habitat in the main Everglades is not only off road, but way off road.

After about 10 minutes, Kalil and Rodriguez emerged like 19th-century bushwhackers from the dense foliage. "All right, let's try it," Kalil said. "We can always back up!" A flicker of chagrin crossed Rodriguez's face, perhaps imagining the repair bill, but he nodded his assent. Unlike Kalil, contract hunting was his full-time job.

As we slowly inched forward, stems, roots, and branches thumped against the undercarriage of Rodriguez's truck, a slow yet distressingly abrasive percussion marking the beat of our frustrating search. "It really is due for a haircut," Kalil said absently. Snakes supposedly can't hear, of course, because they lack eardrums, but the notion of finding a python amid the seismic racket Rodriguez's truck was making seemed as remote as our little geopositional dot in the immense Everglades wilderness.

Why do they do it? Given the mosquitoes, razor-sharp saw grass, poison ivy, vehicular trauma, and alligators, not to mention ornery pythons, they certainly don't do it for the money. When the South Florida Water Management District hired its first 25 contract python hunters in 2017—Kalil was among the few women—the pay was less than minimum wage. At the time, hunters received $13 an hour when they're in the field—"less than somebody working at McDonald's," Kalil observed drily—plus extra for any pythons they catch. It's $50 for the first four feet, then $25 for every foot after that. The 13-footer that Kalil brought in the previous week, for example, netted her $275.

When they complete a catch, the hunters snap a picture on their phones, call it in, report their position, and then usually euthanize the animal with a gunshot on the spot. It is illegal to transport an invasive species in Florida without a permit; Kalil packs a .44-caliber pistol for this unpleasant aspect of the job, and to a person, every hunter with whom I spoke authentically bemoans the need to do it. All the captures are entered into an interactive map with location, date, and time of day so other hunters can see where pythons have been taken—but with a catch. Administrators withhold the exact location for a month, so hunters aren't forced to disclose their latest best hunting grounds. Each contract hunter can bring along four volunteers on each excursion. Kalil invited Toutant to accompany us in the hope that she might become a hunter herself.

When Kalil started in 2017, the contract hunters were

overwhelmingly male, but she was not shy about chastising them on their technique. "Cowboying it"—dragging pythons by the tail along the ground, for example—was painful to the animals, she pointed out, because the scales on their belly catch on the ground. "You think it's a small difference," she said. "But to the snake, it's a big difference!" A lot of the other hunters, she added, tended to emulate Steve Irwin, the flamboyant Australian wildlife TV star who frequently snatched snakes by the tail; Kalil's idol was Bill Haast, a legendary snake celebrity in South Florida who founded the Miami Serpentarium in 1947, played a key role in developing antivenom, was reportedly bitten 172 times by venomous snakes, and lived to be 100. "That's why I catch pythons and all snakes by the head," Kalil said. "Because that's how he would do it. Whereas Steve Irwin? Most of the guys that I'm out here with, the younger guys, that's their idol. And I love him. He did a lot for herpetology, and other animals as well. But unfortunately, he would grab by the tail and do the heroic dance and lift them up and finally get them under control. That's why so many people do catch snakes by the tail, because of the use of him as a reference. That's the unfortunate part about his legacy, because not everybody has the understanding of these animals as well as he did."

The guys also made fun of Kalil when she was the first to outfit her truck with a python bench. She designed the raised platform and handrail behind the cab herself, flanked by a bank of high-powered lights, and commissioned a marine welder to fabricate the setup. Now virtually everyone hunts with a similar rig.

January was not necessarily a bad month to seek pythons; breeding season runs roughly from December through February, and it was likelier to find large, reproductively mature males and females than later in the summer around Python Challenge time, when many of the captures by contrast were relative fingerlings. But as we bumped down one obscure levee road after another, spits of rain came down.

As the prospects of seeing any pythons grew ever dimmer, Kalil talked about the snake that almost got her. She was road cruising in the Shark Valley section of Everglades National Park with two legendary python hunters, Wayne Rassner, then-chairman of the South Florida National Park Trust, and Joe "Snake Man of the Everglades" Wasilewski, when she spotted a python that the men had missed in the water abutting the road. She jumped out of the back seat of the car, charged into two feet of water, grabbed the snake by the neck, and held it up triumphantly. "Wayne goes back to the car for a bag," she said, "Joe goes back to get his camera. My phone rings."

With one hand holding the python's head, Kalil let go of the tail with the other hand and reached for her cell phone. It was her daughter, just checking in.

"Deanna, I just got a python!"

"Congratulations!"

Congratulations were a bit premature. As she stood in the water holding her cell phone, Kalil felt a strange tightness around her neck. "As I was talking with her," Kalil said, "the snake crawled up my arm, backward, because I was holding it, and wrapped around my neck." The tightening began to feel like strangulation. "I'm holding the snake, trying to pull it off. I can't. It's seven feet. It's longer than my arm. So I can't pull it off with one hand. I'm trying to put my phone in; I can't. So now I've got my phone in one hand, the snake in the other wrapped around my neck. All this time, I'm thinking, *I'm a free diver. I can hold my breath for almost two minutes. I'm good...* But that's not the case. The case is that they're putting pressure on your carotid arteries, and that can make you—well, it got me dizzy within 15 seconds." Kalil's almost-last words to her daughter were, "I've got a snake, gotta hang up."

She ran out of the water to the car. By that point, she couldn't speak, so she tapped Rassner on the back. Rassner, recalling the incident, told me, "You could see the muscles of that snake just tightening, and she was turning red, and was having trouble speaking." Despite the constriction, however, pythons are relatively easy to pull off—if you have a free set of hands. "As strong as they are," Rassner said, "you can unwrap them." The whole episode transpired in less than a minute, and Kalil may have been closer to dying of embarrassment than as python prey.

In case you're wondering, the bite of a python "feels like a hundred needles plunged into your hand all at once," she said. This is partly because pythons have two rows of teeth, both of which point backward into their mouths so that prey can't wriggle out. Kalil is reluctant to get specific about how many times she's been bitten, except to say "more than 20." The reason for her reticence is because "it means I made a mistake." But her worst bite ever? Oddly enough, a garter snake, to which she had a strong allergic reaction.

The maze of Holey Land levee roads was, as advertised, so confusing that Kalil and Rodriguez had to pause multiple times to figure out where we were. It was as desolate a landscape as one could imagine, a subtropical version of Bunyan's Slough of Despond. Massive high-voltage power line towers, so ubiquitous as to be Florida's unofficial state flower, pricked the horizon practically to Lake Okeechobee. At one point, we spotted dense black smoke, looking almost like stationary tornadoes, rising up into the sky. It was sugarcane fields intentionally set afire. All the irrigation you could see to the northern horizon, dozens of miles to Lake Okeechobee, accommodated the farming. After farmers harvested the sugarcane, they gathered the leftover chaff—including its chemical residues from fertilizer, Kalil pointed out—and set it on fire. "Yeah, it looks like an atom bomb going off," she said in disgust. "Literally, the cloud comes up just like a mushroom cloud. And they'll do that all day long." It was another perverse reason to admire the pythons—in this incredibly despoiled, inhospitable, and toxic ecosystem, with the visual props of apocalypse, they still managed to survive, thrive, and spread. But like so many creatures, they are also living, breathing diagnostics of a sick planet; the flesh of pythons contains unhealthy levels of mercury.

"I wish the sun would come out," Rodriguez bleated at one point in the afternoon. "I was really excited about today."

By late afternoon, no one had any illusions. We saw tricolored herons and black-headed vultures and ibises and anhingas and limpkins—the same beautiful waterfowl that the pythons now see as prey. We saw a dead alligator floating in one canal and an otter and perhaps a bobcat. We did not see another soul, another vehicle, for hours. And then suddenly, about five hours into the hunt, we began to see cars. A lot of cars. And then we heard gunshots, announcing the presence of duck hunters. As we inched slowly and carefully over the last bit of road in Holey Land and headed back to Miami, we suffered a final indignity: a traffic stop by a Florida Fish and Wildlife inspector, who pulled us over, checked our credentials for being there, and noted that we'd been traversing hunting grounds. It was all resolved amicably, but it added a browbeating bureaucratic coda to the outing.

In two fruitless days of hunting, we didn't see a single python. We knew they were there. Given the nearly zero detection probability, failure was hardly a surprise. But we also knew that, cryptic to our mortal eyes, the snakes continued to inch northward.

The day wasn't a complete wash, herpetologically speaking. Toutant, who had eyes like a hawk, spotted a juvenile black racer (*Coluber constrictor*) in the underbrush by the side of the road and snatched it before it escaped into the weeds. Thin as a pencil, inquisitive as a puppy, the snake remained calm as Toutant handed the reptile off to Kalil. Like the snake whisperer she is reputed to be, Kalil gently ran her finger under the jaw of the racer and cooed. "If you tickle their chin, they're like, 'Oh, okay, this is interesting. What's going on?' They're curious." After a few minutes of interspecies communion, Kalil gently released the snake into the underbrush. The racer seemed in no hurry to move on.

With the clouds of burning sugarcane chaff still sullying the sky, one of the things Kalil had said as we rumbled down a levee road in a stormwater treatment area stuck with me. "All these levees are prime locations to put nests in," Kalil said, shaking her head. "If they weren't here, if the canal systems weren't here, I don't believe the pythons would be here. If we weren't here, they wouldn't be here, either. It is all about us. We've made a nice environment for them."

Kalil invited me to come back later in the spring to try again, and I had intended to do so. But as I read the up-to-the-minute 119-page USGS report on Florida's "invasive" pythons, I realized that not finding them, perversely, may have been the more authentic experience. Not finding them is the single most important fact clouding every baseline assessment of the problem, every control strategy, every government policy moving forward.

Despite the thousands of snakes pulled out of the Greater Everglades Ecosystem over the years by python hunters like Kalil, despite the upbeat propaganda of the annual Python Challenges, despite cutting-edge technologies like scout snakes and surveillance drones and fancy python traps, the reality is that nothing short of an extended ice age is going to get rid of, or even significantly diminish, the Burmese python population in Florida. With climate change, eliminating them appears ever more unlikely.

No one who understands the situation even indulges the fantasy of "eradication" anymore, and scientists are quietly pondering even more radical remedies. Buried in that 2023 government report, close to the end, the 37 authors—invasive species experts all—described the first tentative steps to genetically engineer our way out of this human-made catastrophe. They raised the possibility of using the new gene-editing technology called CRISPR to introduce a genetic flaw in the Florida python population—a lesion designed to disable the X chromosome and thus render all the offspring male, ultimately making the snakes incapable of reproducing. The technology is known as gene drive, and the prospect of introducing it into the Everglades ecosystem reminded me of a remark by Frank Mazzotti about another earlier python control technology—"What could possibly go wrong?"

Until that controversial science matures, Donna Kalil gets up every day, Monday through Friday, checks the weather radar around 5:00 a.m., climbs into her truck, unlocks some nondescript gate on some overgrown levee road, chalky tire tracks barely visible in the underbrush, crawls along at five miles per hour, consults her map to see where pythons have been caught before, and continues this quixotic search. Or maybe not so quixotic. The week after I left Florida, she sent me a picture of herself with a Burmese python, all 15 feet of it, draped over her shoulders; it was one of seven snakes she'd hauled in that week. We know the exact dimensions because Kalil passed the snake on to University of Florida researcher Melissa A. Miller, who measured it, outfitted it with a radio transmitter, and sent it back into the wild to collect more data for the scientists.

There's no question that the main public message attached to Florida's python problem is menace—invasion (with human fingerprints all over it), ecological devastation, predation without consequence, and hope (mostly without foundation). It seems to me that the python "problem" might send a different, equally important message. Snakes find a way to survive. Snakes have *always* found a way to survive. They've been doing it for 100 million years or so. If they have to live in an arid, hot desert, they figure it out. If they have to live in tropical jungles, they figure it out. If they have to live in salt water, they figure it out. If they have to adjust to the colder temperatures of North America, or the ultimately warmer temperatures of climate change, they will figure it out. Without limbs, without metabolically warmed blood, without ears, without eating sometimes for as long as a year, they figure it out, and, in those rare moments when we actually manage to see their cryptic forms, they grace us with a glimpse of resilience and the unearthly, eternal beauty of survival.

Snake Road
Avenue of the Dead, Teotihuacan, Mexico

There is an ancient and hallowed "road" in Mexico—actually, a 40-meter-wide, three-mile-long esplanade, interrupted by staircases and plateaus—called the Avenue of the Dead, which connects some of the most wondrous architectural monuments that survive from antiquity. In ancient times, the road ran for several more miles, but its current southern terminus brings contemporary visitors (all tourists on foot, no cars) to an 1,800-year-old pyramidal temple dedicated to the serpent-god Quetzalcoatl, the Feathered Serpent. To the indigenous people who created this remarkable metropolis, according to modern-day archaeologists, Quetzalcoatl represented wind, rain, clouds, fertility, and the mythological origin of time.

On March 20, 2023, a day ushering in the spring solstice, when the sun is directly over the equator and time is equally divided between day and night, I found myself standing near the lower part of the Avenue of the Dead, in the plaza of the Temple of the Feathered Serpent in Teotihuacan, the Mesoamerican archaeological site about 25 miles northeast of Mexico City. In the 21st century, as in the 4th century CE, when Teotihuacan was among the most populous and advanced metropolitan complexes on earth, solar events are a serious matter. Throngs of visitors garbed all in white, many accented by bright-red headbands and thin red cords encircling their waists, stood in the plaza and raised beseeching hands toward the heavens in a millennium-old ritual, begging to receive energy from the cosmos, while a shaman-like figure in the center exhorted the encircling pilgrims. The same ritual played out on numerous platforms throughout the archaeological site, whose massive Pyramid of the Moon and Pyramid of the Sun, along with the Temple of Quetzalcoatl, are surely among the most magnificent monuments ever fashioned by human hands.

The crowds that descended on Teotihuacan that day in 2023, the estimated 200,000 citizens who may have enacted similar rituals nearly 2,000 years ago, the science that allowed the architects of Teotihuacan to position the Avenue of the Dead and its pyramids and temples so precisely that sunlight would predictably intersect other buildings in the complex on the most important ceremonial days of the year, the hands that began to craft the Temple of Quetzalcoatl around 150 CE, the societal dedication that persuaded three or four generations of citizenry to labor on these massive construction projects for more than a century to create the pyramids, the suasion of the rulers and priests who organized and maintained this urban florescence for a period of at least 600 years—all of it may descend, in a sense, from a communal belief in the power of the snake.

The ruling elites of the city identified with and personified the main deities of the Basin of Mexico. There were four main districts, each associated with an animal power group: birds (eagles, owls, raptors); canids (coyotes, wolves, dogs); felids (jaguars, pumas); and snakes. "The serpents were maybe the original people in Teotihuacan," said Elba Estrada, an archaeologist at the Museo des Murales, referring to the city's origins around 100 CE. "They may have founded the city." Indeed, archaeologist Linda Manzanilla, who has studied the political organization of Teotihuacan for decades, believes that the *serpiente* faction, which represented the southeastern quadrant of the city nearest the Temple of the Feathered Serpent, exercised exceptional power on Teotihuacan's ruling council for perhaps 250 years, including the period when the Temple of the Feathered Serpent was built.

This pyramid—the "most sumptuous monument" in the complex, according to several prominent archaeologists—possessed "a sculptural splendor that was unsurpassed during the following centuries of the city's life." It marks the spot where time began, where time could be measured in calendar-specific divisions (hence the importance of ritual calendars and solstices), where that unique entity that archaeologists describe as "time-destiny" took shape—the Mesoamerican concept of linear time that embraces day-to-day activity, history, social transformations, the worldview of an entire culture over entire epochs. The original temple was ringed by 364 sculptures, half of them massive serpent heads, with one more atop the pyramid, indicating that the pyramid was not only a sacred site but an architectural codex enumerating the 365 days of the year. It was also a fabulous polychrome monument, as a re-creation of the snake facade in the National Museum of Anthropology in Mexico City makes clear. The temple served as a cultural template for the sacred power of the snake, as serpent iconography was adopted (with variation) by almost every succeeding Mesoamerican civilization, including the Toltecs (with their snake wall at Tula), the Maya (with their Kukulcan, a version of the feathered serpent), and the Aztecs, whose Templo Mayor in the heart of present-day Mexico City is a veritable snake ball of monumental serpentine statuary.

When it comes to modern attitudes about serpents, the default cliché is that poor man's duality: the tension between fascination and dread. But perhaps no culture in history has done duality better than the civilizations of Mesoamerica. And this goes back even earlier than Teotihuacan, to the ancient reptilian Earth monster Cipactli, dating to perhaps 500 BCE. As the archaeologist Leonardo Lopez Luhan and colleagues have noted, Cipactli "was cut by two gods transformed into serpents, which formed heaven and earth with the two halves, placing columns between them, thus creating the open space that would be occupied by man."

This Mesoamerican dualistic embrace of the Snake—as organism and as symbol—incorporated a quality conspicuously absent from modern-day attitudes: reverence. When you walk along the Avenue of the Dead, which connects the Pyramid of the Moon to the north with the Temple of the Feathered Serpent in the south, what you feel most is reverence. The earliest settlers in Teotihuacan—their exact identity still remains uncertain—clearly regarded the serpent with awe and respect. The sheer shape of the snake represented an animate symbol of lightning, which in turn signified water, wisdom, knowledge, fertility (in the agricultural sense), and rain. The venom of the snake represented mortality, both its suddenness and its finality. The cryptic behavior of the snake, burrowing underground or hiding beneath rocks, represented its ability to travel back and forth between the world of the living and the world of the dead. The shedding of snakeskin represented rebirth and regeneration. The form of the snake biting its tail, the circular symbol known as the ourobouros, represented the cyclical nature of life. The serpent, in short, embodied nature in all its essential dualities. And the Feathered Serpent (Quetzalcoatl) represented something even more fundamental: the origin of the world and time itself.

Neither snakes nor snake power lives forever. Something went terribly wrong with the serpent elites in what had become one of the most sophisticated, multi-ethnic planned urban centers in the world. Around 350 CE, Manzanilla recently wrote, "One of the strong factions of this council, the 'serpents,' was evicted from Teotihuacan"; the Temple of the Feathered Serpent was destroyed, the "serpent" group was banned from the ruling council, and iconographic images of snakes were eradicated from the city, possibly because the snake group, Manzanilla speculated, had plotted a coup d'état. The snake faction in Teotihuacan disappeared, but the symbolic power of the snake persisted elsewhere in Mesoamerica for more than a millennium.

The Spanish came in 1519, bringing with them Christianity and germs. The germs decimated indigenous populations, and religion suppressed the indigenous reverence for serpents. In the Judeo-Christian tradition, the snake became, overnight in the New World, a symbol of evil, temptation, betrayal, ferocity. The Garden of Eden story, transplanted to a new hemisphere, was almost like an invasive species of thought. "Reptile iconography, particularly involving snakes, characterized much of pre-Hispanic Mesoamerican imagery, perhaps more than any other culture…," wrote David Charles Wright-Carr, a professor of visual arts at the University of Guanajuato. "The first Europeans to experience Mesoamerican images responded with a mixture of admiration and repulsion."

Despite all that (or perhaps because of it), it is amazing how common, and commonplace, serpent imagery asserts itself in the everyday landscape of modern Mexico City. You see it in the massive "Snake Circles" of Chapultepec Park, Mexico City's largest urban park—four enormous circles formed by an undulating, slithering, sinuous curve of dark volcanic rock carved into the shape of a snake, little knobs of white in the tail signifying the rattles of a rattlesnake, the body arriving at a fierce and unmistakably crotaline serpent head. You see serpents on modern murals alongside the highway from the airport to the city center. You see live captive pythons (and axolotls) brandished by locals along the shoreline of the Xochimilco tourist boat excursions. You see them in ancient Mesoamerican murals amidst the rubble of what's left of the last great Aztec monument in the Templo Mayor in the heart of the city, before the temple was ransacked by the Spaniards to salvage stones used to erect the nearby cathedral. You see snake icons in room after room of the magnificent Museo Nacional de Anthropología, beginning with the massive mural by Rufino Tamayo that greets visitors in the entryway, depicting the forces of daylight and goodness, the snake, battling the forces of darkness and danger, the jaguar. Its title? *Dualidad.* Duality.

Snakes are omnipresent in the dazzling and devastatingly anti-capitalist murals of Diego Rivera that decorate the government's Secretariat of Public Education, where half-tone rattlesnakes frame gloriously chromatic indictments of the *Wall Street Banquet*, with its depiction of auto magnate Henry Ford and oil magnate John D. Rockefeller, and *The Capitalist Dinner*. Snakes even inhabit the marginalia of a mural titled *The Writer*, which is flanked by a snake and topped with an image of the ourobouros—a symbol as familiar to ancient Mesoamerican cultures as to one of the tourists I saw in Mexico City with a slender tattoo of it on her wrist. At some level, the omnipresence of snake imagery is a kind of modern assertion of reverence. Given the broad institutional demonization of the serpent, it simultaneously feels like a form of transgression—or, more precisely, transgression as a form of reverence.

Reverence is a capacious yet nuanced word that hints at multiple meanings. In these ancient, "primitive" cultures, in their regard for this "primitive" animal, you can sense a special kind of reverence. A reverence for mystery, a reverence for the unknown, a reverence for the unknowable. That kind of reverence has been either lost in modern religion or distorted by small fanatical modern-day sects that conflate the worship of snakes with the worship of dangerous serpents. To modern sensibilities, it is easy to dismiss ancient forms of reverence as superstitious and mystical, "pagan," grounded in myth. But that overlooks the central element of those early belief systems: The snake was a symbol of Nature in all its colors, habitats, beauty, benevolence, menace, indifference, adaptability, and behavior. The reverence, writ large, was for Nature. The modern loathing of the snake, by extension, may reflect our contempt for any aspect of Nature that discomfits, perturbs, or doesn't merit a postcard.

The visit to Teotihuacan reminded me of why I became interested in writing about snakes in the first place. My guide, a young man named Luis Sarabia, mentioned that the ancient Mesoamerican rulers and priests acquired power because of their knowledge and their ability to predict events, like the spring equinox and its temporal link to the optimal time for planting crops such as corn. The real power, I replied, may have resided in those early Mesoamerican citizen-scientists who, by observing celestial events with such acuity, were able to measure—and predict—the passage of calendric time, including the two times in the year that marked the beginning and end of the ritual calendar. Perhaps it was that knowledge, I suggested, that endowed the rulers with power—a power that in Teotihuacan may have initially been considered "the power of the snake." Luis replied that the rulers and priests had many people working for them in the acquisition of this knowledge and power.

Reverence is not strictly mystical. It can be (and often is) evidence-based. If it is enriched by knowledge, as I believe it is, then many of the scientific observations about serpents in the past decade or so should make us regard these creatures in a new—by which I mean ancient—light. Beyond the simplistic duality of fascination and dread, there is room for reverence. Or, perhaps stated more precisely, the *rediscovery* of reverence for these remarkable creatures. As I considered one more profound environmental paradox, the snake's ancient association with rain and modern Mexico City's dire shortage of water, it seems as though the rediscovery of reverence—for nature, for habitat, for conservation—could not come a moment too soon.

Luis said one other thing that stuck with me during our tour of Teotihuacan. He grew up in a small village just outside the archaeological park, he told me, and visited the site virtually every single day. He recalled frequently seeing snakes, harmless ones, slithering through the dry brush and rocks just a few dozen meters away from where the white-shirted pilgrims were even now raising their hands to the heavens. And he continued to notice them after he became an accredited tour guide for the site.

"But that was six or seven years ago," he said in passing. "You don't see them anymore."

Epilogue

Snake Lectures

O n the evening of April 21, 1923, the art historian Aby Warburg gave a slide-show presentation to his fellow patients of
the Bellevue Sanatorium in Kreuzlingen, Switzerland, describing a
long-ago trip to the United States. Warburg, black-sheep scion of the
famous German banking family, who had been hospitalized with a
diagnosis of incurable schizophrenia several years earlier, devoted
almost the entire lecture to a discussion of ritual ceremonies by the
indigenous people of the American Southwest, whom he visited in
pueblos in New Mexico and Arizona during a nine-month trip in
1895–1896. In the course of a 45-minute presentation, he made reference to snakes and snake-related imagery more than 100 times.
Although barely known outside the field of cultural studies, Warburg's unusual talk—now known as the Snake Lecture—is considered one of the seminal contributions to 20th-century art and
cultural history.

The Snake Lecture achieved this degree of acclaim because
it represented the culmination of a decades-long effort by Warburg to articulate a radical theory in not only art history but also
human evolutionary psychology: the idea that a symbolic visual

vocabulary of humankind's relationship with nature extended deep into antiquity, from prehistory and the pre-Christian reliance on magic as the medium of interaction and Asklepios, through medieval times and Renaissance art to, finally, the technological world that had already begun to encroach on the Native American worldview, with its railroads, its telegraph wires, and what he called "Edison's copper snake." After spending months in the United States, mostly immersed in indigenous observation, not all of it benign, Warburg concluded that the similar visual signature of serpents and lightning, and the relationship of these symbols to meteorology and specifically precipitation, represented *the* fundamental primeval image to the human animal throughout human history. It represented rain, agricultural sustenance, survival. The snake-lightning icon, he said, was "a living (primal) rain-snake-saint in animal form."

Decades later, Warburg's lecture—the cogency of which influenced the decision by doctors to release him from the sanatorium—"would help to justify, and internationally enhance, Warburg's reputation as a pioneer of cultural history and indeed cultural psychology," according to German historian Uwe Fleckner. It is famous in part because it introduced the idea of iconography in art—visual images that signified something universally important.

Given the enormous edifice of theory that Warburg erected upon the foundation of Native American serpent rituals, it is notable that he never actually witnessed the Hopi Snake Dance. It takes place every August, and Warburg ended his American sojourn in May. But he compiled copious notes about the ceremony, and what he especially celebrated was the fact that following the dance, which is the culminating event in a nine-day ritual that essentially functions as a communal plea for rain, the snakes were not killed but rather released back into the surrounding landscape as "messengers" to nature—a kind of spiritual catch and release. "Drought," Warburg said, "teaches magic and prayer."

In 1924, D. H. Lawrence, another writer with pagan sympathies and author of the poem "Snake," took the road that leads to Walpi and Oraibi, the two villages in Arizona that continue to stage the Hopi Snake Dance in alternating years. Even in Warburg's time, indigenous peoples mocked and detested the ethnographic prurience of outsiders who flocked to see live, venomous snakes dangling from the mouths of the dancers. Nowadays, the Hopi people forbid tourists from observing the snake dance, in part because of the invasive manifestations of curiosity exhibited by outsiders, dating all the way back to the time of Warburg's visit. In recent years, the tribes have requested that certain archival images not be shown at cultural institutions, including the Warburg Institute. With uncanny prescience, however, Warburg anticipated the way "civilization" would perturb the fragile relationship between the Hopi people and nature. He clearly saw how technologies—"those ominous destroyers of the sense of distance"—imperiled what he called "the space for devotion, which evolved in turn into the space required for reflection."

The "space required for reflection" might be thought of as a different kind of habitat under threat—a psychic habitat, the place where humans attempt to come to terms with nature in a psychological sense, just as the technologies of civilization might be thought of as agents of a kind of habitat destruction of that same psychological space. To underline this point, as the final image in the Snake Lecture, Warburg displayed his photograph of a dandified and clearly well-to-do Victorian gentleman striding through the civic plaza of San Francisco. He chose to end his talk with this picture, he explained, because of the string of telegraph poles that loom above the head of the man and recede to the vanishing point of the photo. It was, he said, "an example of that primordial condition of which the refinement, transcendence, and replacement are the work of modern culture." To punctuate that thought, Warburg added, "The American of today is no longer afraid of the rattlesnake. He kills it; in any case, he does not worship it. It now faces extermination. The lightning imprisoned in wire—captured electricity—has produced a culture with no use for paganism." Which was another way of saying "no use for snakes."

⌒

Telegraph Road in Michigan, named in honor of that pioneering technology, which is now extinct, runs north out of downtown Detroit, the city that blessed the world with the mass production of automobiles and all the vehicles that have guzzled gasoline in the 125 years since. The road squeezes through an endless series of business parks and cookie-cutter office complexes and commercial outlets before it reaches some of the more affluent suburbs of the Motor City. From the "Tel-Twelve Mall," at the intersection of Telegraph and Twelve Mile Road, practically to Pontiac, it's one long commercial strip of Best Buy, Target, Shell, Five Guys, Aldi's, Wing Stop, Starbucks, Carl's Golfland, Hockey World. And of course acres and acres of parking lots. All that asphalt has largely supplanted the forests and wetlands, with their many lakes and ponds, that glaciers gouged into the landscape 20,000 years ago. The closest you come to a reptile may be the "Rattlesnake Pasta" on the menu of the popular 5th Tavern in Bloomfield Hills (the "bite," as it were, courtesy of the Cajun Alfredo sauce).

As a child, I could hear the rumble of traffic on Telegraph Road as I scouted the ponds and culverts just behind the junior high school, prowling for frogs and turtles and snakes. Our family lived just west of the highway, but nature seemed nearby and accessible. Telegraph Road was only two lanes in each direction back then; now it is three and sometimes four lanes wide on each side. The pond behind the junior high school, where we played ice hockey in winter and caught turtles in the spring, is now developed and inaccessible, surrounded by a dozen or so private homes along Lake Crescent Drive. The hill just above Lone Pine Road, where we sometimes went sledding in winter, is now a gated community called the Hills of Lone Pine, developed and inaccessible, with a guardhouse protecting luxury homes (among its residents was reportedly Aretha Franklin). The Martin farm on Echo Road, perhaps a quarter mile from our postwar suburban Colonial, no longer appears active. Except for some commerce-free sections zoned as much for human privilege as for wildlife protection, Telegraph Road is one long upscale strip mall. "It's all changed," said one non-expert, the young waitress who brought my dinner at the tavern. "Everywhere around here used to be nature and trees," she added. "It's all gone."

I wanted to revisit this little patch of exurbia because it's much easier to see habitat destruction after a long absence—nearly half a century in my case. It is stark yet familiar, not as strange as you expected, yet worse than you imagined. The area remains beautiful, and indeed has been prudently zoned and carefully preserved, with natural set-asides of wetlands and protected nature. But the human footprint falls heavy on the landscape, in both macro and micro ways. The hockey pond that reliably froze in winters past apparently does so much less frequently; a recent study showed that average winter temperatures in the Detroit area have increased three degrees Fahrenheit since the 1980s. The wild empty fields are now well-maintained lawns. It's an ecological cliché that's been scribbled upon the landscape of a zillion towns around the world, but Telegraph Road seemed especially significant because it was the highway that connected corporate Detroit to the well-to-do suburban enclaves of the automotive industry's executive class, whose century-long embrace and promotion of a fossil-fuel economy enabled, knowingly or not, population spread, commercial and residential development, habitat destruction, climate change. There were roads before cars, of course, but cars used roads to turn all of us into invasive species.

As it turns out, I also managed to track down several local naturalists who knew James Fowler Sr., the man who led our little herpetology field trip in the 1960s. He passed away a few years ago, but David Mifsud, a private-sector conservation biologist who manages the state of Michigan's Herpetology Atlas, vividly recalls Fowler lamenting how construction of the Tel-Twelve Mall in 1968 destroyed one of his favorite snake-hunting spots (for all I know, perhaps the spot where I caught that ribbon snake). "He said it was one of the best places in the state of Michigan to see blue racers," Mifsud told me. "And now it's a mall." In an irony almost too rich to mention, another of Fowler's acquaintances told me his daytime job was at the Henry Ford Museum in Dearborn.

Snake biology teaches us again and again that one of the keys to survival is diversity—diversity of physiology, diversity of behavior, diversity of coloration, diversity in adaptation, diversity of diet, diversity of molecular resourcefulness. That's true of nature in general, where every ecological niche is unique, dynamic, complex. Nature celebrates diversity; all those name-brand franchises lining Telegraph Road reminded me that commerce and "civilization" demand conformity, predictability, consistency. Development is a form of predation, and a visit to the exurbs of Detroit was just my modest attempt to see how habitats slowly and inevitably and irreversibly surrender to the number one predator on the planet, and how roads accelerate the process. Mifsud told me that when the M-5 highway extension north of Detroit opened to traffic around the year 2000, so many reptiles got run over that the road surface became slick with their blood, leading to accidents. When everyone stayed home during the pandemic, he saw more reptiles alongside rural roads than he had seen in years.

If, as Thomas Cole's *Expulsion from the Garden of Eden* painting and other artworks of that genre suggest, our human destiny is inevitably linked to the snake's—if we're in it together—what are we to make of this perilous, eye-blink moment in the fate of the Snake? Harry Greene reminded us decades ago, "Although many people believe animals relocate when their habitats are destroyed, most organisms have nowhere to go." Part of the imperial indifference of the human animal is our illusionary belief that there is always plenty of habitat left.

On the other hand, it's cautiously heartening to consider the "futurity" of an animal that has been breaking the rules for millions of years and may well continue to do so despite our well-meaning, well-to-do predations. Rick Shine, the Australian scientist, told me a story about file snakes (*Acrochordus arafurae*) in tropical northern Australia that offered at least a morsel of hope. These snakes have an incredibly low metabolic rate and, ecologically speaking (and in a very profound way), may even possess a biological, almost innate sense of patience. During a series of good wet seasons, females sexually mature in a couple of years and produce litters every year. When the environment is more challenging, the snakes dramatically alter their lifestyles and behaviors. They may not reach sexual maturation for 10 years, and they may produce a litter once every decade. They weren't disappearing. They were doing something very snake-like, hunkering down and adapting to a hostile world.

As Shine put it, "You can just sit around there and wait until the world gets better." Of all of evolution's shrewd adaptations, that may be the most brilliant of all. It is something that snakes have done well for a long time. It's an adaptive biological skill, he might have added, that we humans show no evidence of possessing.

~

While we wait, or just hope, for the world to get better, perhaps we can privately work on the kind of psychological habitat reclamation implicit in Aby Warburg's snake lecture. Perhaps we can burrow into the "fear module" and aerate that internal, psychic habitat, beginning by rethinking what respect for nature, writ large, and respect for snakes, writ small, really mean. This occurred to me, in of all places, when I struggled to pick up a venomous snake at a rattlesnake-handling workshop conducted by Emily Taylor.

Taylor of course loves rattlesnakes, and as part of her Central Coast Snake Services side gig, she tries to educate ordinary citizens to love them, too. On a Saturday in September, she offered one of those workshops to a group of local folks who were dealing with rattlesnakes one way or another—on their jobs, in their backyards, even in their offices. After leading everyone on a morning field trip to Montaña de Oro State Park, a popular recreational site, to look for rattlesnakes in the wild (we saw seven along a popular hiking trail in about 45 minutes, including a pair peacefully sunbathing near a rock on the beach), the group returned to the San Luis Obispo campus to each take a turn, including me, handling a rattlesnake. With tongs.

Walking into the Physiological Ecology of Reptiles Lab was like walking into a wonderful natural history museum, without the glass. The hide of a huge Nile crocodile took up most of one of the tables; two ambassador snakes, Milo the gopher snake and Dorcas the California king snake, hid in their cages; the skeleton of a Malayan tiger prowled atop a cabinet in back; and Taylor propped open the door to the room for latecomers with a basketball-size whale vertebrae. The seriousness of the task became apparent when everyone was asked to sign a release basically stating, in half a dozen different ways, that we understood we might suffer injury or death from snakebite during the course of the workshop and accepted the risk. Not only was this a reminder that the wariness, if not fear, of snakes is a primordial human emotion rooted in real danger, but it also clarified the daunting psychological challenge of converting that fear into respect.

The very first thing Taylor said to us was the point of the entire exercise: "You don't have to kill a snake to be safe. And killing a snake can be unsafe...There are lots of ways to pick up a rattlesnake, and we're going to learn the safest way."

Unlike Aby Warburg's famous lecture, Taylor's "Rattlesnake 101" lecture stuck to biological and ecological facts. She told us that rattlesnakes, in the muscle that powers their rattle, have the second quickest muscle contraction (or "twitch speed") among all animals—90 cycles per second, second only to the swim bladder of a male toadfish. That female rattlesnakes can store sperm in their body up to eight years and still fertilize their eggs. That very few males actually get to mate. That you can see the beating fetal heart of a developing rattlesnake on ultrasound. That "baby newborn rattlesnakes punching through the amniotic sac is the most adorable thing on the planet." That ground squirrels chew rattlesnake skins and rub the masticated cream all over their faces, apparently as a social signal to fellow squirrels. That an adult rattlesnake only needs to eat a single squirrel, with its 500 to 700 calories, to maintain its weight for a year, which is roughly half of a single Big Mac combo meal at McDonald's with its nearly 1,200 calories (adult mountain lions, Taylor noted, need to consume 1.5 million calories to make it through a year). That only 20 percent of the roughly 8,000 annual venomous snakebites in the US are due to rattlesnakes (copperheads account for fully half of all bites, and cottonmouths another 30 percent). That a Boy Scout who was bitten in 2015 at nearby Lake Cachuma required 24 vials of antivenom, at a cost of $150,000 (total medical bill for snakebite treatment: $600,000). That antivenom in Mexico, by contrast, costs nothing, and veterinary costs to treat snakebite, including antivenom, are typically thousands of dollars (about 30,000 dogs in the US suffer snakebites each year).

It was at precisely this point in Taylor's lecture, as she diagrammed the structure of an antivenom antibody molecule on the blackboard, that we heard for the first time from the protagonist of the afternoon's program: a small female Pacific rattlesnake named Fizz. Inside a bucket at the front of the room, she suddenly started rattling. *Fizz* was an apt name, because the sound she made was a prolonged replica of the sound you hear when opening the top of a carbonated drink after shaking the bottle a bit—rapid, evanescent, shrill, earily disquieting.

Taylor picked up Fizz with tongs and deposited her on the floor. You might think that a rattlesnake, or any snake, would make a run for it after being deposited on the linoleum floor of the lab. Instead, Fizz arranged herself into a hypervigilant coil—curled up, as Taylor put it, "like a small cinnamon bun." One by one, everyone in the workshop took turns approaching Fizz, gripping her by the midsection with tongs, lifting her up and placing her in a bucket, closing the lid of the bucket with the tongs (so as not to expose their limbs or digits to a strike), reopening the bucket, reaching in with the tongs, and depositing Fizz back on the floor for the next person. Fizz, surprisingly patient but not amused, did not stop rattling for the better part of an hour.

When it came my turn, I can't say I reminded anyone of Raymond Ditmars or Bill Haast. But I found myself in an edifying psychological no-man's-land as I approached the venomous animal. I knew there was potential danger, of course; the lawyered language of the release form had made that crystal-clear. At the same time, having read about the skeletal fragility of snakes and how pinning venomous species often fractured their bones, I felt very conscious about not wanting to harm Fizz by squeezing the tongs too tightly. And the tongs, after all, were a safe but unnatural intermediary between human and animal, about as far as you could get from the Hopi practice of gripping rattlesnakes by the neck with your teeth. As a result, Fizz kept slithering out of the tongs and I had to start the grasping motion all over again. Soon she became quite annoyed, striking at the tongs every time I maneuvered them close to her. She did it when I tried to pick her up off the floor, and she did it when I tried to extract her from the bottom of the bucket after an embarrassingly long effort.

"You have to be prepared for the unexpected!" Taylor had cheerfully exclaimed during our extended standoff.

It seemed like a throwaway line at the time, but as I think back on the encounter now, Taylor's remark hints at something deep and fundamental, something that perhaps the ancients grasped better than we do. Fizz the rattlesnake was small, composed, and in no way aggressive, although her venom gland, not much larger than a lentil, was still big enough to wreak serious havoc. And that havoc could happen in an instant; that, after all, is why we primates developed our exceptional vision and our exceptionally large brains. But even our big brains can't control the unexpected, the unpredictable, the sudden. Indeed, how appropriate at this moment in time that so much of the snake's symbolic power is connected to meteorology, with its unpredictable and accelerating catastrophes—drought, floods, global warming, food insecurity.

We went out to a brewpub for a bite to eat after the workshop. As we rode to the eatery, Fizz kept up a steady buzz in the back of Taylor's car. It was as if she wouldn't let us forget that respect—for her, for snakes in general, for nature itself—essentially required acceptance of all the things we can't control. That was more terrifying, I realized, than any fear of any snake. Accepting it might ultimately be thought of as a kind of citizenship test for living, as we all ultimately do, in Cottonmouth Country.

Acknowledgments

~~~~~~

Expulsion

The first person I want to thank is someone I have met only once for a couple of hours, haven't seen in more than 40 years, and, until I gave him a call a few months ago, almost certainly wouldn't have remembered me at all. His name is Dale Marcellini and, through no fault of his own, he triggered an episode of ophidiophobia, a fancy word for an intense and often irrational fear of snakes, that altered the trajectory of my career as a writer.

In the early 1980s, Marcellini was curator of reptiles at the National Zoo in Washington, D.C., and I was an editor at a small, quirky magazine in New York called *Attenzione*, which focused on all things Italian and Italian-American. In practical editorial terms, this meant looking for interesting people whose last names ended in a vowel and writing stories about them. I stumbled across a reference to Marcellini in an article and immediately thought he could be the focus of a fascinating magazine feature. He was game, and in the summer of 1981 I traveled down to Washington to interview him. He turned out to be a genial, intelligent, and worldly naturalist with interests well beyond herpetology. (I later learned that he would volunteer as an apprentice chef in one of Washington's best restaurants and ran a vineyard in Napa Valley.) We spent a pleasant couple of hours in the reptile house, I went back and wrote the story, and the magazine dispatched a photographer to take pictures. If *Attenzione* had any claim to fame, it was because of its stylish visual design; the art director, Paul Hardy, won a National Magazine Award that same year. Paul loved the photos of boas and geckoes and other slithery creatures, and he designed an eye-popping six-page technicolor spread of beautiful herpetological specimens. There was more than a little subversion in this scaly feature. The magazine usually celebrated the designs of Missoni, Fendi, and Versace.*

All went well until the executive editor, a product of the Fashion Institute of Technology and former editor of *American Home* magazine, ventured into the art department to look at the final page proofs. The moment she saw photos of snakes, she exploded. "We can't have these ugly creatures in our beautiful magazine," she cried. "Take it out! Kill the story!" For a town as cosmopolitan as New York, it always surprised me how provincial and small-minded editorial offices could be. To make a long story short, the executive editor insisted that the snake story be killed, the editor refused, she fired him, and virtually the entire staff—including me—quit in protest. (Yes, we were all young and naive in those days.)

This unexpected expulsion left me precipitously unemployed, and with a New York rent to pay, I stumbled into a freelance gig organized by *Encyclopedia Britannica*—a science encyclopedia for lay readers. Like the proverbial monkeys chained to typewriters and aspiring to ape Shakespeare, I sat with colleagues like Charles Mann (*1491*), Robert Crease (historian of Brookhaven National

---

* S. Hall, "All Things Weird and Wonderful," *Attenzione* (June 1982).

Laboratory), and other underemployed scribes as we banged out reader-friendly explanations of science at 12 cents a word. The director of the project, a fellow named Charles Van Doren, flew in from Chicago to give us pep talks; no one suspected that a few years later, Ralph Fiennes would portray him in the movie *Quiz Show*. The feedback was encouraging. Somewhere along the line, I stopped writing New Fiction short stories and unpublishable novels at night and started writing about science. I haven't stopped since. Thanks, Dale.

I wish to thank all the scientists, herpetologists, curators, archivists, and peer reviewers who assisted in getting *Slither* into the world; they saved me from serial inaccuracies, although I own any mistakes that may have slithered through: Mathew Allendar; Henry Astley; Ian Bartoszek; Rick Bauer; Patricia Brennan; Gordon Burghardt, Jr.; Sean Bush; Daren Card; Nicholas Casewell; Todd Castoe; Howie Choset; Christopher Cooper; Sheila Coulson; Hayley Crowell; Rachel Davis; Nicholas DePerno; Sylvie Diochot; Eugene Edwards: Elba Estrada; Megan Folwell; Tim Garnett; J. Whitfield Gibbons; Daniel Goldman; Elena Gracheva; Harry Greene; John Guy; Mimi Halpern; James Harding; William Hayes; Matthew Holding; Gunther Hollopeter; David Hu; Margaret E. Hunter; Lynne Isbell; Alan Jaslow; Bruce Jayne; David Julius; Donna Kalil; Michael Kirkland; Rachel Keeffe; Yvonne Kelly; John Kubie; William Lamar; Christopher Lay; Leslie Leinwand; Matthew Lewin; Eric Lingueglia; Lorenzo Lujan Lopez; Robert Lovich; Linda Manzanilla; Hamid Marvi; Robert Mason; Frank Mazzotti; Karin McElhatton; Melina Melfi; Edmund Meltzer; Joseph Mendelson; David Mifsud; James Murphy; M. Rockwell Parker; Wayne Rassner; Denise Readinger; Robert Reed; Marc Rodriguez; Ken Ross, Sr.; Derrick Rossi; Elda Sanchez; Gonzalo Sanchez; Luis Sarabia; Daniela Schiller; Rick Shine; Jake Socha; Emily Taylor; Tamekia Wakefield; "Weazel"; Bronwen Wikkiser; J. D. Willson; Wolfgang Wuster.

Acknowledgments

For research help on the Michigan herpetology episode, I am especially grateful to Deborah Rice, head archivist, Cranbrook Center for Collections and Research, Bloomfield Hills, Michigan; and Elizabeth Parkinson of Cranbrook. John Marshall of the Bloomfield Hills Historical Society was a reliable fount of amazing historical photos and documents. Curtis Morgan of the *Miami Herald* provided python coordinates; Mike Kirkland of the South Florida Water Management District provided logistical help; Donna Kalil and Marc Rodriguez welcomed me on several hunting forays, despite the bad luck I brought. Edward Bleiberg, Katerina Barbash, Gonzalo Sanchez, Edmund Meltzer, and Wendy Golding were all instrumental in describing the Snakebite papyrus; Nicholas Casewell, Matthew Lewin, and Leslie Boyer were especially helpful on the science of venom. Yeshwant Bakhle was a fantastic source on the early discoveries that led to Captopril. Greg Schneider and Hayley Crowell gave me a remarkable tour of the University of Michigan's reptile specimen collection.

To snake people who provided detailed help and guidance, I want to thank Harry Greene, now an adjunct at the University of Texas; Gordon Burghardt of the University of Tennessee; Daren Card of the Broad Institute; Chris Friesen of Woolagong University; Jake Socha of Virginia Tech; Gunther Hollopeter of Cornell University; Henry Astley of the University of Akron; Christopher Cooper of Critter Consulting; and Bronwen Wickkiser of the City University of New York. To non-snake people who were equally helpful, thanks to Rachel Benioff; Maggie Delgado and Bill Sheppard; David Harlan and Michael Czech of the University of Massachusetts-Worcester; and Laura Alonso of Weill-Cornell Medical Center.

Special profuse thanks go to Emily Taylor and Joe Mendelson for above and beyond help.

For their long-term support and encouragement, I want to thank my colleagues at New York University (Dan Fagin, Brooke Borel, Virginia Hughes, Sophia Domokos, Charles Seife, Robin Lloyd, Robin Henig, and Ivan Oransky), at Rockefeller University (Emily Harms, Tim Stearns, Kristen Cullen, and Marta Delgado), and at Cold Spring Harbor Laboratory (Zach Lippman, Alyson Kass-Eisler, Monn-Monn Myat, and Catherine Perez). Students at all three institutions, in magically invisible yet palpable ways, continually astonish me by how much I learn from them.

For opening doors, suggesting sources, providing hospitality, offering critical feedback, or simply lending support (short-term and lifelong), it is a pleasure to thank Tom O'Neill and So-Young; Joseph Helguera; Larry Sulkis and Martineke; Jerry Roberts and Linda Kiefer; Richard Klug and Kate Stearns; Jessica Nicoll and Barry Oreck; Letta Tayler and Herbert Buchsbaum; Nelson Smith and Toko Nagase; Ellen Rudolph; Lisa Shea; Thea Lurie and Joel Kaye; Bill Grueskin and Caryl Yanow; Chris and Jim Wiggins; Kate Doyle and Tim Weiner; Mary Nance and Steve Tager; Steve Warner and Martha Haakmat; Steve and Connie Reddicliffe; Richard McGahey; Connie Casey and Harold Varmus. In Italy, Riccardo Scalera, Ernesto Filippi, Claudio Corvino, and Gianpaolo Montinaro provided invaluable assistance in Cocullo. And I am forever grateful to art historian Ivo Bomba (and Muni) for bringing Aby Warburg's "Snake Lecture" to my attention.

For his enduring friendship and peerless literary wisdom, I especially want to thank Joseph McElroy (and Barbara Ellmann) for everything from recommendations on the best translations of Pindar and Ovid to conversations about D. H. Lawrence's "Snake" poem. Of the countless tidbits sent my way, one of my favorite snake anecdotes, possibly apocryphal, arose out of his bringing little-known St. Spyridon to my attention. According to one legend, this saint gave poor people gold to pay off their debts, then converted the gold into snakes once the debt had been erased. The patron saint of post-capitalism?

I owe special, and long-overdue, gratitude to the Guggenheim Foundation, which funded a fellowship to support several projects that ultimately failed to launch. The support was nonetheless important as a vote of confidence, and I take great pleasure in noting that the very first interview for what ultimately became *Slither* took place during the first year of the Guggenheim fellowship.

At Grand Central, Colin Dickerman's admitted fear of snakes made him an avid early cheerleader of *Slither*, and Maddie Caldwell proved to be not only a relentlessly enthusiastic and supportive editor from day one, but a brilliant fount of editorial suggestions large and small, from overall structure to the minutiae of punctuation; we were on the same page throughout the project (except perhaps on the use of parentheses). Editorial assistant Morgan Spehar found time to make sharp and thoughtful editorial suggestions while deftly fielding every author request. And gratitude to the entire Grand Central production team: designer Albert Tang, who conceived the fetching cover; copy editor Laura Jorstad, whose meticulousness saved me from many a grammatical and factual embarrassment; managing editor Rebecca Holland; interior designer Jeff Stiefel; production editor Bob Castillo; production coordinator Melissa Mathlin; manufacturing coordinator Jennifer Tordy; marketer Theresa DeLucci; and publicist Estefania Acquaviva.

Thanking my long-time agent Melanie Jackson is always more than pro-forma, but especially so for *Slither*. Beyond the wonderfully collegial relationship we've enjoyed over 40+ years, I credit Melanie with, in a sense, launching the entire idea of this book when she suggested, "Why don't you write a book about an animal?" My immediate reply was: "Only if it's an animal that most people hate." It didn't take long to settle on snakes. It sounds perverse, but I thought making an argument about conservation would be much more challenging and interesting when the animal we're trying to save is eternally condemned in the Bible and almost universally loathed.

And then there's family. First and foremost, I want to thank my late parents, Robert and Delores Hall, who didn't live to see this book but surely had a hand in shaping it, from allowing a few snakes in the house (okay, the garage) early on to giving me Harry Greene's *Snakes: The Evolution of Mystery in Nature* for Christmas in 1997. In the familial corner, I'm also lucky to have the Halls of Oregon (Eric, Dawn, Nathan, and Sam) and the Levines of Connecticut (Jedd, Susan, Danielle, Casey, and Jon).

~~~

Closer to home, and to the heart, I am blessed in every way to have three extraordinary people in my inner circle. As always, Micaela and Sandro were precious confederates, reliably supplying emotional support, factual comeuppance, and the occasional well-deserved roasting. And as always, peerless Mindy did it all—astute reader, clear-eyed editor, strategic planner, indefatigable research associate, field coordinator and documentarian, legacy advocate, emotional pillar, hug supplier, and the nearest and dearest shoulder to cry on. As anyone who has written a book knows, the sacrifice of family is a tariff we all regrettably exact. Without them, however, nothing would be possible.

Sources and Notes

INTRODUCTION: "SNAKES, ARDENCY OF"

2 **Mr. Fowler, the elder:** "Herpetology for Juniors," newsletter, Cranbrook Institute of Science, Winter 1965. The course announcement identifies James Fowler as the instructor—an affiliation confirmed by James Harding (below).

2 **a strange bump-bump-bump:** This "aggression" was consistent with the "bitey" behavior of garter and ribbon snakes, per B. B. Bowers, A. E. Bledsoe, and G. M. Burghardt, "Responses to escalating predatory threat in garter and ribbon snakes (*Thamnophis*)," *Journal of Comparative Psychology* 107:25-33 (1993).

3 **"That's a nice little ribbon snake":** J. Alan Holman, James H. Harding, Marvin M. Hensley, and Glenn R. Dudderar, *Michigan Snakes: A Field Guide and Pocket Reference*, Michigan State University Extension Bulletin E-2000, East Lansing, 1999.

3 ***Dainty* is the word:** Raymond L. Ditmars, *Snakes of the World: Where and How They Live*, Pyramid, New York, 1962, p. 64.

3 **Native American trail:** An 1817 survey map of the orchard area, in the archives of the Bloomfield Township Historical Society in Michigan, marks the path of this indigenous trail; the trail appears to cut across the property. John Marshall, personal communication, Bloomfield Township Historical Society, November 6, 2023.

4 **I took that ribbon snake home:** My poor reptile husbandry is apparent from Philip Purser, *Garter & Ribbon Snake Care*, TFH Publications, Neptune City, NJ, 2005.

4 **technically venomous:** Nicholas Casewell, Centre for Snakebite Research & Interventions, Liverpool School of Tropical Medicine, Zoom interview, August 14, 2023.

5 **"snakes, ardency of":** Charles Darwin, *The Descent of Man, and Selection in Relation to Sex*, Penguin, New York, 2004 (based on the 1879 edition), p. 400.

5 **"warmth of feeling":** *Oxford English Dictionary*, compact edition, Oxford University Press, New York, 1971, p. 437.

5 **"spooky, but adorable":** Metal snake glassware: West Elm website, https://www.westelm.com/products/metal-snake-wine-glass-sets-d14343/pip-print.html; accessed February 19, 2024. Serpenti Viper jewelry: Bulgari website, https://www.bulgari.com/en-us/jewelry/necklaces/serpenti-viper-necklace-white-gold-348165; accessed June 19, 2024. On Serpenti being Bulgari's bestselling line, see Kathleen Beckett, "Bulgari's Serpenti turns 75," *New York Times*, December

5, 2022. "Serpenti is our leading line," CEO Jean-Christophe told Beckett. "Sales have gone up 30 times in the past 10 years."

6 *The Autobiography of a Snake*: Drawings by Andy Warhol, Thames & Hudson, New York, 2016. An editor's note next to the title page reads: "Any resemblance of the snake, a creative individual who loves celebrities, to Andy Warhol, a creative individual who loves celebrities, is purely coincidental."

6 **"a rustle in the grass"**: D. H. Lawrence, "*Apocalypse* and the writings on Revelation," edited by Mara Kalnins, Cambridge University Press, Cambridge, UK, 1980, p. 123. Lawrence viewed the dragon and the serpent as among the oldest, and somewhat synonymous, symbols of the human relationship with unpredictable nature. The complete quote reads: "The dragon and serpent symbol goes so deep in every human consciousness, that a rustle in the grass can startle the toughest 'modern' to depths he has no control over."

7 **"We primates uniquely evolved"**: Lynne A. Isbell, *The Fruit, the Tree, and the Serpent: Why We See So Well*, Harvard University Press, Cambridge, MA, 2009, p. 6.

7 **"I think snakes trigger"**: Matthew Holding, personal interview, Ann Arbor, MI, October 24, 2023. The concept of a "preparedness module" in the primate brain was first proposed in the early 1970s by psychologist Martin Seligman, who sought to bridge the gap between evolutionarily shaped innate responses to stimuli (instincts) and acquired responses (learning). He proposed that phobias (intense, irrational fears) in humans derived from this evolutionarily honed preparedness; Arne Öhman and Susan Mineka were among the first scientists to use the term "fear module" in the context of snake phobia. Öhman has argued that the neurological attentiveness humans pay to snakes is comparable only to the intensity with which we perceive angry faces. A. Öhman, "Of snakes and faces: An evolutionary perspective on the psychology of fear," *Scandinavian Journal of Psychology* 50:543-552 (2009). Isbell didn't use the term *fear module*, but her hypothesis is consistent with the idea of a discrete pathway in the brain dedicated specifically to the detection of serpents.

7 **"I've spent hours"**: Emily Taylor, Physiological Ecology of Reptiles Laboratory, Department of Biological Sciences, California Polytechnic State University, San Luis Obispo, Zoom interview, February 11, 2023.

8 **"Snakes seem to do both"**: Daren Card, postdoctoral fellow, Department of Organismic and Evolutionary Biology, Harvard University, Cambridge, MA, Zoom interviews, January 9, 2023, and March 17, 2024 (Card is now at the Broad Institute). On recombination mechanisms, see C. Hoge et al., "Patterns of recombination in snakes reveal a tug-of-war between PRDM9 and promoter-like features," *Science* 383:846 (2024).

8 **"I'm not sure that we know"**: Rick Shine, professor of biology, Macquarie University, Sydney, Australia, Zoom interview, December 14, 2023. Titles of papers are from Shine's CV.

9 **the Madness**: James Murphy, former herpetology curator, Dallas Zoo and National Zoo, telephone interview, September 20, 2022.

10 **It's not just that snakes lack limbs**: Hongyu Yi, "How snakes lost their legs," *Scientific American*, January 2018.

10 **virtually every other terrestrial animal**: Henry Astley, University of Akron, Zoom interview, August 29, 2022.

12 **some 130 to 150 million years**: As in many other areas of vertebrate evolution, current timelines on snake evolution vary between traditional science (based on the fossil record) and molecular science (based on DNA and lineage analysis). According to one recent analysis, the group *Serpentes* emerged about 130 million years ago. See A. Y. Hsiang et al., "The origin of

snakes: Revealing the ecology, behavior, and evolutionary history of early snakes using genomics, phonemics, and the fossil record," *BMC Evolutionary Biology* 15 (2015).

12 **"one or more unknown":** P. O. Title et al., "The macroevolutionary singularity of snakes," *Science* 383:918-923 (2024).

13 **The signs of stress:** N. Cox et al., "A global assessment highlights shared conservation needs of tetrapods," *Nature* 605:285-290 (2022).

13 **"In the late Sixties":** Weazel, personal communication, July 12, 2024.

13 **"The first predictor of extinction":** Weazel, telephone interview, July 25, 2022.

13 **solicitations for donations:** As a couple of examples, recent mailings include pitches from the Nature Conservancy (tokay gecko, barn swallow, hummingbird, fur seal pups, giraffes) and the Natural Resources Defense Council (honeybees). Props, however, to the US Postal Service for including the San Francisco garter snake in its 2022 "Endangered Species" commemorative stamp series.

14 **"the wild creatures":** Edward O. Wilson, *Consilience: The Unity of Knowledge*, Knopf, New York, 1998, p. 78. Wilson drew a distinction between "snakes" (the animals in nature) and "serpents" ("their dream equivalents"). "The slithering bodies and lethal strikes of real snakes make them ideal for magic," Wilson wrote. "Their images evoke blends of emotion that fall on a triangular gradient defined by the three points of fear, revulsion, and reverential awe. Where the real snake frightens, the dream serpent transfixes."

14 **cultural historian Aby Warburg:** Uwe Fleckner, *The Snake and the Lightning: Aby Warburg's American Journey*, Hatje Cantz Verlay, Berlin, 2023. Writing about the Hopi Snake Dance, Warburg noted: "So the snake is not sacrificed in this Snake Dance, but is only, through consecration and influencing dance gestures, turned into a messenger and, having returned to the souls of the dead, is dispatched in the form of lightning to produce storms in the sky," p. 158.

15 **"Birth, not death":** Louise Glück, *Poems 1962–2012*, Farrar, Straus and Giroux, New York, 2012, p. 41. Not everyone views the water moccasin (*Akistrodon piscivores*) as an ominous creature. "You will be surprised how relaxing it is to be around cottonmouths," Harry Greene told writer Constance Casey during a field expedition in Florida. C. Casey, "The smartest snakes: The strange myths and even stranger reality of cottonmouths," *Slate*, September 8, 2014, https://slate.com/technology/2014/09/cottonmouth-natural-history-myths-research-feeding-and-mating-habits-of-of-water-moccasins.html; accessed July 26, 2024.

SNAKE ROAD: CATSKILL, NEW YORK

16 **On April 1, 1841:** "Lecture on American Scenery," in Thomas Cole, *The Collected Essays and Prose Sketches*, edited by Marshall Tymn, John Colet Press, St. Paul, MN, 1980, pp. 197-213. See also "Thomas Cole" in Metropolitan Museum of Art, *American Paradise: The World of the Hudson River School*, Harry N. Abrams, New York, 1987, pp. 119-140.

16 **"an unusual combination":** John K. Howat, *The Hudson River and Its Painters*, Penguin, New York, 1972, p. 35.

16 **"it would be well to cultivate":** Cole, *The Collected Essays and Prose Sketches*, p. 200.

16 **"futurity":** Cole, *The Collected Essays and Prose Sketches*, p. 210.

17 **"I cannot but express":** Cole, *The Collected Essays and Prose Sketches*, p. 210.

17 **Part of the connection:** Cox et al., "A global assessment highlights shared conservation needs of tetrapods."

18 *Expulsion from the Garden of Eden:* According to the Museum of Fine Arts in Boston, "Cole

believed the American wilderness to embody a state of divine grace and lamented that the signs of progress were rapidly approaching." Museum of Fine Arts, https://collections.mfa.org/objects/33060; accessed July 26, 2024.

18 **Words matter:** Translations of Genesis chapter 3 include "more crafty than any other" in *The New Oxford Annotated Bible* (augmented third edition), edited by Michael D. Coogan, Oxford University Press, New York, 2007, pp. 9-16; "cunning" in Robert Alter, *The Five Books of Moses*, Norton, New York, 2004, pp. 24-28; and "shrewdest of all" in *The Torah: A Modern Commentary*, edited by W. Gunther Plaut, Union of American Hebrew Congregations, New York, 1981, pp. 34-37.

19 **"The serpent tricked me":** *New Oxford Annotated Bible*, p. 14.

19 **In the history of Western art:** For a sampling of Garden of Eden artwork, and for a wonderful collection of snake-centric paintings, drawings, sculpture, and art objects, see Marilyn Nissenson and Susan Jonas, *Snake Charm*, Harry N. Abrams, New York, 1995.

20 **"Nature has spread":** Cole, *The Collected Essays and Prose Sketches*, p. 211.

21 **"Among the inhabitants":** Cole, *The Collected Essays and Prose Sketches*, p. 211.

1. FEMALE 21 AND THE BLACK MAMBA

23 **"Maryland Man with 124 Snakes":** NBC News, April 13, 2022, https://www.nbcnews.com/news/us-news/maryland-man-124-snakes-house-died-snake-bite-autopsy-finds-rcna24250; accessed February 13, 2024.

23 **"Snake on a Plane!":** CNN News, January 17, 2024, https://www.cnn.com/travel/snake-on-plane-airasia-thailand-intl-hnk/index.html; accessed February 16, 2024.

23 **"Fatal Black Mamba Bite!!!":** Brian Barczyk, https://www.youtube.com/watch?v=FNBvuCSEqXg. The video focused on the 2018 death of Ryan Vincent Soobrayan of South Africa, who became unconscious and went into anaphylactic shock 50 seconds after getting nipped on the finger by a snake he was milking. According to YouTube, Barczyk's video has had more than 4.5 million views.

24 **"satiating their horror":** Charles Darwin, *The Descent of Man, and Selection in Relation to Sex*, Penguin, New York, 2004 (based on the 1879 edition), pp. 92-93. In this passage, Darwin described informal snake-and-monkey experiments he conducted at the London Zoo.

24 **"Black mamba pet snake":** *Daily Freeman* [Kingston, NY], June 16, 2011.

24 **The tale began:** Reconstruction of the 2011 Putnam Lake, New York, incident is based on Putnam County Sheriff's Office, incident report, June 16, 2011 (obtained by FOIL request); New York State Department of Environmental Conservation, law enforcement complaint, November 30, 2011; "Law Enforcement Complaint Summary," January 25, 2012 (obtained by FOIL request); and interviews with eyewitnesses.

25 **He told officers that there were:** Putnam County Sheriff's Office incident report.

25 **"I'm not pulling any punches":** Nicholas DePerno, Putnam County Sheriff's Office (retired), telephone interview, May 13, 2024.

25 **100 percent lethal:** See D. A. Warrell, "Clinical toxicology of snakebite in Africa and the Middle East/Arabian Peninsula," in *Handbook of Clinical Toxicology of Animal Venoms and Poisons*, edited by J. Meier and J. White, CRC Press, Boca Raton, 1995, pp. 433-492.

26 **"You could see the plexiglass bend":** Ken Ross Sr., chief, Society for the Prevention of Cruelty to Animals, Putnam County, NY, telephone interview, June 29, 2022.

26 **By the time Cooper arrived:** Christopher Cooper, wildlife consultant, Dutchess County, NY,

telephone interviews, June 30, 2022, and April 25, 2024; personal interview, Putnam Lake, NY, August 30, 2022.

27 **"sort of go-to person":** C. Cooper, interview, June 30, 2022.

28 **escaped cobra:** *Washington Post*, July 2, 2021.

28 **"really poisonous" brown snake:** CNN, December 30, 2023.

28 **"Do you know what":** K. Ross, interview, June 29, 2022.

29 **"the most dangerous poisonous snakes":** Herpetologists discourage using the word *poisonous* to describe venomous snakes these days, pointing out that poisons are technically ingested while venoms are injected. Hence the use of *venomous* throughout, except in cases where *poisonous* appears in a direct quote or published text.

29 **The official inventory:** New York State Department of Environmental Conservation, law enforcement complaint, November 30, 2011.

29 **"like the one Cleopatra used":** Despite the persistent repetition of the Cleopatra-asp story, most scholars dismiss this apocryphal tale as the cause of death.

29 **"She was not breathing":** Putnam County Sheriff's Office incident report.

30 **"most feared snake of the African continent":** A. L. Oliveira et al., "The chemistry of snake venom and its medicinal potential," *Nature Reviews Chemistry* 6:451-469 (2022).

30 **"very aggressive when threatened":** Oliveira et al., "The chemistry of snake venom and its medicinal potential."

30 **"the shaft of a traveling":** Ditmars, *Snakes of the World*, p. 180. Of an incident when a mamba at the Bronx Zoo was accidentally poked and escaped a cage, Ditmars wrote, "I have never noted a quicker movement on the part of a snake."

31 **"possible suicide":** Putnam County Sheriff's Office incident report, p. 1. According to later New York State records, the death certificate listed "the immediate cause of death as multiple drug intoxication with a consequence being a poisonous snake bite." New York State Department of Environmental Conservation, narrative entry, January 25, 2012.

31 **In a subsequent, mordantly melodramatic:** *Fatal Attraction: Snake Secrets*, Animal Planet, September 2012.

32 **Soon after, it popped up:** For example, CBS News ("Putnam Lake woman killed by pet snake," June 16, 2011); Reuters ("World's deadliest snake suspected in owner's death," June 16, 2011); ABC News, outlets in New York, Chicago, and Los Angeles ("Bite from pet snake suspected in death of NY woman," June 16, 2011); *Johannesburg Sowetan* ("World's deadliest snake suspected in owner's death," June 17, 2011); *London Daily Mail* ("Did she commit 'suicide by snake'?," June 20, 2011).

32 **The World Health Organization estimates:** WHO Snakebite Fact Sheet.

33 **mostly because they refuse medical attention:** Emily Taylor, lecture, rattlesnake-handling workshop, California Polytechnic State University, September 9, 2023.

33 **doubled to 4 million people:** J. E. Smith, *Stolen World*, Crown, New York, 2011, p. 185.

33 **Animal dealers imported:** J. Guzy et al., US Geological Survey, 2023, p. 32.

33 **there are probably many more:** J. Virata, email to author, December 18, 2023.

33 **When beloved Michigan "reptile influencer":** Soban Deb, "Brian Barczyk, 54, a reptile evangelist who attracted millions on YouTube and TikTok," *New York Times*, January 21, 2024, p. 22.

33 **like the possibly apocryphal:** "Suicide by snake," *Mirror*, November 13, 2017.

33 **fearless snake-handler:** "Freer rattlesnake handler dies from bite at Rattlesnake Roundup," KIII TV, Corpus Christi, TX, May 2, 2022.

34 **the male resident in Charles County:** "Maryland man with 124 snakes," NBC News.

34 **Grace Wiley, onetime curator:** J. B. Murphy and D. E. Jacques, "Death from snakebite: The intertwined stories of Grace Olive Wiley and Wesley H. Dickinson," *Bulletin of the Chicago Herpetological Society*, 2006.

34 **like Karl Schmidt:** D. Pla et al., "What killed Karl Patterson Schmidt? Combined venom gland transcriptomic, venomic and antivenomic analysis of the South African green tree snake (the boomslang), *Dispholidus typus*," *Biochimica et Biophysica Acta* 1861:814-823 (2017). Of note, Schmidt was bitten by a snake he'd been asked to identify by the then-director of Chicago's Lincoln Park Zoo and well-known television animal expert Marlin Perkins.

34 **well-known herpetologist Joe Slowinski:** Jamie James, *The Snake Charmer*, Hyperion, New York, 2008.

34 **"As he was dying":** C. Cooper, interview, August 30, 2022.

35 **"gumption trap":** Robert M. Pirsig, *Zen and the Art of Motorcycle Maintenance: An Inquiry into Values*, William Morrow, New York, 1974, pp. 303-318.

35 **"That's when you have":** Harry Greene, emeritus professor, Cornell University, Rancho Cascabel, Mason County, TX, Zoom interview, March 2, 2023.

35 **his 1997 opus:** Harry W. Greene, *Snakes: The Evolution of Mystery in Nature*, University of California Press, Berkeley, 1997.

35 **his ruminative 2013 memoir:** Harry W. Greene, *Tracks and Shadows: Field Biology as Art*, University of California Press, Berkeley, 2013.

35 **"Everything is weathered":** Greene, *Tracks and Shadows*, p. 85.

36 **"academic gone a bit feral":** H. W. Greene, "A part of apart: Ought nature-lovers ever wear fur?," in *Heart of the Wild: Essays on Nature, Conservation, and the Human Future*, edited by B. A. Minteer and J. B. Lobos, Princeton University Press, Princeton, NJ, 2024, p. 216.

36 **that snakes were "unworthy":** Greene, *Tracks and Shadows*, p. 68.

37 **"I'm a physician":** Hardy's remarks were recalled by Greene in a March 2023 interview. Hardy died in 2016; accounts of his herpetological exploits are in R. A. Villa, "Sonoran herpetologist dedication to the memory of David Loop Hardy, Sr., MD," *Sonoran Herpetologist* 29:32-43 (2016).

37 **"He was, like me, a military brat":** H. Greene, Zoom interview, June 17, 2024. Greene related that Hardy once found a bushmaster egg in Costa Rica, hid it in his pocket while returning to the US, hatched it in Arizona, and kept the snake in his home.

38 **In those early technological days:** Greene credited Howard Reinert (College of New Jersey) and David Cundall (Lehigh University) with the key 1982 innovation in this technology. H. K. Reinert and D. Cundall, "An improved surgical implantation method for radio-tracking snakes," *Copeia* 1982(3):702-705 (1982).

38 **hospitable to snakes in general:** Chiricahua National Monument, National Park Service, https://www.nps.gov/chir/learn/nature/reptiles.htm.

40 **Hardly anyone had suggested:** The prevailing view during the 20th century was that maternal behavior in pit vipers remained unproven and was considered unlikely, but as Hardy and Greene pointed out, there were several chance, one-time observations of parental attention in vipers described in the literature, including an entry in James Audubon's 1849–1850 journal recording an early example in an unspecified species, probably western diamondbacks.

40 **first hint at much more complex:** D. L. Hardy Sr. and H. W. Greene, "Borderland Blacktails: Radiotelemetry, Natural History, and Living with Venomous Snakes," in *USDA Forest Service Proceedings RMRS-P-10*, pp. 117-121, 1999. Around the same time as Hardy and Greene's

field study in Arizona, Terence Farrell at Stetson University in Florida and colleagues conducted experiments with captive pygmy rattlesnakes (*Sistrurus miliarius*) suggesting maternal recognition and protection of neonates in a laboratory setting; the two groups pooled their findings in the book chapter cited below.

40 **"Harry Greene was the first"**: Emily Taylor, Zoom interview, February 11, 2023.

41 **"you can fail to see"**: Pirsig, *Zen and the Art of Motorcycle Maintenance*, p. 311.

43 **Fate was not quite as kind:** D. L. Hardy Sr. and K. R. Zamudio, "Compartment syndrome, fasciotomy, and neuropathy after a rattlesnake envenomation: Aspects of monitoring and diagnosis," *Wilderness and Environmental Medicine* 17:36-40 (2006).

43 **They boiled down 15 years:** The most detailed description of the Chiricahua field observations appears in Harry W. Greene, Peter G. May, David L. Hardy Sr., Jolie M. Sciturro, and Terence M. Farrell, "Parental behavior by vipers," in *Biology of the Vipers*, Eagle Mountain Publishing, Eagle Mountain, UT, 2002, pp. 179-205. According to this account, the researchers first captured, anesthetized, and implanted a radio transmitter in Female 21 in 1994, and the excited phone call from Hardy to Greene likely occurred in July 1995.

44 **"especially relished"**: Greene, *Tracks and Shadows*, p. 7 (Greene's encounters with Female 21 are described in pp. 6-8, 162-168).

44 **"The blacktails often seemed"**: Greene, *Tracks and Shadows*, p. 164.

44 **"ignorance of their lives"**: Greene, *Tracks and Shadows*, p. 162.

44 **"Natural historians"**: Greene, *Tracks and Shadows*, p. 9.

SNAKE ROAD: STATE HIGHWAY 254, EL DORADO, KANSAS

46 **In the summer of 1987:** W. M. Langley et al., "Responses of Kansas motorists to snake models on a rural highway," *Transactions of the Kansas Academy of Science* 92:43-48 (1989). According to one academic herpetologist, "A later study following that one was accepted scientifically, but it's against the law to do that. So the paper got rejected."

48 **Several well-known snake scientists:** G. M. Burghardt, J. B. Murphy, D. Chiszar, and M. Hutchins, "Combating ophiophobia: Origins, treatment, education, and conservation tools," in *Snakes: Ecology and Conservation*, edited by S. J. Mullin and R. A. Seigel, Comstock Publishing, Ithaca, NY, 2009.

48 **Israeli neuroscientists not long ago:** Uri Nili et al., "Fear thou not: Activity of frontal and temporal circuits in moments of real-life courage," *Neuron* 66:949-962 (2010).

49 **"revealed an interesting dissociation"**: D. Schiller, "Snakes in the MRI machine: A study of courage," *Scientific American*, July 20, 2010.

49 **As Darwin described:** Darwin, *The Descent of Man*, pp. 92-93.

50 **primatologists discovered that newborn vervet monkeys:** Burghardt et al., *Snakes: Ecology and Conservation*, p. 268.

50 **"We can conclude"**: Burghardt et al., *Snakes: Ecology and Conservation*, p. 268.

50 **"fear module"**: The original observation came from the work of Öhman and Mineka in the early 2000s. "The fear module," they wrote, "is a relatively independent behavioral, mental, and neural system that has evolved to assist mammals in defending against threats such as snakes" (in A. Öhman and S. Mineka, "The malicious serpent: Snakes as a prototypical stimulus for an evolved module of fear," *Current Directions in Psychological Science* 12:5-9 (2003).

50 **"In Africa, where hominids evolved"**: Burghardt et al., *Snakes: Ecology and Conservation*, p. 266.

51 **Snake Detection Theory:** Lynne A. Isbell, *The Fruit, the Tree and the Serpent*, Harvard University Press, Cambridge, MA, 2009.

51 **"an object in the environment":** L. Isbell, telephone interview, April 11, 2024.

52 **"It doesn't take long":** L. Isbell, telephone interview, April 11, 2024.

53 **One of Harry Greene's most interesting:** T. N. Headland and H. W. Greene, "Hunter-gatherers and other primates as prey, predators, and competitors of snakes," *Proceedings of the National Academy of Sciences* 108:20865-20869 (2011).

53 **One of the oldest human ritual sites:** S. Coulson, S. Staurset, and N. Walker, "Ritualized behavior in the Middle Stone Age: Evidence from Rhino Cave, Tsodilo Hills, Botswana," *PaleoAnthropology* 2011:18-61 (2011); and S. Coulson, personal communication, July 2, 2024. *ScienceDaily* referred to it as the "world's oldest ritual" (November 30, 2006). Although the *PaleoAnthropology* paper does not emphasize a serpent connection, Coulson and Staurset told an interviewer in 2011 that a local San chieftain "felt the carving represented a powerful snake" and that earlier generations of local people participated "in male rituals involving the rubbing of wooden spears, coated in snake fat, along the grooves of the panel [the python-like rock] to gain the power of the snake" ["Middle Stone Age ritual at Rhino Cave, Botswana," *Radical Anthropology*, November 2011, pp. 12-17].

54 **"These foul and loathsome":** Quoted in Burghardt et al., *Snakes: Ecology and Conservation*, p. 265.

55 **"We are still reacting":** Desmond Morris and Ramona Morris, *Men and Snakes*, McGraw Hill, New York, 1965, p. 215.

56 **"Before Adam and Eve":** Rabbi Dovid Rosenfeld, "Judaism and snakes," and "What was the serpent?," The Jewish Website, https://aish.com/judaism-and-snakes, January 1, 2021.

56 **the Talmud, which generally forbids:** In true Talmudic fashion, this exemption has been extensively debated by scholars.

56 **"how certain ideas":** Elaine Pagels, *Adam, Eve, and the Serpent*, Random House, New York, 1988, p. xxviii.

57 **"they contain true divine light":** Burghardt et al., *Snakes: Ecology and Conservation*, p. 278.

57 **"brutally suppressed":** Burghardt et al., *Snakes: Ecology and Conservation*, p. 278.

57 **"the worship of Asclepius":** Emma J. Edelstein and Ludwig Edelstein, *Asclepius: A Collection and Interpretation of the Testimonies*, Ayer, Salem, NH, 1988 (reprint of the original 1945 edition, Johns Hopkins Press), p. vii.

58 **"It's not innate!":** J. Whitfield Gibbons, professor emeritus, University of Georgia, Zoom interview, July 26, 2022.

58 **"Snake":** The text of the poem is available at the Poetry Foundation website, https://www.poetryfoundation.org/poems/148471/snake-5bec57d7bfa17. The Lawrence poem, from his 1923 collection *Birds, Beasts, and Flowers*, has inspired much critical assessment among English literature academics, far fewer among herpetologists. Lawrence regarded the snake as a symbol of impulsive human passion, "the fluid, rapid, startling movement of life within us" that is "swift and surprising as a serpent" (*Apocalypse*, p. 123). Of paganism, with its affinity for snakes, he wrote: "The instinctive policy of Christianity towards all true pagan evidence has been and is still: suppress it, destroy it, deny it" (*Apocalypse*, p.87).

59 **"the conservation of snakes":** Burghardt et al., *Snakes: Ecology and Conservation*, p. 262.

59 **"The most important question":** Burghardt et al., *Snakes: Ecology and Conservation*, p. 276.

2. SNAKE GUYS

60 **the architecture of the building:** For historical details on the Bronx Zoo's World of Reptiles, see Madeleine Thompson, "Bronx Zoo's Reptile House with Italian foundations," Wildlife Conservation Society blog, https://blog.wcs.org/photo/2016/01/14/bronx-zoos-reptile-house-with-italian-foundations; accessed August 11, 2024. Author observations of the Reptile House are based on personal visits June 30, 2022, and February 28, 2024.

61 **"It was extremely realistic":** H. Greene, interview, March 2, 2023.

62 **yearned to be:** David Quammen, personal communication, May 24, 2024.

62 **"Give us a huge snake":** "Rumors of a snake," in David Quammen, *Natural Acts: A Sidelong View of Science and Nature*, Dell, New York, 1985, p. 41.

62 **"the deadliest and most treacherous":** For this and many other biographical details about Ditmars, I have consulted Dan Eatherley, *Bushmaster: Raymond Ditmars and the Hunt for the World's Largest Viper*, Arcade, New York, 2015, p. 250.

62 **"most massive":** "Conservancy captures most massive Burmese python in Florida," Conservancy of Southwest Florida press release, June 22, 2022, https://conservancy.org/conservancy-captures-most-massive-burmese-python-in-florida; accessed August 10, 2024.

63 **"The major challenge for the zoo herpetologist":** Burghardt et al., *Snakes: Ecology and Conservation*, 2009, p. 274.

63 **an estimated 100,000-person bump:** Eatherley, *Bushmaster*, p. 252.

63 **when the Bronx Zoo put a spitting cobra:** Raymond L. Ditmars, *Snakes of the World: Where and How They Live*, Pyramid, New York, 1962, p. 174. Ditmars wrote, "It was necessary to remove the cobras every five or six days as the glass was so showered during their 'spitting' at visitors it was impossible to see through it."

63 **the hidden genetic history of spitting cobras:** T. D. Kazandjian et al., "Convergent evolution of pain-inducing defensive venom components in spitting cobras," *Science* 371:386-390 (2021). Among the most fascinating observations in this paper was that all three lineages of spitting cobras independently evolved venoms that specifically caused excruciating pain in the eyes of adversaries ("ocular toxicity"); their targets were believed to be bipedal apes and early hominins.

64 **beginning with Aristotle:** Aristotle, *The History of Animals*, book 2, translation by D'Arcy Wentworth Thompson, available online at http://classics.mit.edu/Aristotle/history_anim.1.i.html.

64 **Benjamin Franklin was a snake guy:** For Franklin's rattlesnake essays, see *A Benjamin Franklin Reader*, edited and annotated by Walter Isaacson, Simon & Schuster, New York, 2003.

65 **"Rattlesnakes for Felons":** *A Benjamin Franklin Reader*, pp. 149-150.

65 **Franklin's modest proposal was adapted:** On the Synanon assassination attempt, see C. Risen, "Paul Morantz, 77, California lawyer who took on cults and gurus," *New York Times*, November 10, 2023, p. B-11. The *Times* obituary noted of the attempted assassination at Morantz's LA home: "As he walked in the door, he reached his left hand into his mailbox. As he did, he noticed a dark, lumpy shape. He didn't have time to pull back before the object, a four-and-a-half-foot diamondback rattlesnake, bit him on his wrist." Morantz was hospitalized for six days; according to the *Times*, "Doctors said he was lucky to survive."

66 **"Don't mess with us":** Synanon founder Charles Dederich, quoted in Christopher Goffard, "A cult targeted a crusading L.A. lawyer. The weapon: A rattlesnake," *Los Angeles Times*, May 29, 2024.

66 **persists as an all-purpose political insult:** See M. Flegenheimer, "Pence struggles trying to warn yet lure voters," *New York Times*, August 5, 2024, p. A-1. This insult was especially ironic as the Pence family kept a pet snake.

67 **"The Rattlesnake as America's Symbol":** *A Benjamin Franklin Reader*, pp. 263-266. See also Bob Ruppert, "The rattlesnake tells the story," *Journal of the American Revolution*, 2015, which reproduces a good collection of 18th-century snake imagery in political publications and cartoons.

68 **Christopher Gadsden created a flag:** Laurence Klauber, *Rattlesnakes: Their Habits, Life Histories, and Influence on Mankind*, University of California Press, Berkeley, 1997 (two-volume set). Volume 2, pp. 1241-1242, has an especially detailed account of the early history of the Gadsden flag that makes clear it was employed as a signaling device on a Revolutionary-era warship. The Gadsden flag may even have undone the magic of Franklin's metaphor and inspired public scorn of snakes, according to a recent *Washington Post* article. Six days after the first shots of the Civil War were fired at Fort Sumter, in April 1861, a merchant vessel out of Savannah, Georgia, sailed into Boston Harbor flying a white version of the "Don't Tread on Me!" flag; because the Gadsden flag had become associated with Southern secessionists, an angry mob of Northerners gathered in protest, seized the flag, and destroyed it. "As images of stomped, stabbed and eaten snakes proliferated in Union illustrations," the *Post* reported, "Confederates gradually relinquished their serpent." Laura Brodie, "The disgraced Confederate history of the 'Don't Tread on Me' flag," *Washington Post*, June 14, 2023.

69 **timber rattlesnakes "have been eliminated":** South Carolina Department of Natural Resources, "Venomous snakes," https://www.dnr.sc.gov/wildlife/snakes/snakes11.html; accessed August 10, 2024.

69 **posted a brief, four-paragraph announcement:** C. Beach and L. Smith, "Black Lives Matter," website, American Society of Ichthyologists and Herpetologists, June 18, 2020. For a history of the name change, see "Journal name change," American Society of Ichthyologists and Herpetologists, https://www.asih.org/ichsandherps/journal-name-change; accessed September 28, 2024.

70 **arguably the most prolific snake collector:** For a detailed account of E. D. Cope's life, see Henry Fairfield Osborn, "Biographical memoir of Edward Drinker Cope, 1840–1897," presented at the annual meeting of the National Academy of Sciences, 1929, https://www.nasonline.org/wp-content/uploads/2024/06/cope-edward.pdf; accessed August 10, 2024.

70 **"militant paleontologist":** Osborn, "Biographical memoir," p. 170.

71 **"the evolutionary significance of human character":** "The evolutionary significance of human character," in E. D. Cope, *The Origin of the Fittest: Essays on Evolution*, D. Appleton and Company, New York, 1887.

71 **"inferior mental co-ordination":** In E. D. Cope, "The relation of the sexes to government," *Popular Science Monthly*, October 1888, quoted in Osborn, "Biographical memoir," pp. 165-166.

71 **"They would vote through emotional suasion":** Osborn, "Biographical memoir," p. 166.

72 **"As every anatomist knows":** E. D. Cope, "Two perils of the Indo-European," *The Open Court*, January 23, 1890, pp. 2052-2054. Although Cope's racist views were hardly unique in 19th-century culture, what makes them particularly pernicious is the glaze of scientific authority with which they were advanced. He justified his opposition to mixed-race parentage, for example, by claiming that "hybrid" matings inevitably produced inferior offspring.

72 **"It appears to the writer":** Cope, "Two perils of the Indo-European," pp. 2052-2054.

72 **"When we think more deeply":** Osborn, "Biographical memoir," p. 169.

73 **It began in the early 1990s:** For this account of the *Copeia* name change, I have consulted Tamekia Wakefield, MD, New York, telephone interview, May 15, 2024; Alan Jaslow, professor emeritus, Rhodes College, Memphis, Zoom interview, November 20, 2023; M. Rockwell Parker, Department of Biology, James Madison University, Harrison, VA, Zoom interview, August 15, 2023; and Emily Taylor, past president, American Society of Ichthyologists and Herpetologists, Zoom interview, July 20, 2023.

73 **"Wow, that's messed up":** Wakefield, telephone interview, May 15, 2024.

74 **"The suggestion was seen as laughable":** Alan Jaslow, email sent to DEI Committee, American Society of Ichthyologists and Herpetologists, June 15, 2020.

74 **"told me it was hurtful":** Alan Jaslow, personal communication to author, November 20, 2023.

74 **"Finding out about Cope":** Parker interview. For another account, see M. R. Parker, "Refusing to cope with a name," *Ichthyology & Herpetology* 109:3-4 (2021).

75 **To people who know and love snakes:** The main sources on Ditmars's life and career are Ditmars's own books and Eatherley, *Bushmaster*. It is unclear if snakes still inhabit New York's Central Park; neither the Central Park Conservancy nor the Central Park Rangers keep track nowadays of reports of snakes in the park.

75 **how Ditmars strapped a caged king cobra:** Eatherley, *Bushmaster*, p. 122.

77 **"He seemed to have":** Eatherley, *Bushmaster*, p. 29.

77 **"I could feel it":** Eatherley, *Bushmaster*, p. 34.

78 **"accidental means":** *New York Times*, May 13, 1942.

79 **"wasn't Ditmars's finest":** Eatherley, *Bushmaster*, p. 288.

79 **In an iron-horse prequel:** Eatherley, *Bushmaster*, pp. 134-135.

79 **a spitting cobra launched itself:** Eatherley, *Bushmaster*, pp. 183-185.

79 **anatomical anomaly:** Eatherley, *Bushmaster*, p. 156.

79 *Killing the Killer*: Eatherley, *Bushmaster*, pp. 200-201.

79 *The Jungle Circus*: Eatherley, *Bushmaster*, p. 192.

79 **he sometimes dressed up orangutans:** Eatherley, *Bushmaster*, pp. 174-178.

80 **"Ditmars's showmanship":** William Lamar, quoted in Eatherley, *Bushmaster*, p. 235.

80 **"Ditmars Fails":** Eatherley, *Bushmaster*, p. 233.

80 **"sort of freemasonry":** Eatherley, *Bushmaster*, p. 135.

80 **In a tiny bit of ephemera:** R. Ditmars, letter to Grace Wiley, May 29, 1925, reproduced in James B. Murphy and David E. Jacques, "Death from snakebite: The entwined histories of Grace Olive Wiley and Wesley H. Dickinson," *Bulletin of the Chicago Herpetological Society: Special Supplement* (2006), p. 1. After stating that the Bronx Zoo had never observed rattlesnakes breeding in captivity in the 25 years of the Reptile House, Ditmars, almost grudgingly, told Wiley, "It is possible that you are the first to breed captive rattlers. I cannot remember other records."

81 **"Our efforts of half a century":** Roger Conant, quoted in J. B. Murphy, "Zoos and aquariums: What is next," *Herpetological Review* 45:532-534 (2016). Eatherley, *Bushmaster*, quoted former Bronx Zoo curator Peter Brazaitis as saying of the Reptile House: "A leaky roof, cold weather, and an insufficient coal-fired heating system had, over the years, cost the lives of many warm-loving reptiles," p. 96. Among other instances of mortality, Eatherley (pp. 116-123) noted that the Bronx Zoo lost 500 snakes to a tick infestation in 1919 and that many snakes suffered head and neck injuries during force-feeding or venom milking, although in zoo publications, officials claimed force-feeding "appears to be as perfect as when it is brought about by more natural processes," pp. 116-117. Snake mortality in zoos was common at the time.

82 **"What aren't you noticing?":** Joseph Mendelson III, director of research, Zoo Atlanta, Zoom interview, September 14, 2022.

82 **The market for exotic reptiles:** For a summary of wildlife trafficking laws and illegal smuggling, see F. Hierink et al., "Forty-four years of global trade in CITES-listed snakes: Trends and implications for conservation and public health," *Biological Conservation* 248:108601 (2020). Among other things, this paper reported that the main exporting nations for live snakes are Ghana, Indonesia, Togo, and Benin, while the main importing nations are China and the United States. "Despite an increasing shift towards captive breeding of pythons," the report concluded, "a large proportion of traded snakes are still harvested from the wild, with potential implications for snake conservation."

83 **Among the zoos exposed:** Jennie Erin Smith, *Stolen World: A Tale of Reptiles, Smugglers, and Skulduggery*, Crown, New York, 2011, pp. 168, 178.

83 **Murphy, now retired:** James B. Murphy, "Biographical sketch and bibliography of James B. Murphy," *Smithsonian Herpetological Information Service* 148 (2016). In addition, Murphy published a series of recollections titled "Portrait of a herpetologist as a young man" in the *Bulletin of the Chicago Herpetological Society*, beginning in 2021, that vividly conveyed the Madness.

85 **"We tried that":** J. Murphy, retired curator, National Zoo, telephone interviews, September 20, 2022, and May 8, 2024.

85 **In 1988, Marcellini led a study:** Dale L. Marcellini and Thomas A. Jenssen, "Visitor behavior in the National Zoo's reptile house," *Zoo Biology* 7:329-338 (1988).

86 **"Enrichment has become":** Murphy, "Zoos and aquariums," p. 533.

87 **"Snake guys," he told me:** D. Marcellini, retired curator of reptiles, National Zoo, Berkeley, CA, telephone interview, May 20, 2024.

87 **"basically an ode":** E. Taylor, Zoom interviews, February 11, 2023; July 20, 2023; September 8, 2023; and January 24, 2024.

87 **The good news was that:** Katelyn N. Rock et al., "Quantifying the gender gap in authorship in herpetology," *Herpetologica* 77:1-13 (2021).

89 **You could almost count the prominent female snake scientists:** Catherine Cooper Hopley, *Snakes: Curiosities and Wonders of Serpent Life*, Dutton, 1882. There were obviously more than three; for a global update, see *Women in Herpetology: 50 Stories from Around the World*, edited by Umilaela Arifin, Itzue Wendolin Caviedes Solis, and Sinlan Poo, Global Women in Herpetology Project, 2023.

89 **Helen Thompson Gaige:** Alison Davis Rabosky, "Building a tangled bank: Today's Michigan herpetology community," in *Letters from Michigan Herpetology*, edited by Greg Schneider and Linda Trueb, Special Publication 3, University of Michigan Museum of Zoology, Ann Arbor, 2021.

89 **the checkered career:** J. B. Murphy and D. E. Jacques, "Grace Olive Wiley: Zoo curator with safety issues," *Herpetological Review* 36(4):365-367 (2005).

90 **"Project RattleCam" channel:** For livestreams of sites in California and Colorado, see https://rattlecam.org.

91 **she and Amarello organized:** For examples of handwritten letters sent by children seeking to make the Sweetwater, Texas, Rattlesnake Roundup a no-kill event, see https://www.rattlesnakeroundups.com/love-letters.

91 **In 1970, only about 1 in 10:** Rock et al., "Quantifying the gender gap in authorship."

92 **"far more interesting":** Clara Ditmars, quoted in Eatherley, *Bushmaster*, p. 134.

SNAKE ROAD: INTERSTATE 78, HAMBURG, PENNSYLVANIA

93 **Reptile shows date back:** James B. Murphy and Ken McCloud, "The evolution of keeping captive reptiles and amphibians," *Herpetological Review* 41(2):134-142 (2010). Murphy and McCloud note that the "Exeter Exchange" in London, which sold animals from the 1770s to 1829, "specialized in the sale of exotic reptiles."

93 **a famous 1898 exposition:** See Eatherley, *Bushmaster*, pp. 74-75.

94 **"When my brother first put on":** Denise Readinger, personal interview, Hamburg, PA, October 15, 2022.

94 **"emotional support" boa constrictor:** Matt Cohen, "TSA finds 4-foot 'emotional support' boa constrictor in woman's carryon at Tampa airport," *Tampa Bay Times*, January 9, 2023.

94 **"males with tattoos":** C. Cooper, personal interview, August 30, 2022.

95 **"advanced hobbyist":** "Discover venomous reptiles at Hamburg Reptile Show," https://hamburgreptileshow.com/?from=192.168.1.19:9090; accessed August 11, 2024. The show boasts "the largest selection of venomous reptiles on the East Coast!"

95 *Conservation thru Commercialization*: Smith, *Stolen World*, p. 151. Smith attributed the slogan to Florida reptile dealer (and smuggler) Tom Crutchfield, who apparently began using it around 1990. "The slogan," Smith wrote, "borrowed the self-serving ideology long in vogue at the zoos: Grab the animals while you can because they're going extinct anyway."

95 **global CITES treaty:** The 1973 Convention on International Trade in Endangered Species of Wild Flora and Fauna regulates international trade in wild animals and plants.

95 **Several protagonists in her tale:** Molt was sentenced to nine months in jail in 1981; T. O'Toole, "U.S. cracks down on rare species imports," *Washington Post*, August 14, 1981. Crutchfield was sentenced to 30 months in prison for smuggling rare reptiles into the United States; US Department of Justice press release, April 16, 1999.

96 **Major US breeders of ball pythons:** Virginia-based breeder Trooper Walsh described the satellite-phone anecdote in J. B. Murphy and K. McCloud, "Reptile Dealers and Their Price Lists," *Herpetological Review* 41:266-281 (2010).

97 **"Genetic Wizard":** See the World of Ball Pythons website, https://www.worldofballpythons.com/wizard/; accessed September 28, 2024.

97 **"leucistic" ball python:** Rebecca Giggs, "Skin in the game," *New Yorker*, February 26, 2024, pp. 20-26.

97 **echo of the 19th-century "pigeon fancy":** C. Darwin, *On the Origin of Species*, 1859, pp. 553-568. In a prescient comment about fanciers (or hoarders) of any stripe, Darwin noted that the pigeon breeder "perceives extremely small differences, and it is in human nature to value any novelty, however slight, in one's own possessions" (p. 568).

97 **"There really has been a sea change":** J. Mendelson, Zoom interview, September 14, 2022.

97 **"Having the latest new wild-caught":** J. Virata, personal communication, December 18, 2023.

98 **The total value of the reptile pet business:** "Must-know reptile industry statistics," Gitnux Marketdata Report, December 16, 2023.

98 **"There's a lot of people":** E. Taylor, Zoom interview, February 11, 2023.

99 **"They won't let us":** "Tom," vendor, Mid-Hudson Reptile Exposition, Poughkeepsie, NY, September 18, 2022.

99 **One of the reasons they won't let them:** James Murphy, interview, September 20, 2023.

100 **Hayley Crowell got her first corn snake:** H. Crowell, Zoom interview, November 13, 2023.

100 **"I grew up with snakes":** D. Rossi, telephone interview, January 9, 2024.

3: A PANDEMONIUM OF MOLECULES

101 **Karin McElhatton knew she was in trouble:** K. McElhatton, Studio Animal Services, telephone interview, July 22, 2022, and personal interview, Studio Animal Services ranch, Castaic, CA, September 8, 2023.

102 **"dogs of every demeanor":** Studio Animal Services website, https://studioanimalservices.com; accessed August 11, 2023. Although McElhatton usually doesn't handle movie snakes, the company did collaborate on a serpent scene with Cher in *The Witches of Eastwick*.

102 **there's a separate Hollywood subculture:** "It had to be snakes…" is probably the most famous serpent-related line of dialogue in the history of Hollywood, uttered by Harrison Ford in *Indiana Jones and the Raiders of the Lost Ark* (1977). From Indiana Jones to *Anaconda* to *Snakes on a Plane*, serpents have been a regular and oft-deployed scare component in movies; less well known is that roughly half a dozen Hollywood-affiliated animal handlers specialize in snakes, from making sure that outdoor, on-location shoots do not have any venomous snakes lurking on the premises to providing snakes for scenes.

104 **"pandemonium of molecules":** A. Alagón, Autonomous University of Mexico, quoted in C. Arnold, "The snakebite fight," *Nature* 537:26-28 (2016).

104 **McElhatton had no way of knowing:** On the severity of lower-extremity defensive bites, see W. K. Hayes, "The snake venom-metering controversy: Levels of analysis, assumptions, and evidence," in *The Biology of Rattlesnakes*, Loma Linda University Press, Loma Linda, CA, 2008, p. 211, and C. E. Person et al., "Paradoxical exception to island tameness: Increased defensiveness in an insular population of rattlesnakes," abstract, *Toxicon* 119:375-376 (2016). Although the data is modest, Hayes said the latter study provides "compelling evidence" that rattlesnakes "can deliver more venom when defensive" (William Hayes, personal communication, August 27, 2023).

104 **"Defensive bites are more severe":** Sean Bush, MD, Duke University Hospital, Durham, NC, telephone interview, August 27, 2022.

105 **Not that it's any consolation:** J. M. Alves-Nunes et al., "Study of defensive behavior of a venomous snake as a new approach to understand snakebite," *Scientific Reports* 14:10230 (2024). Alves-Nunes wore special safety boots for this study and stepped on the head, midsection, and tail of each viper 10 times; none of the more than 40,000 provoked bites broke through, but Alves-Nunes did suffer an envenomation from a rattlesnake bite, requiring a hospital visit and antivenom treatment.

105 **"intrinsically ecological":** N. R. Casewell et al., "Causes and consequences of snake venom variation," *Trends in Pharmacological Sciences* 41:570-581 (2020).

105 **The venom of the Mojave rattlesnake:** J. L. Strickland et al., "Evidence for divergent patterns of local selection driving venom variation in Mojave rattlesnakes (*Crotalus scutulatus*)," *Scientific Reports* 8:17622 (2018); D. J. Massey et al., "Venom variability and envenoming severity outcomes of the *Crotalus scutulatus scutulatus* (Mojave rattlesnake) from Southern California," *Journal of Proteomics* 75:2576-2587 (2012); and G. Zancolli et al., "When one phenotype is not enough: Divergent evolutionary trajectories govern venom variation in a widespread rattlesnake species," *Proceedings of the Royal Society B* 286:1-10 (2019).

105 **a dose as small as 15 milligrams:** See the VenomousReptiles.org website, http://www.venomousreptiles.org/pages/venchart; accessed October 4, 2024.

105 **"I can show you roads":** Wolfgang Wüster, department of zoology, Bangor University, Zoom interview, January 24, 2024.

106 **"Every bite is dynamic":** E. Taylor, Zoom interview, January 24, 2024.

108 **"It sounds like she got bit"**: William Hayes, Loma Linda University, Loma Linda, CA, Zoom interview, November 14, 2023. The victim of a northern Pacific rattlesnake bite included in Klauber reported immediate pain at the site of the bite, "followed shortly by a general tingling sensation all over the body"; swelling at 4 minutes; a "woozy sensation" at 15 minutes; loss of ability to walk at 30 minutes, followed by "heavy, gasping breathing." Klauber, *Rattlesnakes: Their Habits, Life Histories, and Influence on Mankind*, pp. 865-866.

109 **Each year, an estimated 138,000:** For US snakebite fatalities, S. A. Seifert, J. O. Armitrage, and E. E. Sanchez, "Snake envenomation," *New England Journal of Medicine* 386:68-78 (2022).

109 **An estimated 2.7 million:** "Snakebite envenoming," World Health Organization fact sheet, https://www.who.int/news-room/fact-sheets/detail/snakebite-envenoming; accessed August 11, 2024.

109 **"disease of poverty":** D. J. Williams et al., "Strategy for a globally coordinated response to a priority neglected tropical disease: Snakebite envenoming," *PLoS Neglected Tropical Diseases* 13(2):e0007059 (2019).

109 **"Snakebite is the most important tropical disease":** Williams et al., "Strategy for a globally coordinated response."

109 **Over the three decades ending in 2018:** S. C. Greene et al., "Epidemiology of fatal snakebites in the United States 1989–2018," *American Journal of Emergency Medicine* 45:309-316 (2021).

109 **By contrast, in recent years an average of 28 people:** Comparative mortality statistics for lightning strikes, Centers for Disease Control and Prevention, https://www.cdc.gov/disasters /lightning/docs/LightningDeaths_09132022-H.pdf; for bee and hornet stings, CDC, https:// www.cdc.gov/mmwr/volumes/72/wr/pdfs/mm7227a6-H.pdf; for food poisoning, CDC, https://www.cdc.gov/foodborneburden/index.html; for pedestrian fatalities, Governors Highway Safety Association, https://www.ghsa.org/resources/Pedestrians24; for deaths from unintentional falls, motor vehicle traffic accidents, and unintentional poisoning, National Center for Health Statistics, CDC, https://www.cdc.gov/nchs/fastats/accidental-injury.htm (all accessed July 6, 2024).

110 **"Almost no one in America":** Sean Bush, telephone interview, 2022.

110 **one day in June 2006:** R. F. Raney, "Southern Californians see a rise in venomous snakes," *New York Times*, June 29, 2006.

111 **"Provoked bites":** Seifert et al., "Snake envenomation."

111 **The priests possessed a kind of medical field guide:** For physical details and dating of the Egyptian document, see "Snakebite Papyrus," Brooklyn Museum website, https://www .brooklynmuseum.org/opencollection/print/object/60690; accessed December 11, 2023. For recent English-language translations, see the note in the "Eastern Parkway" section. See also W. Golding, "The *Brooklyn Papyrus* snakebite and medicinal treatments' magico-religious context," *Religions* 14:1300 (2023); and G. Sanchez and W. B. Harer Jr., "Toxicology in ancient Egypt," in *Toxicology in Antiquity*, volume 1, Elsevier Science & Technology, 2018.

112 **In 1895, Albert Calmette:** On the development of antivenom, see M. B. Pucca et al., "History of envenoming therapy and current perspectives," *Frontiers in Immunology* 10:1598 (2019).

112 **"They are probably the most expensive":** Nicholas Casewell, Centre for Snakebite Research & Interventions, Liverpool School of Tropical Medicine, Liverpool, UK, Zoom interview, August 14, 2023.

113 **"That's a fucking krait!":** In Jamie James, *The Snake Charmer: A Life and Death in Pursuit of Knowledge*, Hyperion, New York, 2008, p. 8. James assembled a wealth of detail about this ill-starred expedition, which I have consulted for this account.

114 **one of the "Big Four":** The other three are the saw-scaled viper, Russell's viper, and Indian cobra.

114 **"I am the king":** James, *The Snake Charmer*, p. 208.

114 **always been an occupational hazard:** See James, *The Snake Charmer*, p. 22, for Schmidt and Mertens; B. Murphy, "William 'Marty' Martin, authority on eastern timber rattlesnakes, dies at 80," *Washington Post*, August 10, 2022; and J. B. Murphy and D. E. Jacques, "Grace Olive Wiley: Zoo curator with safety issues," *Herpetological Review* 36(4):365-367 (2005), for Wiley.

115 **"Tales of his reckless":** James, *The Snake Charmer*, p. 238.

115 **"a natural fit":** Matthew Lewin's career and early snakebite research from M. Lewin, founder, Ophirex, Inc., Corte Madera, CA, Zoom interviews, August 8, 2022, and September 20, 2023; and Lewin CV. In addition to interviews with Lewin and Casewell, I have summarized the Ophirex origin story based on interviews and print accounts, including D. Perlman, "Local doctor on trail of new snakebite treatment," *San Francisco Chronicle*, August 8, 2013; J. Robbins, "An ER doctor's search for a snakebite drug might lead to a COVID-19 treatment," Kaiser Health News, *Los Angeles Times*, November 6, 2020; and J. Ditzler, "'Brink of a major revolution': Pentagon-funded drug trial may end venomous snakebite scourge," *Stars and Stripes*, July 28, 2022.

117 **"keeping to myself in the kitchen":** M. Lewin, personal communication, September 21, 2024.

118 **Neostigmine had been used:** M. R. Lewin et al., "Reversal of experimental paralysis in a human by intranasal neostigmine aerosol suggests a novel approach to the early treatment of neurotoxic envenomation," *Clinical Case Reports* 1:7-15 (2013).

119 **This fantastic pharmacological story:** M. Rocha e Silva et al., "Bradykinin, a hypotensive and smooth muscle stimulating factor released from plasma globulin by snake venoms and trypsin," *American Journal of Physiology* 156:261-273 (1949).

120 **High blood pressure (hypertension) affects:** "Hypertension," World Health Organization fact sheet.

121 **That paradox puzzled Sérgio Ferreira:** For this account of the discovery of ACE inhibitors and captopril, I have used Y. S. Bakhle and B. R. Ferreira, "Sérgio Ferreira and *Bothrops jararaca* at the Royal College of Surgeons, London," *Toxins* 15:522 (2023); Y. S. Bakhle and R. J. Flower, "Sergio Henrique Ferreira (1934–2016)," *British Journal of Pharmacology* 174:341-342 (2017); J. Bryan, "From snake venom to ACE inhibitor—the discovery and rise of captopril," *Pharmaceutical Journal*, April 2019; C. G. Smith and J. R. Vane, "The discovery of captopril," *FASEB Journal* 17:788-789 (2003); D. W. Cushman and M. A. Ondetti, "History of the design of captopril and related inhibitors of angiotensin converting enzyme," *Hypertension* 17:589-592 (1991); and L. M. Opie and H. Kowalik, "The discovery of captopril: From large animals to small molecules," *Cardiovascular Research* 30:18-25 (1995).

121 **"much more potent":** P. Downey, "Profile of Sergio Ferreira," *Proceedings of the National Academy of Sciences* 105:19035-19037.

121 **"the world's only sample":** Bakhle and Ferreira, "Sérgio Ferreira and *Bothrops jararaca*."

121 **changed plans at the last minute:** Y. S. Bakhle, personal communications, March 21, 2024, April 2, 2024, and July 20, 2024.

122 **"one of the greatest pharmacologists":** R. Flowers, "Obituary: Sir John Vane," *Nature Reviews Drug Discovery* 4:10 (2005).

122 **"so we did not have to milk":** Y. S. Bakhle, personal communication, July 20, 2024.

122 **"now underlies the successful treatment":** Y. S. Bakhle, "How ACE inhibitors transformed the renin-angiotensin system," *British Journal of Pharmacology* 177:2657-2665 (2020). Of note in this account, Bakhle wrote that David Cushman, one of the two principal Squibb scientists assigned to the project, was clearing out his files after the project had been shut down when

he came across a paper that gave him and colleague Miguel Ondetti the idea to design the synthetic version of the snake-derived molecule in a different way. Among these new formulations, *after* the project had been closed, was the small molecule that became captopril.

122 **"The progress from snake venom":** Y. S. Bakhle, personal communication, July 20, 2024.

123 **ACE inhibitors are still one of the most lucrative:** As an indication of the popularity of antihypertensive drugs based on snake venom, pharma analysts estimate that the number of patients taking a single ACE inhibitor (lisinopril) in a single country (United States) in a single year (2022) numbered 20.3 million. "Lisinopril—drug usage statistics," ClinCalc DrugStats Database, https://clincalc.com/Drugstats/Drugs/Lisinopril; accessed August 12, 2024. Current estimates of annual global sales of ACE inhibitors, nearly half a century after FDA approval, is roughly $7 billion.

123 **"a rich playground":** A. L. Oliveira et al., "The chemistry of snake venom and its medical potential," *Nature Reviews Chemistry* 6:451-469, 2022.

123 **snake venoms have proven to be exceptionally:** For mambalgins, see S. Diochot et al., "Black mamba venom peptides target acid-sensing ion channels to abolish pain," *Nature* 490:552-555 (2012). On mambalgins, see S. Diochot and E. Lingueglossa, Zoom interview, July 7, 2022. For venom research, D. R. Schield et al., "The roles of balancing selection and recombination in the evolution of rattlesnake venom," *Nature Ecology & Evolution* 6(9):1367-1380 (2022); and C. F. Smith, et al., "Snakes on a plain: Biotic and abiotic factors determine venom compositional variation in a wide-ranging generalist rattlesnake," *BMC Biology* 21:136 (2023). For the ecological niche modeling based on the Brooklyn Papyrus, see E. McBride, I. C. Winder, and W. Wüster, "What bit the ancient Egyptians? Niche modelling to identify the snakes described in the Brooklyn Medical Papyrus," *Environmental Archaeology*, 2023.

125 **"You know a beautiful":** J. Mendelson, interview, September 14, 2022.

126 **"They're technically venomous":** N. Casewell, interview, 2023.

127 **Casewell, who has published extensively:** N. R. Casewell et al., "Causes and consequences of snake venom variation," *Trends in Pharmacological Sciences* 41:570-581 (2020); N. R. Casewell et al., "Complex cocktails: The evolutionary novelty of venoms," *Trends in Ecology and Evolution* 28:219-229 (2013); and B. A. Fry et al., "Evolution of an arsenal," *Molecular and Cellular Proteomics* 7.2:215-246 (2008).

128 **"How does this enzyme":** M. Lewin, interview, 2022.

128 **"an elongated middle finger":** Oliveira et al., "The chemistry of snake venom."

130 **"The card game analogy":** M. Holding, Zoom interview, January 16, 2024.

130 **Casewell cited the current situation in India:** R. R. Senji Laxme et al., "Biogeographic venom variation in Russell's viper (*Daboia russelii*) and the preclinical inefficacy of antivenin therapy in snakebite hotspots," *PLoS Neglected Tropical Diseases*, March 25, 2021. The discrepancy between official and unofficial snakebite mortality figures in India is striking. According to government National Health Profile data for 2018, there were a total of 885 snakebite deaths, J. K. Chakma et al., "White paper on venomous snakebite in India," *Indian Journal of Medical Research* 152: 568-574 (2020); according to a recent analysis, an average of 58,000 Indians died of snakebite every year during the first two decades of this century, W. Suraweera et al., "Trends in snakebite deaths in India from 2000 to 2019 in a nationally representative mortality study," *eLife* 9e54076 (2020).

131 **experts in antibody engineering:** A. H. Laustsen, "Recombinant snake antivenins get closer to the clinic," *Trends in Immunology* 45:225-227 (2024).

132 **That tinkering eventually led:** For the principal varespladib findings, see M. Lewin et al., "Varespladib (LY315920) appears to be a potent, broad-spectrum, inhibitor of snake venom

phospholipase A2 and a possible pre-referral treatment for envenomation," *Toxins* 8:248 (2016); M. Lewin et al., "Delayed oral LY333013 rescues mice from highly neurotoxic, lethal doses of Papuan taipan (*Oxyuranus scutellatus*) venom," *Toxins* 10:380 (2018); and M. R. Lewin et al., "Delayed LY333013 (oral) and LY315920 (intravenous) reverse severe neurotoxicity and rescue juvenile pigs from lethal doses of *Micrurus fulvius* (eastern coral snake) venom," *Toxins* 10:479 (2018).

134 **"He solved the one insurmountable problem"**: Leslie V. Boyer, University of Arizona, Zoom interview, October 4, 2024.

134 **"Quite frankly"**: D. Rossi, interview, 2024.

136 **"It couldn't have gone"**: Malcolm Chandler, quoted in K. Cataudella, "A new pill being tested at Duke could change snake bite treatment worldwide," *Raleigh News & Observer*, September 22, 2023.

136 **Chandler's doctor, Charles Gerardo:** Charles J. Geraldo, "Broad spectrum antidote varespladib oral for snakebite envenoming: The BRAVO trial," poster, "Venom Week" meeting, Durham, NC, June 7, 2024; the Phase II clinical trial results were published as C. J. Gerardo, et al., BMJ Global Health 2024; 9:e015985.

136 **"While the study did not meet"**: Timothy Platts-Mills, chief medical officer, Ophiirex, Inc., written statement, September 21, 2024.

137 **Animal Rule:** FDA, "Product guidance under the animal rule, 2015," http://www.fda.gov/media/88625/download; accessed August 12, 2024.

137 **without describing details, Lewin confirmed:** M. Lewin, telephone interview, September 6, 2024.

138 **The Liverpool group also foresees:** L-O. Albulescu et al., "A therapeutic combination of two small molecule toxin inhibitors provides broad preclinical efficacy against viper snakebite," *Nature Communications* 11:6094 (2020).

138 **never-seen-in-nature proteins:** S. V. Torres, et al., "De novo designed proteins neutralize lethal snake venom toxings," Research Square preprint, May 2024.

SNAKE ROAD: EASTERN PARKWAY, BROOKLYN, NEW YORK

140 **animals with souls:** In addition to personal visits to view the Egyptian galleries at the Brooklyn Museum, I consulted Edward Bleiberg, Yekaterina Barbash, and Lisa Bruno, *Soulful Creatures: Animal Mummies in Ancient Egypt*, Brooklyn Museum in association with D. Giles Ltd., Brooklyn, 2013.

140 **museum curators used X-rays:** Bleiberg et al., *Soulful Creatures*, pp. 116, 124.

140 **the so-called Pyramid Texts:** *The Ancient Egyptian Pyramid Texts*, translation by R. O. Faulkner, Oxford University Press, London, 1969. There are dozens of these snake-directed exhortations; Utterance 241, a typical example, reads, "O you expectoration of a wall, you vomit of a brick, what comes out of your mouth is turned back against yourself" (p. 57). These utterances, according to Faulkner, are "the oldest corpus of Egyptian religious and funerary literature now extant" (p. v).

141 **"The One Who Loves Silence"**: Meretseger, sandstone sculpture, ca. 1479–1400 BCE, Brooklyn Museum.

141 **"The Pyramid Texts are probably the first"**: Edmund Meltzer, telephone interview, September 20, 2024.

141 **"Brooklyn Snakebite Papyrus"**: Brooklyn Museum website, https://www.brooklynmuseum.org/opencollection/objects/60690; accessed August 14, 2024.

142 **"Incidentally," Barbash said:** Yekaterina Barbash, curator of Egyptian, classical, and ancient Near Eastern art, Brooklyn Museum, Zoom interview, January 29, 2024, and personal interview, March 15, 2024.

142 **many letters are represented by snake figures:** Gonzalo Sanchez, personal communication, August 18, 2023.

142 **"Like many cultures":** Bleiberg et al., *Soulful Creatures*, p. 55.

143 **Charles Edwin Wilbour, a journalist:** John M. Adams, "A bad dream of New York: The rise, fall, and redemption of Charles E. Wilbour," available on academia.edu at https://www.academia.edu/6990369/A_BAD_DREAM_OF_NEW_YORK_The_Rise_Fall_and_Redemption_of_Charles_E_Wilbour; and Mark Rose, "Wilbour's legacy," *Archaeology*, August 2005. According to Adams's account, Wilbour "fled to Europe to avoid joining Tweed in jail…," p. 2.

143 **For many years, the lone translation:** For English-language translations of the Snakebite Papyrus, I have consulted G. M. Sanchez, E. S. Meltzer, W. Wüster, N. R. Casewell, and G. W. Schuett, *Snake Identification in the Ancient Egyptian Brooklyn Medical Papyrus: A New Study of the Twenty-Four Extant Registers of the "Snakebite Papyrus,"* Lockwood Press, Columbus, GA (2024); W.R.J. Golding, "The Brooklyn Papyrus (47.218.48 and 47.218.85) and its snakebite treatments," PhD dissertation, University of South Africa, Pretoria, 2020 (pdf available on academia.edu); and Montgomery Q. Stewart, "Ancient snakebite literature: The Brooklyn Medical Papyrus and Nicander's Theriaca," CUNY Academic Works, 2020, https://academicworks.cuny.edu/cgi/viewcontent.cgi?article=1287&context=bc_pubs; accessed August 14, 2024.

144 **"If it bites a man":** Sanchez et al., *Snake Identification in the Ancient Egyptian Brooklyn Medical Papyrus*, p. 59 (attributed to the Egyptian cobra).

144 **"If he vomits":** Sanchez et al., *Snake Identification in the Ancient Egyptian Brooklyn Medical Papyrus*, p. 83 (possibly Field's horned viper).

144 **to suggest the possible identity:** As both Sanchez et al. and Golding noted, nearly half a dozen scholars have taken a shot at identifying the precise species indicated in each register, or chapter, of the Snakebite Papyrus.

145 **"Hail to you, Onion":** Golding, "The Brooklyn Papyrus," p. 219.

145 **"The treatments are essentially":** G. Sanchez, telephone interview, August 18, 2023.

145 **one coveted herbal ingredient, silphium:** M. Q. Stewart, "Ancient snakebite literature," p. 8.

146 **"Pyramid Texts" date back:** "Utterance 298," in the Faulkner translation of *The Ancient Egyptian Pyramid Texts*, apparently refers to a venomous snake and describes the use of cords from a sandal to "draw off your poison"—perhaps one of the earliest reports of milking a venomous snake, p. 89.

4: DREAMS OF HEALING

147 **The story of the Greek demigod Asklepios:** For background on the myth and cult of Asklepios, I have used Carl Kerenyi, *Asklepios: Archetypal Image of the Physician's Existence*, translation by Ralph Manheim, Pantheon, New York, 1959; Ludwig Edelstein and Emma Edelstein, *Asclepius: Collection and Interpretation of Testimonials*, Johns Hopkins University Press, Baltimore, 1945; Milena Melfi, *I Santuari di Asclepio in Grecia*, "L'Erma" di Bretschneider, Rome, 2007; Robin Mitchell-Boyask, *Plague and the Athenian Imagination: Drama, History and the Cult of Asclepius*, Cambridge University Press, Cambridge, UK, 2008; David B. Morris, "Un-forgetting Asclepius: An erotics of illness," *New Literary History* 38:419-441 (2007); H. Christopoulou-Aletra, A. Toglia, and C. Varlami, "The 'smart' Asclepieion: A total healing

environment," *Archives of Hellenic Medicine* 27(2):259-263 (2010); Bronwen L. Wickkiser, *Asklepios, Medicine and the Politics of Healing in Fifth-Century Greece: Between Craft and Cult*, Johns Hopkins University Press, Baltimore, 2008; and Jacques Jouanna, *Greek Medicine from Hippocrates to Galen*, translation by Neil Allies, Brill, Leiden, 2012.

149 **"that unfailing healer"**: In Homer, *The Iliad*, translated by Robert Fagles, Penguin, New York, 1990, book 4, line 220.

149 **"the craftsman of mild remedies"**: In Pindar, *The Odes*, translation by Andrew M. Miller, University of California Press, Oakland, 2019, p. 122.

149 **"put an end to grief"**: In *The Metamorphoses of Ovid*, translation by Allen Mandelbaum, Harvest/Harcourt, San Diego, 1993, p. 543.

150 **Arrayed on couches:** The use of psychedelics prior to incubation and the presence of free-ranging snakes in the *abaton* overnight is not universally accepted. The assertion about psychedelics appears in Christopoulou-Aletra et al., "The 'smart' Asclepieion," p. 259; another account of the temple sleep is in Jeffrey B. Pettis, *The Sleeper's Dream: Asclepius Ritual and Early Christian Discourse*, Gorgias Press, Piscataway, NJ, 2015, pp. 39, 43. Wickkiser believes "much of this is speculative," personal communication, June 15, 2024.

150 **"Snakes were encountered"**: Kerenyi, *Asklepios: Archetypal Image of the Physician's Existence*, p. 35.

151 **"In the case of Asklepios"**: Melina Melfi, Department of Archaeology, Oxford University, Oxford, UK, Zoom interview, April 19, 2023.

151 **A German archaeologist:** J. W. Riethmuller, *Asklepios: Heiligtumer und Kulte* (two volumes), Verily Archaologie und Geschichte, Heidelberg, 2005.

151 **"The cult became so popular"**: Wickkiser, *Asklepios, Medicine and the Politics of Healing*, p. 2; and Bronwen Wickkiser, Solomon Bluhm Professor of Ancient History, Hunter College, New York, Zoom interview, December 7, 2023.

151 **a curious, circular building:** Kerenyi, a colleague of Jung's, argued that the Thymele—the circular building at Epidauros—was the place where the snakes were kept. "I regard it as certain that the snakes were raised in the circular structures of the Asklepieia and that the form of these buildings was adapted to their habits." Kerenyi, *Asklepios: Archetypal Image of the Physician's Existence*, p. 104. There is, however, no scholarly consensus on this point.

152 **"the oldest and most sacred"**: Wickkiser, *Asklepios, Medicine and the Politics of Healing*, p. 78.

152 **"total healing environment"**: Christopoulou-Aletra et al., "The 'smart' Asclepieion," p. 259.

152 **"fourth-century promotional campaign"**: Lynn R. LiDonnici, *The Epidaurian Miracle Inscriptions: Text, Translation and Commentary*, Scholars Press, Atlanta, 1995, p. 1.

153 **"slept in the sanctuary"**: Kerenyi, *Asklepios: Archetypal Image of the Physician's Existence*, p. 41.

153 **a painful, debilitating abscess:** Kerenyi, *Asklepios: Archetypal Image of the Physician's Existence*, p. 33.

153 **"Terrified, she cried"**: Kerenyi, *Asklepios: Archetypal Image of the Physician's Existence*, p. 34. LiDonnici, *The Epidaurian Miracle Inscriptions*, has somewhat different translations of these testimonials, but calls these inscriptions "a campaign that clearly combined the concepts of bodily health, divine favor, and political wholeness and strength," p. 1.

153 **"Krito, we owe a cock"**: D. Kamen, "The manumission of Socrates," *Classical Antiquity* 32:78-100 (2013), p. 78.

153 **"It would take a new cult"**: Wickkiser, *Asklepios, Medicine and the Politics of Healing*, p. 2.

154 **"medical reptile"**: Wickkiser, *Asklepios, Medicine and the Politics of Healing*, quoted on p. 3.

154 **In October 1998:** Stephen M. Secor and Jared Diamond, "A vertebrate model of extreme physiological regulation," *Nature* 395:659-662 (1998).

157 **"absolutely the impetus"**: Leslie Leinwand, Department of Molecular, Cellular and Developmental Biology, University of Colorado, Boulder, telephone interview, July 25, 2013.

157 **Her group published an article:** Cecilia A. Riquelme et al., "Fatty acids identified in the Burmese python promote beneficial cardiac growth," *Science* 334:528-531 (2011).

158 **"And the day may come"**: Lawrence Altman, "Snakes' feat may inspire heart drugs," *New York Times*, October 27, 2011.

158 **"If I had realized how hard"**: L. Leinwand, interview, July 25, 2013. As of fall 2024, Hiberna had apparently not tested a potential pharmaceutical product of the python research.

158 **In October 2011:** T. A. Castoe et al., "Report from the First Snake Genomics and Integrative Biology Meeting," *Standards in Genomic Science* 7:1 (2012). The Consortium for Snake Genomics, led by Pollock, Castoe, and Secor, laid out the rationale for the python genome project in T. A. Castoe et al., "Sequencing the genome of the Burmese python (*Python molurus bivittatus*) as a model for studying extreme adaptations in snakes," *Genome Biology* 12:406 (2011).

159 **"nothing crazy, about average"**: Todd A. Castoe, Department of Biology, University of Texas–Arlington, Zoom interviews, September 7, 2022; June 22, 2023; and April 30, 2024. David Pollock did not respond to multiple interview requests.

163 **The first evidence emerged in 2013:** T. A. Castoe et al., "The Burmese python genome reveals the molecular basis for extreme adaptation in snakes," *Proceedings of the National Academy of Sciences* 110:20645-20650 (2013). The key follow-up papers were: A. L. Andrew et al., "Rapid changes in gene expression direct rapid shifts in intestinal form and function on the Burmese python after feeding," *Physiological Genomics* 47(5):147-157 (2015); A. L. Andrew et al., "Growth and stress response mechanisms underlying post-feeding regenerative organ growth in the Burmese python," *BMC Genomics* 18:338 (2017); B. W. Perry et al., "Multi-species comparisons of snakes identify coordinated signaling networks underlying post-feeding intestinal regeneration," *Proceedings of the Royal Society B* 286 (2019); A. K. Westfall et al., "Identification of an integrated stress and growth response signaling switch that directs vertebrate intestinal regeneration," *BMC Genomics* 23:6 (2022); and A. K. Westfall, et al., "Single-cell resolution of intestinal regeneration in pythons without crypts illuminates conserved vertebrate regenerative mechanisms," *PNAS* 2024 Oct. 22; 121(43)e2405463121.

164 **frenzied dissections:** For an account of the Tuscaloosa dissection procedure, see Carl Zimmer, "Eat rat, make new body: Easy stuff for pythons," *New York Times*, May 12, 2020.

170 **"An animal able to take"**: Clifford H. Pope, *The Giant Snakes*, Knopf, New York, 1973, p. 99. (The lone mention of "metabolism" is on p. 106.)

173 **molecule called NRF-2:** F. He et al., "NRF2, a transcription factor for stress response and beyond," *International Journal of Molecular Science* 21:4777 (2020).

175 **These cells are known as Best4+ cells:** T. Malonga et al., "BEST4+ cells in the intestinal epithelium," *American Journal of Physiology—Cell Physiology* 326:C1345-1352 (2024).

177 **this striking constellation:** According to a 2020 NIH-sponsored study of 9,710 patients with type 2 diabetes who had undergone bariatric surgery, roughly 85 percent had seen a remission in their diabetes at five years after surgery. See K. M. McTigue et al., "Comparing the 5-year diabetes outcomes of sleeve gastrectomy and gastric bypass: The National Patient-Centered Clinical Research Network (PCORNet) Bariatric Study," *JAMA Surgery*, May 1, 2020.

178 **"Todd Castoe is the most"**: J. Mendelson, Zoom interview, September 14, 2022.

180 **"Only the gods"**: B. Wickkiser, Zoom interview, December 7, 2023.

SNAKE ROAD: ISOLA TIBERINA, ROME, ITALY

181 **This is the only fragment:** For accounts of the Asklepian sanctuary on Isola Tiberina, see *The Metamorphoses of Ovid*, Allen Mandelbaum translation, pp. 538-543; Augustus J. C. Hare, *Walks in Rome*, Kegan Paul, Trench, Turner & Co., London, 1925, pp. 584-588; Georgina Masson, *The Companion Guide to Rome*, Collins, London, 1965, pp. 118-120 ("one of the most charming and picturesque places in the city"); Robert Hughes, *Rome: A Cultural, Visual, and Personal History*, Knopf, New York, 2011, p. 173; and H. V. Morton, *A Traveller in Rome*, Methuen, London, 1957, pp. 413-416.

181 **"polluted by a fatal":** *The Metamorphoses of Ovid*, Allen Mandelbaum translation, p. 538.

182 **"The Greek response":** *The Metamorphoses of Ovid*, Allen Mandelbaum translation, p. 539.

182 **"The god / lifted himself":** *The Metamorphoses of Ovid*, Allen Mandelbaum translation, p. 543.

182 **"put an end to grief":** *The Metamorphoses of Ovid*, Allen Mandelbaum translation, p. 543.

183 **To this day:** The two hospitals, Ospedale Fatebenefratelli and the Ospedale Israelitico Poliambulatorio, postdate a hospice for pilgrims during the Middle Ages; based on personal visits in April 2023 and June 2024.

183 **one of Caravaggio's greatest masterpieces:** For background on *Madonna dei Palafrenieri*, see the Galleria Borghese website, https://www.collezionegalleriaborghese.it/en/opere/madonna -and-child-with-saint-anne-madonna-dei-palafrenieri; accessed July 10, 2024.

5. THE WASABI CONNECTION

185 **The first thing Elena Gracheva noticed:** The descriptions of the Kingsville experiments and dissections are based on Elena Gracheva, Department of Cellular and Molecular Physiology, Yale University, New Haven, CT, Zoom interview, May 31, 2022, and personal interview, April 17, 2023; Elda Sanchez, director, National Natural Toxins Research Center, Kingsville, TX, Zoom interview, June 28, 2022; David Julius, Department of Physiology, University of California–San Francisco, Zoom interview, June 28, 2022; and Gunther Hollopeter, Department of Molecular Medicine, Cornell University, Ithaca, NY, Zoom interview, August 27, 2022, and telephone interview, May 18, 2024.

186 **Harry Greene mentioned a couple:** H. Greene, "Rattlesnakes, gravestones, and rural realities," *Mason County News*, August 11, 2021.

186 **Of all the satellite outposts:** For background on National Natural Toxins Research Center, see the NNTRC website at https://www.tamuk.edu/agriculture/institutes-and-other-units/nntrc /index.html; posted venom prices were accessed June 1, 2022.

188 **Snakes with pit organs have an ability:** E. O. Gracheva et al., "Molecular basis of infrared detection by snakes," *Nature* 464:1006-1011 (2010).

189 **life-threatening envenomations:** Klauber, *Rattlesnakes*, pp. 974-975. Klauber calls these "illegitimate bites."

190 **"inconvenient subjects":** Gracheva et al., "Molecular basis of infrared detection by snakes," p. 1007.

190 **"If you want new ideas":** Gordon M. Burghardt, "Insights found in century-old writings on animal behavior and some cautions for today," *Animal Behaviour* 164:241-249 (2020).

190 **"Our anthropocentric way":** Uexkull, quoted in Burghardt, *Animal Behaviour*, p. 243.

191 **Burghardt tested the feeding preferences:** G. M. Burghardt, "Chemical preference studies on newborn snakes of three sympatric species of *Natrix*," *Copeia* 1968:732-737 (1968).

191 **"Just still the most mind-blowing":** H. Greene, interview, March 2, 2023.

192 **this chemosensory processing center:** On Jacobson's organ, see H. T. Roper-Hall, "Jacobson's organ," *Proceedings of the Royal Society of Medicine* 38:41-50 (1944). Roper-Hall noted that the organ was "fully described by Jacobson (a Danish anatomist) in 1811; although Ruysch, in 1703, showed interest in it." A preeminent 18th-century anatomist known primarily for his cabinets of curiosity, Ruysch and his work were described in Mike Jay, "The melancholy of anatomy," *New York Review of Books*, December 8, 2022, pp. 38-40.

193 **two separate tines:** Described in H. B. Lillywhite, *How Snakes Work: Structure, Function and Behavior of the World's Snakes*, Oxford University Press, New York, 2014, p. 172.

193 **"Smelling in stereo":** Kurt Schwenk, "Smelling in stereo: The real reason snakes have flicking, forked tongues," *The Conversation*, June 17, 2021, https://theconversation.com/smelling-in-stereo-the-real-reason-snakes-have-flicking-forked-tongues-142363.

193 **"Snakes are the most chemically sensitive":** R. Mason, Zoom interview, June 26, 2023.

193 **"We now know":** Mason lab website, https://masonlab.ib.oregonstate.edu/robert-t-mason; accessed August 6, 2024. For a good general discussion of the role of semiochemicals in snake behavior, see R. T. Mason and M. R. Parker, "Social behavior and pheromonal communication in reptiles," *Journal of Comparative Physiology A* 196:729-749 (2010).

193 **"I really wanted to understand":** Mimi Halpern, Department of Anatomy and Cell Biology (retired), State University of New York–Downstate, Brooklyn, telephone interview, July 13, 2023; and John Kubie, Department of Cell Biology, SUNY-Downstate, Brooklyn, telephone interview, February 20, 2023. For a summary of this early work on the vomeronasal system, see T. G. Schulterbrandt et al., "Patterns of tongue-flicking by garter snakes (*Thamnophis sirtalis*) during presentation of chemicals under varying conditions," in *Chemical Signals in Vertebrates* 11, edited by J. L. Hurst et al., Springer, New York, 2008, pp. 345-356.

194 **"It wasn't high-level training":** J. Kubie, telephone interview, February 20, 2023.

195 **That idea received a harrowing test:** James B. Murphy and David E. Jacques, "Death from snakebite: The entwined histories of Grace Olive Wiley and Wesley H. Dickinson," *Bulletin of the Chicago Herpetological Society: Special Supplement*, 2006. Wiley tested the idea while operating a roadside reptile exhibit in California after losing her zoo job.

196 **"He really didn't bite me":** Murphy and Jacques, "Death from snakebite," p. 1.

196 **"kind of brilliant":** Robert Mason, Department of Integrative Biology, Oregon State University, Corvallis, Zoom interview, June 26, 2023.

196 **"I think it's entirely plausible":** Harry Greene, Zoom interview, March 3, 2023.

196 **Burghardt proposed "vomodor":** W. E. Cooper Jr. and G. M. Burghardt, "Vomerolfaction and vomodor," *Journal of Chemical Ecology* 16:103-105 (1990).

197 **The molecular basis of smell:** L. Buck and R. Axel, "A novel multigene family may encode odorant receptors: A molecular basis for odor recognition," *Cell* 65:175-187 (1991).

198 **If Julius's name is faintly familiar:** M. J. Caterina et al., "The capsaicin receptor: A heat-activated ion channel in the pain pathway," *Nature* 389:816-824 (1997).

198 **A few years later:** S. E. Jordt et al., "Mustard oils and cannabinoids excite sensory nerve fibres through the TRP channel ANKTM1," *Nature* 427:260-265 (2004).

199 **"My kid was little":** D. Julius, interview, June 28, 2022.

202 **a marvelous technique called RNA seq:** This new technology emerged in the mid-2000s; Jonathan Weissman, one of the leaders in its development, was a co-author on the pit organ paper.

205 **a kind of ecological thermostat:** A. R. Krochmal and G. S. Bakken, "Thermoregulation is the pits: Use of thermal regulation for retreat site selection by rattlesnakes," *Journal of Experimental Biology* 206:2539-2545 (2003).

206 **"the infrared and visual information":** H. Lillywhite, *How Snakes Work*, p. 175.

SNAKE ROAD: VIA ORTO MAGLIOCCO, COCULLO, ITALY

209 **the itinerant 10th-century Benedictine monk:** According to historical sources, San Domenico, born in 951 CE in Foligno (Umbria), established numerous hermitages in central Italy during the 10th and 11th centuries, and died in Sora (Lazio) in 1031. He stopped in Cocullo for seven years.

209 **the real action is around the corner:** The snake registration scene in the Cocullo municipal building and "Mostra Dei Serpenti" are based on a visit on April 30, 2023.

209 **cervone:** For names and characteristics of Italian snake species, I have used M.A.L. Zuffi, S. Scali, and E. Filippi, "I serpenti d'Italia: Specie, distribuzione, nova e acquisition recenti," *Gazzetta Ambiente*, 2016, pp. 9-20.

210 **has persisted, according to some experts:** This claim appears in J. A. Mendoza-Roldan et al., "Parasites and microorganisms associated with the snakes collected for the 'Festa Dei Serpari' in Cocullo, Italy," *PLoS Neglected Tropical Diseases* 18(2) (2024).

210 **The modern festival survives:** Century-old accounts of the Cocullo snake festival are remarkably similar to the present-day ceremony. See Estella Canziani, *Through the Apennines and the Lands of the Abruzzi: Landscape and Peasant Life*, Houghton Mifflin, Boston, 1928; and Linda Clarke Smith, "A survival of an ancient cult in the Abruzzi," in *Studi e Materiali di Storia delle Religioni*, volume 4, Anomina Romana Editoriale, Rome, 1928. See also Emanuele Pecoraro, "La festa dei separi a Cocullo tra antropologia, arte religione e mito," *Quaderni de Villa Sandra* 29:29-32 (2018). One of the best accounts of the festival, oddly enough, appears in a cookbook: Carol Fields's *Celebrating Italy: The Tastes and Traditions of Italy Revealed Through Its Feasts, Festivals, and Sumptuous Foods*, Morrow, New York, 1990, pp. 21-35. Fields includes a local recipe from Cocullo for Cervone, a rum-and-spice cake in the shape of a snake.

211 **"It's so cool":** Nicole Szafranski, graduate student, University of Tennessee, interview, April 30, 2024.

212 **"So there was a problem":** Gianpaolo Montinaro, Field Study Department, Regulatory Affairs Infobrokerage Faunistics, Hirschberg, Germany, Zoom interview, May 23, 2023.

212 **The event had been the academic fief:** The most important anthropological study is Alfonso di Nola, *Gli aspetti magico-religiosi di una cultura subalterna italiana*, Boringhieri, Torino, 2001.

213 **"The Marsi were a hardy":** *Encyclopaedia Britannica*, 11th edition, volume 17, p. 774. The Britannica entry also includes a brief history of Angizia.

213 **Chief among those magicians:** See Pecoraro, "La festa dei separi a Cocullo tra antropologia," for background on Angizia.

215 **"I will give you":** Gianpaolo Montinaro, May 1, 2023.

216 **As they have done for at least a century:** Archival black-and-white film of the 1928 Cocullo snake festival can be viewed at https://patrimonio.archivioluce.com/luce-web/detail /IL3000051107/1/-457.html.

216 **"San Domenico did not choose":** Claudio Corvino, personal interview, May 1, 2023, and personal communication, July 15, 2024.

216 **That was the theory of Alfonso di Nola:** di Nola, *Gli aspetti magico-religiosi di una cultura subalterna italiana*, concedes that "there must certainly have existed in the Abruzzo region a permanent tradition of anti-ophidic magic linked to the clan of Marsa," but he suggests that by the 17th and 18th centuries, these traditions took the form of itinerant snake charmers, which in turn created a rift between the ecclesiastical authorities, who continued to demonize serpents, and the rural, subordinate class whose reverence for snakes persisted despite what di Nola called "the accusation of imposture and charlatanism," p. 144.

217 **geophysicists from the Max Planck Institute:** Gianpaolo Montinaro, personal communication, August 23, 2024.

218 **"generate conservation policies":** Mendoza-Roldan et al., "Parasites and microorganisms associated with the snakes."

6. THE EVOLUTION OF PLEASURE

219 **"Good morning, Carmen!":** Patricia Brennan, associate professor, Mount Holyoke College, Mount Holyoke, MA, personal interview, April 18, 2023. Unless otherwise noted, all Brennan quotes are from this interview. For background on Brennan's career, in addition to the interview, I have also consulted "Patricia Brennan, C.V.," and "About Me" at www.pattybrennan .com; and C. Fitzgerald, "Mount Holyoke professor Patty Brennan receives lifetime honor for genital morphology research," *Mount Holyoke News*, February 24, 2023.

219 **shifty:** P. Brennan, email to author, July 16, 2023.

220 **"Oh, I should have asked":** Rachel Keeffe, postdoctoral fellow, Mount Holyoke College, personal interview, April 18, 2023.

221 **The procedure resembled:** On Plaster Caster history, see Randall Roberts, "Cynthia Plaster Caster, artist known for rock-star penis sculptures, dies at 74," *Los Angeles Times*, April 22, 2022. In addition to Jimi Hendrix, Plaster Caster reportedly took casts of members of the Beach Boys, the Kinks, the Lovin' Spoonful, and many others.

222 **Brennan was one of four:** Megan J. Folwell, Kate L. Sanders, Patricia L. R. Brennan, and Jenna M. Crowe-Riddell, "First evidence of hemiclitoris in snakes," *Proceedings of the Royal Society B* 289, December 21, 2022.

222 **"scientifically unstoppable":** Richard O. Prum, *The Evolution of Beauty: How Darwin's Forgotten Theory of Mate Choice Shapes the Animal World—and Us*, Doubleday, New York, 2017, p. 162.

223 **snake sex is a little bizarre:** Intersexuality in snakes has been known for at least half a century; see Laurence M. Hardy, "Intersexuality in a Mexican colubrid snake (*Pseudoficimia*)," *Herpetologica* 26:336-343 (1970). On the ability of male snakes to emit female pheromones, see R. Shine et al., "Facultative pheromonal mimicry in snakes: 'She-males' attract courtship only when it is useful," *Behavioral Ecology and Sociobiology* 66:691-695 (2012).

225 **One day early in the summer of 2021:** Based on a timeline of the project provided by Folwell, personal communication, June 29, 2023.

225 **many of the claims left her "confused":** Megan Folwell, PhD student, University of Adelaide, Australia, Zoom interview, June 12, 2023.

226 **The volume is a compendium:** J. Sean Doody, Vladimir Dinets, and Gordon M. Burghardt, *The Secret Social Lives of Reptiles*, Johns Hopkins University Press, Baltimore, 2021.

226 **Male and female cottonmouths:** H. B. Lillywhite and C. M. Sheehy III, "The unique insular population of cottonmouth snakes at Seahorse Key," in *Islands and Snakes: Isolation and Adaptive Evolution*, edited by Harvey B. Lillywhite, Oxford University Press, Oxford, UK, and New York, 2019, pp. 227-230.

226 **physically moved male rattlesnakes:** M. L. Holding et al., "Experimentally altered navigational demands induce changes in the cortical forebrain of free-ranging northern Pacific rattlesnakes (*Crotalus o. oreganus*)," *Brain, Behavior and Evolution* 79:144-154 (2012).

227 **YouTube snake porn:** There are dozens of videos under "snakes copulating." On Escher's last woodcut, see "Snakes," Escher in het Paleis, https://www.escherinhetpaleis.nl/showpiece /snakes/?lang=en.

227 **"They're copperheads":** "Copperheads mating," YouTube, https://www.youtube.com
/watch?v=texmLWBpumI; accessed June 19, 2022 (no longer available).

227 **Greene wrote of a pair of western diamondbacks:** Greene, *Snakes: The Evolution of Mystery in
Nature*, p. 127.

227 **In a facetiously titled paper:** R. Shine et al., "Are snakes right-handed? Asymmetry in
hemipenis size and usage in garter snakes (*Thamnophis sirtalis*)," *Behavioral Ecology* 11:411-415
(2000). Among the findings reported in this paper were that the hemipenis on the right side
of the snake's body was larger, and produced a larger copulatory plug, than the left side, which
likely improved odds of fertilization.

228 **Snake sex can be downright:** Brahminy blind snake, Lillywhite, *How Snakes Work*, p. 195.

228 **University of Florida herpetologist:** Lillywhite, *How Snakes Work*, p. 208.

229 **one of her first scientific papers:** R. L. Pitman et al., "Sightings and possible identity of a
bottlenose whale in the tropical Indopacific: *Indopacetus pacificus*?," *Marine Mammal Science*
15:531-549 (1999).

230 **an ancient species of neotropical bird:** P. Brennan, "About Me," http://www.pattybrennan
.com/aboutme; accessed March 29, 2023. In describing how she developed her interest in
sexual selection and evolution during this PhD project, she wrote, "I also witnessed tinamou
mating in the field and saw my first bird penis…Since I was not even aware that birds had
penises, I decided to investigate this further for my post-doctoral project. As they say, the rest
is history."

231 **She went to a farm near Sheffield:** P.L.R. Brennan et al., "Coevolution of male and female
genetical morphology in waterfowl," *PLoS One* 2(5) (May 2007). For typical media coverage,
see B. J. King, "Ducks do it differently, and science wants you to know about it," National Public Radio, July 10, 2014; and J. Thomson, "The secret sex lives of ducks," *Salon*, July 25, 2021,
https://www.salon.com/2021/07/25/the-secret-sex-lives-of-ducks.

231 **Using a high-speed video camera:** P.L.R. Brennan, C. Clark, and R. Prum, "Explosive eversion and functional morphology of waterfowl supports sexual conflict in genitalia," *Proceedings
of the Royal Society B* 277:1309-1314 (2010). An eleven-second video on YouTube has been
viewed 2.9 million times, https://www.youtube.com/watch?v=qwjEeI2SmiU; accessed July 15,
2024.

232 **"Don't we really need":** The Hannity-Carlson exchange is recounted in Prum, *The Evolution
of Beauty*, pp. 174-177, and R. O. Prum, "Duck sex and the patriarchy," *New Yorker*, May 17,
2017.

232 **she refused to back down:** P. Brennan, "Why I study duck genitalia," *Slate*, April 2, 2013; and
P.L.R. Brennan, R. W. Clark, and D. W. Mock, "Time to step up: Defending basic science and
animal behaviour," *Animal Behaviour* 94:101-105 (2014).

233 **"She was in my ear":** Christopher Friesen, School of Earth, Atmospheric and Life Sciences,
University of Wollongong, Australia, personal communications, July 19, 2023, and August 16,
2024.

234 **The frigid prairie of central Manitoba:** The key papers in the garter snake research are C. R.
Friesen et al., "Sexual conflict over mating in red-sided garter snakes (*Thamnophis sirtalis*)
as indicated by experimental manipulation of genitalia," *Proceedings of the Royal Society B*
281:1774 (2013); and C. R. Friesen et al., "Female behavior and the interaction of male and
female genital traits mediate sperm transfer during mating," *Journal of Evolutionary Biology*
29:952-964 (2016). For background on the Manitoba garter snake site, in addition to the
Brennan interview, I consulted Robert T. Mason, professor, Oregon State University, Corvallis,

Zoom interviews, June 26, 2023, and September 21, 2023; Rick Shine, professor, Macquarie University, Australia, Zoom interview, December 13, 2023; and M. Rockwell Parker, Biology Department, James Madison University, Harrisonburg, VA, Zoom interview, August 15, 2023.

236 **"small males":** Friesen et al., "Sexual conflict over mating in red-sided garter snakes."

238 **an obscure medical paper:** P. B. Pendergrass et al., "A technique for vaginal casting utilizing vinyl polysiloxane dental impression material," *Gynecologic and Obstetric Investigation* 32:121-122 (1991). The difficulties Pendergrass faced are recounted in Rose Eveleth's reappraisal, "The failed vagina story," in *The Last Word on Nothing*, July 28, 2016, https://www.lastwordonnothing.com/2016/07/28/the-failed-vagina-story; accessed July 15, 2024.

239 **In an early test of the technique:** D. Orbach et al., "Biomechanical properties of female dolphin reproductive tissue," *Acta Biomaterialia* 86:117-124 (2019); and P.L.R. Brennan et al., "Functional morphology of the dolphin clitoris," *Current Biology* 31:R1-R3 (2022).

240 **The 3-D shapes began to tell:** J. F. Greenwood et al., "Divergent Genital Morphologies and Female-Male Covariation in Watersnakes," symposium at the annual meeting of the Society for Integrative and Comparative Biology, January 6, 2022.

241 **"disagree on the details":** Fitzgerald, "Mount Holyoke professor Patty Brennan."

242 **It is not exactly clear:** On early descriptions of snake hemipenes, see H. Gadow, "Remarks on the cloaca and on the copulatory organs of the Amniota," *Proceedings of the Royal Society of London* 40:266-267 (1886); E. D. Cope, "Prodomus of a new system of the non-venomous snakes," *The American Naturalist* 27:477-484 (1893); and E. D. Cope, "On the hemipenes of the Sauria," *Proceedings of the Academy of Natural Sciences of Philadelphia* 48:461-467 (1896).

242 **"Morphologie Der Hemipenes":** Use of the term *hemipenis* in an 1833 publication was cited by C. Schmidt in a 2000 book (M. Folwell, personal communication, January 23, 2024).

242 **"brilliant attempt":** H. G. Dowling and J. M. Savage, "A guide to the snake hemipenis: A survey of basic structure and systematic characteristics," *Zoologica: Scientific Contributions of the New York Zoological Society* 45:17-28 (1960).

242 **"snakes are *not* straightforward":** M. Folwell, Zoom interview, June 12, 2023.

243 **"conspicuously overlooked":** M. Folwell, K. Sanders, and J. Crowe-Riddell, "The squamate clitoris: A review and directions for future research," *Integrative and Comparative Biology* 62:559-568 (2022). As Folwell et al. tartly observed, "Hemipenes have piqued the interest of reproductive, morphological, and evolutionary researchers for centuries, and provided many insights into the mating systems of squamates but this is only half the story."

243 **no less than 10 mistaken:** Folwell et al., "The squamate clitoris."

244 **The variety of snake species native to Australia:** Although neither Folwell nor Brennan mentioned it in their papers, the enormous mythological importance that snakes have among Australia's aboriginal population also has some morphological parallels. In the spectacularly beautiful aboriginal "dream paintings," for example, snakes often represented sexual danger and were depicted encircling or devouring women. According to one myth that surprisingly echoes the emerging snake genitalia story, the so-called Snake Man Yirrbardbard possessed a barbed penis, transformed himself into a serpent, and fatally bit a woman who refused his advances. *Dreamings: The Art of Aboriginal Australia*, edited by Peter Sutton, Asia Society Galleries and George Braziller, New York, 1988, p. 45.

245 **For decades, vertebrate biologists:** M. Folwell, personal communication, January 23, 2024.

248 **"A lot of guy herpetologists":** H. Greene, Zoom interview, June 17, 2024.

248 **"People were basically saying":** P. Brennan, Zoom interview, August 23, 2024.

248 **The internet blew up:** For media accounts of the hemiclitoris research, see K. J. Wu, "Surprise!

Snakes have clitorises," *The Atlantic*, December 13, 2022; and Alex Fox, "Scientists overlooked the snake clitoris, until now," *New York Times*, December 13, 2022, https://www.nytimes .com/2022/12/13/science/snakes-clitoris-hemiclitores.html.

SNAKE ROAD: FIFTH AVENUE, NEW YORK CITY

250 **When a dissolute young prince:** Sources for the Gautama narrative include John S. Strong, *The Experience of Buddhism: Sources and Interpretations*, 2nd edition, Wadsworth, Belmont, CA, 2002, pp. 3-18; Karl Jaspers, *Socrates, Buddha, Confucius, Jesus,* Harcourt, Brace, San Diego, 1962; and Karen Armstrong, *Buddha,* Viking Penguin, New York, 2001. The 60,000 wives, seven walls, seven moats, 500 guards, and ringing bells come from an ancient Sanskrit text recounting the "Great Departure" reproduced in Strong, *The Experience of Buddhism*, pp. 9-12.

251 **according to an ancient sutta:** For the Mucalinda myth, see "Muccalinda Sutta: About Muccalinda," www.ancient-buddhist-texts.net; accessed November 13, 2023. For a recent interpretation of this sutta and the Mucalinda story, see J. Johns and J. R. Nag, "Muchalinda Buddha: An interdisciplinary approach to reinterpret the depiction of the Buddha with Muchalinda Naga," *Journal of Archaeological Studies in India* 1:140-157 (2021).

252 **"a permission-giving faith":** Holland Cotter, "Buddhist art from India: Where the natural meets the supernatural," *New York Times*, July 21, 2023, p. C-1.

252 **That sandstone sculpture of Mucalinda:** John Guy, *Tree & Serpent: Early Buddhist Art in India*, Metropolitan Museum of Art / Yale University Press, New Haven, CT, 2023, pp. 29-30. Snake deities were not solely a feature of Buddhist art; in a 2019 essay about serpent symbols in medieval India, Guy made the point that "by the 1st–2nd century CE, male and female personified *naga* deities, identified by their arching snake-hood canopies, became an established presence in the divine repertoire of all Indic faiths, Brahmanism, Jainism and Buddhism." John Guy, "Snakes, crocodiles and lizards: Protective goddesses in medieval India," in *Indology's Pulse: Arts in Context*, edited by C. Wessels-Mevissen and G.J.R. Mevissen, Aryan Books International, New Delhi, 2019.

253 **"Snake (*naga*) shrines":** Exhibition sign, *Tree & Serpent: Early Buddhist Art in India, 200 BCE– 400 CE*, Metropolitan Museum of Art, New York, July 21–November 13, 2023.

253 **In a piece of limestone sculpture:** Guy, *Tree & Serpent*, pp. 33-35.

253 **"Two iconographic devices":** Guy, *Tree & Serpent*, p. 8.

253 **"The teachings of the Buddha":** Exhibition signage, *Tree & Serpent*.

254 **"the three origins of all religions":** Burghardt, interview, January 18, 2023. Forlong's 1883 opus *Rivers of Life* traces the history of human religious practice from 10,000 years ago to the end of the 19th century.

254 **"Scientists believe that this tolerance":** BBC, *Planet Earth III*, "A highly venomous cobra hunts a toad through a village in India," https://www.bbc.co.uk/programmes/p0gwn0by; accessed July 16, 2024. I am grateful to Rick Shine for bringing this report to my attention.

7. NO LEGS? NO PROBLEM

256 **In the spring of 2014:** H. Marvi et al., "Sidewinding with minimal slip: Snake and robot ascent of sandy slopes," *Science* 346:224-229 (2014).

257 **"if you look too long":** Daniel Goldman, quoted in E. Pennisi, "Sidewinder robots slither like snakes," *Science*, October 9, 2014.

257 **But the initial drama:** Background on the 2014 experiment comes from H. Marvi, School for Engineering of Matter, Transport and Energy, Arizona State University, Tempe, Zoom interview, July 28, 2022; Joseph Mendelson III, director of research, Zoo Atlanta, Zoom interview, September 14, 2022; Howie Choset, co-director, Robotics Institute, Carnegie Mellon University, Pittsburgh, Zoom interview, August 5, 2022, and telephone interview, January 3, 2024; Daniel Goldman, Department of Physics, Georgia Institute of Technology, Atlanta, Zoom interview, November 22, 2023; and Henry Astley, Department of Biology, University of Akron, Zoom interview, August 29, 2022, and personal interview, October 31, 2023.

258 **the epic failure:** E. Pennisi, "Sidewinder robots slither like snakes"; E. Yong, "Cave-exploring snake robot gets inspiration from sidewinders," *National Geographic*, October 9, 2014, https://www.nationalgeographic.com/science/article/cave-exploring-snake-robot-gets-inspiration-from-sidewinders; accessed August 15, 2024; and B. Dorminey, "Robotic snakes slither their way into ancient archaeology," *Forbes*, September 2013.

259 **"Failing miserably":** J. J. Socha, "Of snakes and robots," *Science* 346:160-161 (2014), p. 160.

259 **"Their limbless robots were *terrible*":** D. Goldman, telephone interview, November 23, 2023.

260 **"Having no limbs":** Socha, "Of snakes and robots," p. 160.

261 **11 distinct and unique:** Bruce C. Jayne, "What defines different modes of snake locomotion?," *Integrative and Comparative Biology* 60:156-170 (2020).

261 **One of those modes:** As Jake Socha points out, "slithering" is not a technical term. "I think that most people consider any form of a snake moving on the ground (or even in the air) as 'slithering,'" he says (J. Socha, personal communication, August 21, 2024).

261 **"theoretical study":** D. L. Hu et al., "The mechanics of slithering locomotion," *Proceedings of the National Academy of Sciences* 106:10081-10085 (2009).

262 **roughly 85 million years ago:** Hongyu Yi, "How snakes came to slither," *Scientific American* 318:70-75 (2017).

262 **You can still see vestigial:** Hayley Crowell, University of Michigan, personal interview, October 31, 2023.

263 **"I think one way":** Bruce Jayne, Department of Biological Sciences, University of Cincinnati, Zoom interview, July 25, 2022. Jayne's YouTube channel (https://www.youtube.com/@jaynebc1/videos) features many examples of snake locomotion.

264 **"soft machine":** Shigeo Hirose, oral history conducted by Selma Sabanovic and Matthew R. Francisco, IEEE:https://ethworg/Oral-History:Shigeo_Hirose. The ETHW has a collection of more than 800 oral histories in electrical and computer technology which can be accessed via http:ethw.org/Oral History:List_of_all_Oral_Histories.

265 **"snake dishes":** The restaurant story is in "Shigeo Hirose—relentless passion for the creation of robots," interview with Tokyo Institute of Technology, March 2013, https://www.titech.ac.jp/english/public-relations/research/stories/shigeo-hirose; accessed July 17, 2024.

265 **Snake Center Cafe:** *Time Out*, March 28, 2023, https://www.timeout.com/tokyo/restaurants/tokyo-snake-center; accessed July 17, 2024.

266 **"What makes them interesting":** H. Choset, interview, August 5, 2022.

268 **"properly coordinated sequence":** B. Chong et al., "Coordinating tiny limbs and long bodies: Geometric mechanics of terrestrial swimming," *PNAS* 119: (2022).

272 **On September 19, 2017:** This account of the Mexico City earthquake expedition is based on J. Whitman, N. Zevallos, M. Travers, and H. Choset, "Snake robot urban search after the Mexico City earthquake," IEEE International Symposium on Safety, Security, and Rescue Robotics (SSRR), 2018; M. Hutson, "Searching for survivors of the Mexico earthquake—with snake robots," *Science*, October 4, 2017; "Carnegie Mellon snake robot used in search for Mexico quake

survivors," Carnegie Mellon University press release, September 27, 2017, https://www.cmu .edu/news/stories/archives/2017/september/snakebot-mexico.html; accessed July 19, 2024; E. Ackerman, "What CMU's snake robot team learned while searching for Mexican earthquake survivors," *IEEE Spectrum*, October 13, 2017; and Nico Zevallos, personal communication, February 23, 2023..

273 **"which could not contain survivors":** J. Whitman et al., "Snake robot urban search after the Mexico City earthquake."

273 **"We waited for":** J. Whitman et al., "Snake robot urban search after the Mexico City earthquake."

274 **"We need to do":** H. Choset, quoted in Hutson, "Searching for survivors of the Mexico earthquake."

274 **Roughly 1,500 heart disease:** Choset, interview, 2022.

274 **a tiny, snake-inspired robot:** For details on the surgical snake-bot, see E. Strickland, "The future of robotic surgery: Snake-like bots that glide into orifices," *IEEE Spectrum*, July 26, 2016.

274 **But the company lost:** S. Whooley, "Medrobotics is selling off assets amid Chapter 7 bankruptcy," *MassDevice*, May 7, 2024, https://www.massdevice.com/medrobotics-surgical-robotics -ip-for-sale; accessed August 8, 2024.

274 **"Is it a success":** H. Choset, interview, 2022.

275 **"on its last legs":** H. Choset, quoted in a Carnegie Mellon press release, 2017.

275 **how the scales added a microscopically small:** J. M. Rieser et al., "Functional consequences of convergently evolved microscopic skin features on snake locomotion," *PNAS* 118 (2021).

275 **How the snakes simplified:** H. C. Astley et al., "Surprising simplicities and syntheses in limbless self-propulsion in sand," *Journal of Experimental Biology* 223 (2020).

276 **perfectly adapted?:** B. Chong et al., "Moving sidewinding forward: Optimizing contact patterns for limbless robots via geometric mechanics," Robotics: Science and Systems (conference), July 12–16, 2021.

277 **"She discovered an absolutely gorgeous":** P. E. Schiebel et al., "Mechanical diffraction reveals the role of passive dynamics in a slithering snake," *Proceedings of the National Academy of Sciences* 116:4798-4803 (2019).

277 **"I noticed that it was often":** Henry Astley, personal communication, February 10, 2024.

278 **"It really is a fully":** H. Astley, personal interview, October 31, 2023.

SNAKE ROAD: "B ROAD," NEAR AIKEN, SOUTH CAROLINA

279 **The experiment, conceived by:** J. Whitfield Gibbons, professor emeritus, University of Georgia, Zoom interview, July 26, 2022.

279 **Savannah River Site:** Savannah River Site website, https://www.srs.gov/general/news/fact sheets/srs_overview.pdf; accessed August 15, 2024.

279 **Why does a snake cross:** K. M. Andrews and J. W. Gibbons, "How do highways influence snake movement? Behavioral responses to roads and vehicles," *Copeia* 2005(4):772-782 (2005).

279 **known as road ecology:** *Roads and Ecological Infrastructure: Concepts and Applications for Small Animals*, edited by S. P. Riley, P. Nanjappa, and K. M. Andrews, Johns Hopkins University Press, Baltimore, 2015. A recent general-interest history of the field is Ben Goldfarb, *Crossings: How Road Ecology Is Shaping the Future of Our Planet*, W.W. Norton, New York, 2023; Goldfarb suggests that road ecology as a discipline had its origins only in the mid-1990s, p. 5.

Sources and Notes

282 **Roads even altered the trajectory:** L. M. Klauber, "Night collecting on the desert with ecological statistics," *Bulletin of the Zoological Society of San Diego* 14:7-64 (1939).

282 **"new technique":** Klauber, "Night collecting on the desert," p. 8. Klauber reported that he began these systematic collections in 1927 and extended them during nighttime drives as a power company executive to inspect the Hoover Dam. L. M. Klauber, "Amphibians and reptiles observed enroute to Hoover Dam," *Copeia* 1932:118-128 (1932).

282 **A study that came out:** T. K. Miller et al., "Wildlife rehabilitation records reveal impacts of anthropogenic activities on wildlife health," *Biological Conservation*, 2023, https://doi .org/10.1016/j.biocon.2023.110295. Vehicle collisions were the main cause of injury, according to the study.

282 **"Reptiles suffered the highest":** E. Blakemore, "Humans have 'large, negative impact on wildlife,' researchers find," *Washington Post*, December 10, 2023.

283 **In Australia, Rick Shine uncovered:** R. Shine, interview, December 13, 2023; and G. P. Brown et al., "The ecological impact of invasive cane toads on tropical snakes: Field data do not support laboratory-based predictions," *Ecology* 92:422-431 (2011).

284 **notably including the collaboration:** R. Shine et al., "Why did the snake cross the road? Effects of roads on movement and location of mates by garter snakes (*Thamnophis sirtalis parietalis*)," *Ecology and Society* 9 (2004).

284 **some of the most hospitable and protected:** For a summary of Department of Defense conservation efforts on behalf of snakes, see R. E. Lovich, C. Petersen, and A. Dalsimer, "Department of Defense Natural Resources Program. Strategic Plan for Amphibian and Reptile Conservation and Management on Department of Defense Lands" (2015); and DoD Fact Sheet, https://www.denix.osd.mil/dodparc/denix-files/sites/36/2023/01/DoD-PARC-Fact -Sheet_2020_Final_508-1-1.pdf; accessed July 20, 2024.

284 **amount to 26 million acres:** S. Sicard, "How much land does the military really own?" *Military Times*, August 15, 2022.

284 **"more rare, threatened, and endangered":** Lovich et al., "Department of Defense Natural Resources Program," p. 6.

284 **Sikes Act:** Enacted in 1960, the Sikes Act "directs the Secretary of Defense, in cooperation with the U.S. Fish and Wildlife Service and state fish and wildlife agencies, to carry out a program for the conservation and rehabilitation of natural resources on military installations." US Fish and Wildlife Service website, https://www.fws.gov/law/sikes-act.

285 **"We unwittingly became the stewards":** Rob Lovich, co-director, Partners in Amphibian and Reptile Conservation (PARC), Department of Defense, Zoom interview, August 7, 2023.

285 **While monitoring a population:** S. Tetzlaff et al., "First report of snake fungal disease from Michigan, USA involving massasaugas, *Sistrurus catenatus* (Rafinesque 1818)," *Herpetology Notes* 8:31-33 (2015). By 2020, a survey of 657 snakes tested in 56 military bases in 31 states revealed that 17.2 percent of the serpents tested positive for snake fungal disease. M. C. Allender et al., "Ophidiomycosis, an emerging fungal disease of snakes: Targeted surveillance on military lands and detection in the western US and Puerto Rico," *PLoS One* 15(10):e0240415 (2020).

285 **severe wasting and desiccation:** Matthew Allender, College of Veterinary Medicine, University of Illinois, Urbana, Zoom interview, January 12, 2024.

286 **In 1923, the industrial barons:** S. Glassman, "Blazing the Tamiami Trail," *South Florida History Magazine*, Winter 1989, pp. 3-5, 12-13.

286 **roughly a million Ford Model T's:** Historical vehicle and mileage statistics are from Richard

381

F. Weingroff, "Promoting the road during a war," US Department of Transportation, https://highways.dot.gov/sites/fhwa.dot.gov/files/not3.pdf; accessed July 20, 2024. Current mileage and percent of paved/unpaved are from "Public road length—2020," Federal Highway Administration, US Department of Transportation, October 26, 2021, https://www.fhwa.dot.gov/policyinformation/statistics/2020/pdf/hm16.pdf; accessed July 20, 2024. As of December 2022, the Federal Highway Administration reported that there were 283,400,986 registered private and commercial vehicles in the US ("United States number of registered vehicles," CEIC Data, https://www.ceicdata.com/en/indicator/united-states/number-of-registered-vehicles; accessed August 8, 2024.

286 **"animal-vehicle conflicts":** Thomas E. S. Langton, "A history of small animal road ecology," in Riley et al. (eds.), *Roads and Ecological Infrastructure*, p. 8. Apropos of snakes, Langton pointed out that as early as the 1980s, Manitoba installed road warning signs for red-sided garter snakes, "which are killed in large numbers on roads following mass emergence from hibernation dens," p. 9.

287 **The Tamiami Trail might be Exhibit A:** Antonia Florio, "Removing the cork in the bottle: Reconstructing Tamiami Trail to restore water flow to Everglades National Park," https://www.nps.gov/articles/000/removing-the-cork-in-the-bottle-reconstructing-tamiami-trail-to-restore-water-flow-to-everglades-national-park.htm; accessed August 15, 2024. Even this National Park Service article stated that construction of the highway was "disastrous for the Everglades ecosytem."

288 **"It's also clearly true":** R. Shine, interview, December 13, 2023.

8. THE PYTHON QUEEN OF SOUTH FLORIDA

290 **The irony of searching:** Donna Kalil, contract hunter, South Florida Water Management District, personal interview, January 26, 2023 (Florida Everglades), and January 27, 2023 (Holey Land / Rotenberger Wildlife Refuge Areas). Also accompanying and commenting on these expeditions were Marc Rodriguez (contract hunter) and Tanya Toutant (volunteer).

290 **Tamiami Trail:** S. Glassman, "Blazing the Tamiami Trail," *South Florida History Magazine*, Winter 1989, pp. 3-5, 12-13, is a splendid account of the road's early history, including a fascinating snake anecdote directly related to a pivotal early event in the highway's construction. In 1923, a convoy of mostly Ford Model T sedans—saluted upon their departure from Fort Myers by Henry Ford himself and Thomas Edison—set out to drive the unpaved east–west portion of the route on what they thought would be at most a four-day trip to Miami. Mired in mud and running out of food after six days on the barely passable trail, the convoy's two Native American guides threatened to abandon the group in the middle of the Everglades because one of the motorists had killed two snakes in a pond near Turner River Strand. The pond teemed with dozens of water moccasins, according to Glassman's account, and "both the Indians had threatened to abandon the expedition because of the wanton killing. Finally, Indian agent Hanson convinced the Miccosukees to stay, and Conapatchee [one of the guides] told the whites that if they minded their business, the snakes would mind theirs. To prove his assertion, he sashayed barefoot through the throng of snakes. The moccasins…merely wiggled out of his way," p. 5.

291 **had already yanked 735 pythons:** As of end of August 2024, the number was 992. "I am striving to get at least 1001 so I can say I caught more than a thousand pythons, and then I will go for 2000!," D. Kalil, personal communication, August 28, 2024.

291 **Python-Hunting Queen:** For profiles of Kalil, see Allie Conti, "Meet Donna Kalil: The python-hunting queen of South Florida," *Field & Stream*, February 8, 2021; Brittany Shammas, "One of Florida's only female python hunters risks it all for the Everglades," *Miami New Times*, July 10, 2018; and Susanne Rust, "Meet the women hunting giant pythons 'eating everything' in the Everglades," *Los Angeles Times*, September 12, 2022.

291 **A famous 2005 photograph:** C. Morgan, "It's Alien versus Predator in Glades creature clash," *Miami Herald*, October 5, 2005. The 13-foot snake and the 6-foot alligator were, according to Morgan, "locked so gruesomely it is hard to make head, tails or another other body part out of either."

292 **"To spot a python":** Michael Kirkland Sr., invasive animal biologist, South Florida Water Management District, West Palm Beach, Zoom interview, January 19, 2023.

292 **"This woman is incredible":** Marc Rodriguez, contract hunter, South Florida Water Management District, interview, January 26, 2023.

293 **In a 2011 book:** Michael E. Dorcas and John D. Willson, *Invasive Pythons in the United States: Ecology of an Introduced Predator*, University of Georgia Press, Athens, GA, and London, 2011.

293 **"may very well number":** Dorcas and Willson, *Invasive Pythons in the United States*, p. 112.

293 **a little-known experiment:** Dorcas and Willson, *Invasive Pythons in the United States*, p. 106. A subsequent, more detailed account appears in M. E. Dorcas and John D. Willson, "Hidden giants: Problems associated with studying secretive invasive pythons," in *Reptiles in Research*, edited by William I. Lutterschmidt, Nova Science Publishers, Hauppauge, NY, 2013.

294 **"Many areas of South Florida":** Dorcas and Willson, "Hidden giants," p. 380.

296 **The South Carolina enclosure:** The Florida Fish and Wildlife Commission reported that there have been 20,467 pythons removed since the year 2000; of that total, 13,000 have been the result of contract hunting begun in 2017. M. Kirkland, personal communication, December 7, 2023.

297 **The full origin story:** Jacquelyn C. Guzy et al., "Burmese pythons in Florida: A synthesis of biology, impacts, and management tools," *NeoBiota* 80:1-119 (2023), hereafter "Guzy et al." This is the most comprehensive recent overview of the python situation in Florida; among the co-authors of the report also mentioned in this chapter are John David Willson, Robert Reed, Ian Bartoszek, Margaret Hunter, Frank Mazzotti, Melissa Miller, and M. Rockwell Parker. Other useful overviews include Rebecca G. Harvey et al., "Burmese Pythons in South Florida: Scientific support for invasive species management," Institute of Food and Agricultural Sciences, University of Florida, IFAS Publication WEC-242span, July 2008; and Dorcas and Willson, *Invasive Pythons in the United States*.

298 **"most plausible" scenario:** Guzy et al., p. 34.

298 ***Tampa Daily Times* reported:** Guzy et al., p. 32.

298 **Nonetheless, when Burmese pythons were observed:** W. E. Meshaka Jr., W. F. Loftus, and T. Steiner, "The herpetofauna of Everglades National Park," *Florida Scientist* 63(2):84-103 (2000).

298 **"These observations suggest":** Guzy et al., p. 33.

299 **at least 294,000 Burmese pythons:** Guzy et al., p. 32.

299 **"It wasn't an act of God":** Frank J. Mazzotti, Department of Wildlife Ecology and Conservation, Fort Lauderdale Research and Education Center, University of Florida, telephone interview, July 8, 2022.

299 **There is a persistent suggestion:** On media stories suggesting that the explosion of the Florida python population came in the aftermath of Hurricane Andrew, see for example "Swamp things," *New Yorker*, April 13, 2009, https://www.newyorker.com/magazine/2009/04/20/swamp-things; accessed August 8, 2024.

300 **outright conspiracy theories:** Dorcas and Willson, *Invasive Pythons in the United States*,

p. 64. On the European conspiracy theory about government dissemination of vipers, Riccardo Scalera, personal communication, December 19, 2023. Scalera pointed out that Paolo Toselli, an Italian writer, published a book on popular urban myths called *La Famosa Invasione Delle Vipere Volanti* (The Famous Invasion of the Flying Vipers), Sonzogno, Milan, 1994.

301 **"So it's coming":** Harry Greene, Zoom interview, March 2, 2023.

301 **"may provide the Florida population":** Guzy et al., p. 8.

301 **Beginning in the late 1880s:** On the history of Florida's canal and levee construction, Dorcas and Willson, *Invasive Pythons in the United States*, p. 60.

302 **"That's another reason":** M. Rockwell Parker, interview, August 15, 2023.

302 **astonishing homing abilities:** Harvey et al., "Burmese Pythons in South Florida," described in detail two experiments where telemetered pythons were captured, were released in distant sites, and made their way back over the course of four to six months to the original point of capture; one traveled 48 miles, the other 35.

303 **staggering 200 miles:** M. Kirkland, Zoom interview, January 19, 2023.

303 **"Burmese pythons are capable":** Guzy et al., p. 46.

303 **By 2003, pythons were found:** The Guzy et al. report, p. 35, includes a map that shows the sequential spread of the python population in South Florida.

303 **by 2014:** Margaret E. Hunter, US Geological Survey, Wetland and Aquatic Research Center, Gainesville, FL, Zoom interview, February 23, 2023.

304 **"detection probability":** Guzy et al., p. 25.

304 **"Pythons have essentially zero":** Guzy et al., p. 31.

304 **officials claim that more than 20,000:** M. Kirkland, personal communication, December 7, 2023.

304 **"You hear all kinds":** M. Kirkland, Zoom interview, January 19, 2023.

304 **"There are no reliable":** Guzy et al., p. 37.

304 **"So, the Number":** M. Hunter, interview, February 23, 2023.

304 **"Python Challenge":** According to South Florida Water Management District board member Ron Bergeron, "Under Governor Ron DeSantis' leadership, the state has taken unprecedented action to protect the Everglades and eliminate invasive Burmese pythons from across the landscape." State of Florida press release, August 4, 2023.

305 **academic battle royale:** The original paper is G. H. Rodda, C. S. Jarnevich, and R. N. Reed, "What parts of the US mainland are climatically suitable for invasive alien pythons spreading from Everglades National Park?" *Biological Invasions* 11:241-252 (2009). The strong rebuttal, which was actually published prior to Rodda et al. (because of early online release), was R. A. Pyron, F. T. Burbrink, and T. J. Guiher, "Claims of potential expansion throughout the U.S. by invasive python species are contradicted by ecological niche models," *PLoS One* 3 (2008). For the counter-counterargument, see G. H. Rodda et al., "Challenges in identifying sites climatically matched to native ranges of animal invaders," *PLoS One* 6 (2011).

   The basic controversy involves the accuracy of models that rely on the native range of an invasive species to predict its potential range in the invaded territory. Rodda et al., "What parts of the US mainland are climatically suitable," predicted that pythons in Florida could conceivably find habitable niches throughout the southeastern United States, up the East Coast as far as New Jersey, in the Midwest as far north as Illinois, and along the West Coast up to the Canadian border. Pyron et al., "Claims of potential expansion throughout the U.S.," concluded that the pythons would barely exceed their present range in South Florida, possibly extending to South Texas. Ironically, the two rival models concurred on a seemingly unlikely locale: western Oregon and Washington.

306 **In early January 2010:** For an account of the freeze and its aftermath, see Guzy et al., pp. 12,

22-23; F. J. Mazzotti et al., "Cold-induced mortality of invasive Burmese pythons in South Florida," *Biological Invasions* 13:143-151 (2011); and contemporaneous news reports.

306 **"maladaptive behavior"**: Mazzotti et al., "Cold-induced mortality," p. 143. For the effects of cold weather on the South Carolina python enclosure and the death of the snakes, see Dorcas and Willson, *Invasive Pythons in the United States*.

306 **Python "removals"**: Guzy et al., p. 38.

307 **"Some portion of the southern Florida"**: Guzy et al., p. 23.

307 **Beginning around 2015**: D. C. Card et al., "Novel ecological and climatic conditions drive rapid adaptation in invasive Florida Burmese pythons," *Molecular Ecology* 27:4744-4757 (2018).

307 **"We saw a lot of things"**: Daren Card, postdoctoral fellow, Harvard University, Zoom interview, June 9, 2023.

309 **"If you've got a good amount"**: Todd Castoe, University of Texas–Arlington, Zoom interview, June 22, 2023.

310 **the long-term effect of the 2010 freeze**: Guzy et al., p. 44. The preponderance of evidence suggests, they note, that "we may at least expect that pythons can tolerate climatic conditions farther north than where the population is currently established."

310 **another new genetic technology**: Guzy et al., p. 43; M. E. Hunter, interview, February 23, 2023. As an example of the applications of eDNA, Hunter and colleagues used the technique to detect large populations of Burmese pythons on tree islands in the central Everglades where wading birds breed. S.C.M. Orzechowski et al., "Environmental DNA sampling reveals high occupancy rates of invasive Burmese pythons at wading bird breeding aggregations in the central Everglades," *PLoS One* 14(4) (April 2019). According to unpublished data described at a scientific meeting in 2023, according to Mike Kirkland of the South Florida Water Management District, "There's definitely been positive hits for python eDNA north of Lake Okeechobee and in the lower part of the Kissimmee Basin." But those findings, Kirkland said, "disagree with the consensus that there's been no expansion north of Lake Okeechobee." M. Kirkland, telephone interview, December 12, 2023.

311 **"What I find interesting"**: Gordon Burghardt, University of Tennessee, Zoom interview, January 18, 2023.

312 **Holey Land Wildlife Management Area**: Florida Fish and Wildlife Conservation Commission website, https://myfwc.com/recreation/lead/holey-land/.

312 **"confusing"**: Florida Birding Trail, https://floridabirdingtrail.com/trail/trail-sections/south -section/holey-land-rotenberger-wma; accessed August 8, 2024.

313 **97 percent of python habitat**: According to experts, levee road cruising at best affords access of up to one kilometer (slightly more than half a mile) on either side of the road (Guzy et al., p. 65), which means that roughly 97 percent of the Everglades National Park that is habitable to pythons is inaccessible to hunters using the levee roads.

314 **Snakes supposedly can't hear**: There is some recent research, largely by Bruce Young, suggesting that snakes can in fact hear rather well despite the lack of eardrums.

314 **pay was less than minimum wage**: D. Kalil interview, January 27, 2023.

314 **When they complete a catch**: Of the more than 13,000 snakes removed from the Greater Everglades Ecosystem as of December 31, 2021, according to the Guzy et al. report, the vast majority were "predominantly immature pythons," not the larger and reproductively active adults, p. 65-66.

317 **"You could see the muscles"**: Wayne Rassner, former chairman, South Florida National Park Trust (now known as the Alliance for Florida's National Parks), telephone interview, August 16, 2023; Joe Wasilewski, personal communication, August 13, 2023.

317 **we spotted dense black smoke:** For agricultural contamination of Lake Okeechobee, see D. Egan, "It's toxic slime time in Florida's Lake Okeechobee," *New York Times,* July 9, 2023.

320 **a lesion designed to disable:** Guzy et al., pp. 82-85; and M. E. Hunter, interview, February 23, 2023.

SNAKE ROAD: AVENUE OF THE DEAD, TEOTIHUACAN, MEXICO

322 **Avenue of the Dead:** For background on the archaeological site at Teotihuacan, I have consulted Matthew H. Robb, *Teotihuacan: City of Water, City of Fire*, University of California Press, Berkeley, 2017, and *Teotihuacan: History, Art and Monuments*, Monclem, Mexico City, 2023; as well as additional sources cited below. Personal visits were on March 17 and March 20, 2023.

323 **the science that allowed:** Saburo Sugiyama, "Teotihuacan: Planned City with Cosmic Pyramids," in Robb, *Teotihuacan: City of Water, City of Fire*. Sugiyama argues that cosmology was integrated into the urban layout of Teotihuacan, noting that the central north–south axis of the city (Avenue of the Dead) was roughly 15.5 degrees east of true north, which allowed the setting sun to align with the east–west constructions on August 12 and April 29—the most important ritual calendar dates in Mesoamerica (p. 29).

323 **"The serpents were maybe":** Elba Estrada, Instituto Nacional de Antropologia e Historia, Teotihuacan, director of Museo de Murales Teotihuacanos Beatriz de la Fuente, personal interview, March 17, 2023. Estrada made the comment while pointing out features in an ancient mural in the Museo de Murales. "The control of the serpent lineage was from its foundation (circa 100 AD), until approximately 250 AD." E. Estrada, personal communication, May 31, 2024.

323 **archaeologist Linda Manzanilla:** L. R. Manzanilla, "Corporate societies with exclusionary social components: The Teotihuacan metropolis," in *Origini: Prehistory and Protohistory of Ancient Civilizations*, Gangemi Editori, Rome, 2018, pp. 211-225; and L. R. Manzanilla, "Teotihuacan in Central Mexico: An exceptional metropolis," in *The "City" Across Time*, Bardi Edizioni, Rome, 2023, pp. 141-158.

324 **the "most sumptuous monument":** Alfredo Lopez Austin, Leonardo Lopez Lujan, and Saburo Sugiyama, "The Temple of Quetzalcoatl at Teotihuacan: Its possible ideological significance," *Ancient Mesoamerica* 2:93–105 (1991).

324 **the Mesoamerican concept of linear time:** Alfredo Lopez Austin, "Ecumene time, anecumene time: Proposal of a paradigm," in *The Measure and Meaning of Time in Mesoamerica and the Andes* (conference proceedings), edited by Anthony F. Aveni, 2015.

325 **"was cut by two gods":** Lopez Austin et al., "The Temple of Quetzalcoatl at Teotihuacan: Its possible ideological significance," *Ancient Mesoamerica* 2:93-105 (1991).

325 **The earliest settlers in Teotihuacan:** Alberto Ruy-Sanchez Lacy, "The dreams of the serpent," in "Serpiente en el Arte Prehispanico," *Artes de Mexico* 32 (2002). The entire issue of this magazine is devoted to images and essays about serpents in pre-Hispanic Mexican art.

326 **"One of the strong factions":** L. R. Manzanilla, "Teotihuacan in Central Mexico, p. 151.

326 **"Reptile iconography":** David Charles Wright-Carr, "Sacred reptiles and native worldview: Enactive aesthetics in ancient Mesoamerica," Proceedings of A Body of Knowledge—Embodied Cognition and the Arts Conference, CTSA UCI (2016).

327 ***Dualidad*:** The Tamayo mural and its symbolic elements are described in Felipe Solis, *National Museum of Anthropology*, Monclem, Mexico City, pp. 8-9.

328 **My guide, a young man:** Luis Sarabia, personal interview, March 20, 2023.

EPILOGUE: SNAKE LECTURES

330 **On the evening of April 21:** Background details on the Warburg talk are discussed in Uwe Fleckner, *The Snake and the Lightning: Aby Warburg's American Journey*, translation by Kevin Cook, Hatje Cantz Verlag, Berlin, 2023, pp. 153-161.

330 **one of the seminal contributions:** Fleckner, *The Snake and the Lightning*, p. 11. As Fleckner put it, Warburg's 1923 lecture "has had an almost inestimable impact on art and cultural history research throughout the world." Kenneth Clark called Warburg "without doubt the most original thinker on art-history of our time."

331 **"Edison's copper snake":** Fleckner, *The Snake and the Lightning*, p. 164.

331 **not all of it benign:** Fleckner's account in *The Snake and the Lightning* made clear that Warburg engaged in problematic anthropological behavior during his southwestern sojourn, including inappropriately touching Native Americans during a ritual ceremony (p. 147) and bullying a Zuni child into selling him a native toy with snake-like markings (p. 106).

331 **"a living (primal) rain-snake-saint":** Aby M. Warburg, *Images from the Region of the Pueblo Indians of North America*, translated with an interpretive essay by Michael P. Steinberg, Cornell University Press, Ithaca, NY, 1995, p. 36. Steinberg offered a full translation of the 1923 Snake Lecture.

331 **"would help to justify":** Fleckner, *The Snake and the Lightning*, p. 7.

331 **the idea of iconography in art:** Warburg's theory, for example, deeply influenced E. H. Gombrich's *The Story of Art* (Phaidon, 1950), arguably the most widely read book about art history in the 20th century. Gombrich later wrote an "intellectual biography" of Warburg.

332 **"messengers" to nature:** Warburg, quoted in Steinberg, *Images from the Region of the Pueblo Indians*, p. 38.

332 **"teaches magic and prayer":** Steinberg, *Images from the Region of the Pueblo Indians*, p. 3.

332 **"those ominous destroyers":** Steinberg, *Images from the Region of the Pueblo Indians*, p. 54.

333 **"an example of that primordial":** Steinberg, *Images from the Region of the Pueblo Indians*, p. 53.

333 **"The American of today":** Steinberg, *Images from the Region of the Pueblo Indians*, p. 54. The telegraph pole image is also discussed in Fleckner, *The Snake and the Lightning*, pp. 162-165.

333 **Telegraph Road in Michigan:** I revisited the area on October 22–23, 2023. On the expansion of the highway, J. Marshall, Bloomfield Hills Historical Society, personal communication, November 8, 2023.

334 **average winter temperatures in the Detroit area:** H. Stevens, "Winter is warming almost everywhere," *Washington Post*, March 1, 2024.

335 **"He said it was one of the best places":** David Mifsud, Herpetological Resource and Management, Chelsea, MI, Zoom interview, January 17, 2024, and personal communication, July 30. 2024.

335 **his daytime job was at the Henry Ford:** James Harding, telephone interview, January 22, 2024.

336 **"Although many people":** Harry W. Greene, *Tracks and Shadows: Field Biology as Art*, University of California Press, Berkeley, 2013, p. 3.

337 **"You can just sit around":** Rick Shine, interview, December 13, 2023.

338 **"Rattlesnake 101":** Snake Safety and Handling Workshop, San Luis Obispo, CA, September 9, 2023.

Index

Index

Index

Index

Index

Index

About the Author

Stephen S. Hall has been reporting and writing about the intersection of science and society for more than 40 years. In addition to numerous cover stories in the *New York Times Magazine*, where he also served as a story editor and contributing writer, his work has appeared in *The New Yorker, The Atlantic Monthly, National Geographic, New York Magazine, Wired, Science, Nature, Scientific American, Discover, The Sciences, Hippocrates, Smithsonian*, and more. He is also the author of six critically acclaimed nonfiction books about contemporary science. Among other honors, he has received the Walter Sullivan Award for Excellence in Science Journalism from the American Geophysical Union (2011); the Best Magazine Story of the Year from the American Association for the Advancement of Science—Kavli Foundation in 2017; and an honorary doctorate from Cold Spring Harbor Laboratory in 2022. He also received a Guggenheim Fellowship in 2012.

Since 2007, Hall has served as an adjunct professor of journalism at New York University, where he taught a core-curriculum graduate school seminar in science writing at NYU's Science, Health, and Environmental Reporting Program (SHERP) for 10 years and currently teaches science communication to scientists. He previously taught graduate seminars in science writing and explanatory journalism at Columbia University. Since 2009, he has also conducted hundreds of Science Communication Workshop sessions for scientists and doctors at NYU, Rockefeller University, Mount Sinai School of Medicine, and Cold Spring Harbor Laboratory.

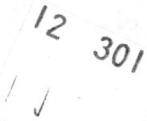